Engineering Science

Engineering Science

Fourth Edition

W. Bolton

Newnes

OXFORD AUCKLAND BOSTON JOHANNESBURG MELBOURNE NEW DELHI

Newnes
An imprint of Butterworth-Heinemann
Linacre House, Jordan Hill, Oxford OX2 8DP
225 Wildwood Avenue, Woburn, MA 01801-2041
A division of Reed Educational and Professional Publishing Ltd

 A member of the Reed Elsevier plc group

First published 1990
Reprinted 1990, 1991
Revised and reprinted 1992
Reprinted 1993, 1994
Second edition 1994
Third edition 1998
Fourth edition 2001

British Library Cataloguing in Publication Data
A catalogue record for this book is available from the British Library

ISBN 0-7506-5259-4

Printed and bound in Great Britain by Martins the Printers Ltd, Berwick upon Tweed

Contents

Preface

Aims　This book aims to provide a comprehensive grounding in science relevant to engineers. As such it is ideal for a wide range of voational courses and foundation programmes at degree level.

The content has been carefully matched to cover the latest UK syllabuses, in particular the new specifications for BTEC National from Edexcel, commencing in 2001/2, and the relevant units of the Vocational A-level (AVCE) from Edexcel and AQA.

The units covered for these courses include:

Science for Technicians
Applied Science in Engineering
Electrical and Electronic Principles
Further Electrical Principles
Mechanical Principles
Further Mechanical Principles

Changes from the third edition　For the fourth edition, the book:

- Has been completely reorganised and reset to more closely match the National and VCE units.

- Includes extra material required to give the comprehensive coverage of all the science and principles at this level.

- Includes more discussion and explanation of basic principles.

- Includes more worked examples.

- Includes more problems.

- Includes suggestions for practical activities

Structure of the book　The book has been designed to give a clear exposition and guide readers through the scientific principles, reviewing background principles where necessary. Each chapter includes numerous worked examples and problems. Answers are supplied to all the problems.

W. Bolton

1 Basics

1.1 Introduction

Engineers make measurements to enable theorics to be tested, relationships to be determined, values to be determined in order to predict how components might behave when in use and answers obtained to questions of the form – 'What happens if?'. Thus, there might be measurements of the current through a resistor and the voltage across it in order to determine the resistance or, perhaps for some device such as a thermistor, to determine the relationship between the current and the voltage. In this chapter, there is a discussion of the measurement and collection of data, errors associated with the results of measurements, the use of graphs, units and a review of some basic terms.

1.2 Measuring and collecting data

In making measurements, it is necessary to select the appropriate instrument for the task, taking into account the limitations of instruments and the accuracy with which it gives readings. Thus if you need to measure the mass of an object to a fraction of a gram then it is pointless using a spring balance since such an instrument cannot give readings to this degree of precision. The spring balance might have a scale which you think you can interpolate between scale markings to give a reading of a fraction of a gram, but it is unlikely that the calibration of the instrument is accurate enough for such interpolations to have any great significance.

With instruments there is a specification of the accuracy with which it gives readings. The term *accuracy* is used for the extent to which a result might depart from the true value, i.e. the errors it might have. *Error* is defined as:

error = measured value − true value

Accuracy is usually quoted as being plus or minus some quantity, e.g. ±1 g. This indicates that the error associated with a reading of that instrument is such that true value might be expected to be within plus or minus 1 g of the indicated value. The more accurate the measurement the smaller will be the error range associated with a measurement.

In some situations the error is specified in the form of:

$$\text{percentage error} = \frac{\text{error in quantity}}{\text{size of quantity}} \times 100$$

Thus, for example, a mass quoted as 2.0 ± 0.2 g might have its error quoted as ±10%. In the case of some instruments, the error is often quoted as a percentage of the full-scale-reading that is possible with an instrument.

Example

An ammeter is quoted by the manufacturer as having an accuracy of ±4% f.s.d. on the 0 to 2 A scale. What will the error be in a reading of 1.2 A on that scale?

The accuracy is ±4% of the full scale reading of 2 A and is thus ±0.08 A. Hence the reading is 1.2 ± 0.08 A.

1.2.1 Sources of error

Common sources of error with measurements are:

1   *Instrument construction errors*
    These result from such causes as tolerances on the dimensions of components and the values of electrical components used in instruments and are inherent in the manufacture of an instrument and the accuracy to which the manufacturer has calibrated it. The specification supplied by the manufacturer for an instrument will give the accuracy that might be expected under specified operating conditions.

2   *Non-linearity errors*
    In the design of many instruments a linear relationship between two quantities is often assumed, e.g. a spring balance assumes a linear relationship between force and extension. This may be an approximation or may be restricted to a narrow range of values. Thus an instrument may have errors due to a component not having a perfectly linear relationship. Thus in the specification supplied by a manufacturer for, say, a temperature sensor you might find a statement of a non-linearity error.

3   *Operating errors*
    These can occur for a variety of reasons and include errors due to:

(i)   Errors in reading the position of a pointer on a scale. If the scale and the pointer are not in the same plane then the reading obtained depends on the angle at which the pointer is viewed against the scale (Figure 1.1). These are called *parallax errors*. To reduce the chance of such errors occurring, some instruments incorporate a mirror alongside the scale. Positioning the eye so that the pointer and its image are in line guarantees that the pointer is being viewed at the right angle. Digital instruments, where the reading is displayed as a series of numbers, avoid this problem of parallax.

(ii) Errors may also occur due to the limited resolution of an instrument and the ability to read a scale. Such errors are termed *reading errors*. When the pointer of an instrument falls

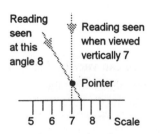

Figure 1.1   *Parallax error*

between two scale markings there is some degree of uncertainty as to what the reading should be quoted as. The worse the reading error could be is that the value indicated by a pointer is anywhere between two successive markings on the scale. In such circumstances the reading error can be stated as a value ± half the scale interval. For example, a rule might have scale markings every 1 mm. Thus when measuring a length using the rule, the result might be quoted as 23.4 ± 0.5 mm. However, it is often the case that we can be more certain about the reading and indicate a smaller error. With digital displays there is no uncertainty regarding the value displayed but there is still an error associated with the reading. This is because the reading of the instrument goes up in jumps, a whole digit at a time. We cannot tell where between two successive digits the actual value really is. Thus the degree of uncertainty is ± the smallest digit.

(iii) In some measurements the insertion of the instrument into the position to measure a quantity can affect its value. These are called *insertion errors* or *loading errors*. For example, inserting an ammeter into a circuit to measure the current can affect the value of the current due to the ammeter's own resistance. Similarly, putting a cold thermometer into a hot liquid can cool the liquid and so change the temperature being measured.

4   *Environmental errors*

Errors can arise as a result of environmental effects. For example, when making measurements with a steel rule, the temperature when the measurement is made might not be the same as that for which the rule was calibrated. Another example might be the presence of draughts affecting the readings given by a balance.

Example

An ammeter has a scale with markings at intervals of 50 mA. What will be the reading error that can be quoted with a reading of 400 mA?

The reading error is generally quoted as ± half the scale interval. Thus the reading error is ±25 mA and the reading can be quoted as 400 ± 25 mA.

1.3 Random errors   The term *random errors* is used for errors which can vary in a random manner between successive readings of the same quantity. This may be due to personal fluctuations by the person making the measurements, e.g. varying reaction times in timing events, applying varying pressures when using a micrometer screw gauge, parallax errors, etc., or perhaps due to random electronic fluctuations (termed noise) in the instruments or circuits used, or perhaps varying frictional effects.

Random errors mean that sometimes the error will give a reading that is too high, sometimes a reading that is too low. The error can be reduced by repeated readings being taken and calculating the mean (or average) value. The *mean* or *average* of a set of n readings is given by:

$$\text{mean } \bar{x} = \frac{x_1 + x_2 + \ldots x_n}{n}$$

where x_1 is the first reading, x_2 the second reading, ... x_n the nth reading. The more readings we take the more likely it will be that we can cancel out the random variations that occur between readings. The *true value* might thus be regarded as the value given by the mean of a very large number of readings.

Example

Five measurements of the time have been taken for the resistance of a resistor: 20.1, 20.0, 20.2, 20.1, 20.1 Ω. Determine the mean value.

The mean value is obtained using the equation given above as:

$$\text{mean} = \frac{20.1 + 20.0 + 20.2 + 20.1 + 20.1}{5} = \frac{100.5}{5} = 20.1$$

1.4 Significant figures

When we write down the result of a measurement we should only write it to the number of figures the accuracy will allow, these being termed the *significant figures*. If we write 12.0 g for the mass of some object then there are three *significant figures*. However, if we quoted the number as 12, there are only two significant figures and the mass is less accurately known.

If we have a number such as 0.001 04 then the number of significant figures is 3 since we only include the number of figures between the first non-zero figure and the last figure. This becomes more obvious if we write the number in scientific notation as 1.04×10^{-3}. If we have a number written as 104 000 then we have to assume that it is written to six significant figures, the last 0 being significant. If we only wanted three significant figures then we should write the number as 1.04×10^5.

When multiplying or dividing two numbers, the result should only be given to the same number of significant figures as the number with the least number of significant figures.

When adding or subtracting numbers, the result should only be given to the same number of decimal places as the number in the calculation with the least number of decimal places.

When the result of a calculation produces a number which has more figures than are significant, we need to reduce it to the required number of significant figures. This process is termed *rounding*. For example, if we have 2.05 divided by 1.30, then using a calculator we obtain 1.5769231. We need to reduce this to three significant figures. This is done by considering the fourth figure, i.e. the 6. If that figure is 5 or greater, the third figure is rounded up. If that figure is less than 5, it is rounded down. In this case the figure of 6 is greater than 5 and so we round up and write the result as 1.58. In any calculation which involves a number of arithmetic steps, do not round numbers until all the calculations have been completed. The rounding process carried out at each stage can considerably affect the number emerging as the final answer.

Example

In an experiment involving weighing a number of items the results obtained were 1.4134 g, 5.156 g and 131.5 g. Quote, to the appropriate number of significant figures, the result obtained by adding the weighings.

1.4134 + 5.156 + 131.5 = 138.0694. But one of the results is only quoted to one decimal place. Thus the answer should be quoted as 138.1 g, the second decimal place figure of 6 rounding the first figure up.

Example

The result of two measurements gave figures of 14.0 and 23.15. If we then have to determine a result by working out 14.0 divided by 23.15, what is the result to the appropriate number of significant figures?

14.0 ÷ 23.15 = 0.6047516. But the result with the least number of significant figures has just three. Hence the result to three significant figures is 0.605, the third figure having been rounded up because the fourth figure is 7.

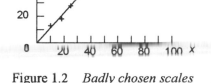

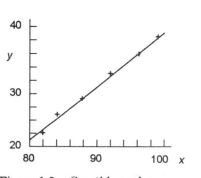

Figure 1.2   *Badly chosen scales*

1.5 Graphs   In plotting graphs it is necessary to consider what quantities to plot along which axis and the scales to be used. Each of the axes will have a scale. When selecting a scale:

1   The scale should be chosen so that the points to be plotted occupy the full range of the axes used for the graph. There is no point in having a graph with scales from 0 to 100 if all the data points have values between 0 and 50 (Figure 1.2).

2   The scales should not start at zero if starting at zero produces an accumulation of points within a small area of the graph. Thus if all the points have values between 80 and 100, then a scale from 0 to 100 means all the points are concentrated in just the end zone of the scale. It is better, in this situation, to have a scale running from 80 to 100 (Figure 1.3).

3   Scales should be chosen so that the location of the points between scale marks is made easy. Thus with a graph paper subdivided into large squares with each having 10 small squares, it is easy to locate a point of 0.2 if one large square corresponds to 1 but much more difficult if one large square corresponds to 3.

Figure 1.3   *Sensible scales*

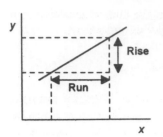

Figure 1.4   *Gradient = rise/run*

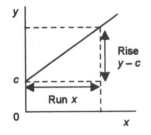

Figure 1.5   *Straight line graph*

4   The axes should be labelled with the quantities they represent and their units.

1.5.1 Equation for the straight line graph

The straight line graph (Figure 1.4) is given with many relationships. The *gradient*, i.e. slope, of the graph is how much the line rises for a particular horizontal run and is m = rise/run. If we consider the rise and run from a point where the graph line cuts the $x = 0$ axis (Figure 1.5):

$$\text{gradient } m = \frac{\text{rise}}{\text{run}} = \frac{y-c}{x}$$

$$y = mx + c$$

This is the general equation used to describe a straight line.

Example

Determine the equation of the straight line graph in Figure 1.6.

Gradient = rise/run = $(20 - 5)/4 = 3.75$ m/s. There is an intercept with the y axis of 5 m. Thus, with s as the distance in metres and t the time in seconds, the equation is $s = 3.75t + 5$.

1.6 Units   The *International System* (SI) of units has the seven basic units:

Length	metre	m
Mass	kilogram	kg
Time	second	s
Electric current	ampere	A
Temperature	kelvin	K
Luminous intensity	candela	cd
Amount of substance	mole	mol

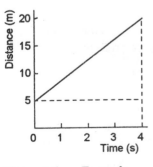

Figure 1.6   *Example*

Also there are two supplementary units, the radian and the steradian.

Since in any equation we must have the units on one side of the equals sign balancing those on the other, we can form the SI units for other physical quantities from the base units via the equation defining the quantity concerned. Thus, for example, since volume is defined by the equation volume = length cubed then the unit of volume is that of unit of length unit cubed and so the metre cubed, i.e. m^3. Since density is defined by the equation density = mass/volume, the unit of density is the unit of mass divided by the unit of volume and thus kg/m^3. Since velocity is defined by the equation velocity = change in displacement in a straight line/time taken, the unit of velocity is unit of distance/unit of time and so is metres/second, i.e. m/s. Since acceleration is defined by the equation acceleration = change in velocity/time taken, the unit of acceleration is unit of velocity/unit of time and thus metres per second/second, i.e. m/s².

$$\text{unit of acceleration} = \frac{m/s}{s} = \frac{m}{s \times s} = m/s^2$$

Some of the derived units are given special names. Thus, for example, force is defined by the equation force = mass × acceleration and thus the unit of force = unit of mass × unit of acceleration and is kg m/s² or kg m s⁻². This unit is given the name newton (N). Thus 1 N is 1 kg m/s². The unit of pressure is given by the defining equation pressure = force/area and is thus N/m². This unit is given the name pascal (Pa). Thus 1 Pa = 1 N/m².

Certain quantities are defined as the ratio of two comparable quantities. Thus, for example, strain is defined as change in length/length. It thus is expressed as a pure number with no units because the derived unit would be m/m. Note that $\sin \theta$, $\cos \theta$, $\tan \theta$, etc. are trigonometric ratios, i.e. θ is a ratio of two sides of a triangle. Thus θ has no units.

Example

Determine the unit of the tensile modulus E when it is defined by the equation E = stress/strain if stress has the unit of Pa and strain is a ratio with no units.

Unit of E = (unit of stress)/(unit of strain) = (Pa)/(no unit) = Pa

1.6.1 Powers of ten notation

The term *scientific notation* or *standard notation* is often used to express large and small numbers as the product of two factors, one of them being a multiple of ten. For example, a voltage of 1500 V can be expressed as 1.5×10^3 V and a current of 0.0020 A as 2.0×10^{-3} A. The number 3 or −3 is termed the *exponent* or *power*. To write a number in powers we have to consider what power of ten number is used to multiply or divide it. Thus, for the voltage 1500 = 1.5 × 1000 = 1.5×10^3 and for the current 0.0020 = 20/1000 = 2.0×10^{-3}.

1.6.2 Unit prefixes

Standard prefixes are used for multiples and submultiples of units, the SI preferred ones being multiples of 1000, i.e. 10^3, or division by multiples of 1000. Table 1.1 shows commonly used standard prefixes.

Example

Express the capacitance of 8.0×10^{-11} F in pF.

1 pF = 10^{-12} F. Hence, since $8.0 \times 10^{-11} = 80 \times 10^{-12}$, the capacitance can be written as 80 pF.

Table 1.1 *Standard unit prefixes*

Multiplication factor	Prefix	
1 000 000 000 = 10^9	giga	G
1 000 000 = 10^6	mega	M
1 000 = 10^3	kilo	k
100 = 10^2	hecto	h
10 = 10	deca	da
0.1 = 10^{-1}	deci	d
0.01 = 10^{-2}	centi	c
0.001 = 10^{-3}	milli	m
0.000 001 = 10^{-6}	micro	μ
0.000 000 001 = 10^{-9}	nano	n
0.000 000 000 001 = 10^{-12}	pico	p

Example

Express the tensile modulus of 210 GPa in Pa without the unit prefix.

Since 1 GPa = 10^9 P, then 210 GPa = 210×10^9 Pa.

1.7 Basic terms   The following are some basic terms:

1   *Density*
    If a body has a mass m and volume V, its density $\rho = m/V$ and has the SI unit of kg/m³.

2   *Relative density*
    Relative density is defined as (density of a material)/(density of water). Since the relative density is a ratio of two quantities in the same units, it is purely a number and has no units.

Example

The density of water at about 20°C is 1000 kg/m³ and the density of copper is 8900 kg/m³. What is the relative density of copper?

Relative density of copper = 8900/1000 = 8.9.

3   *Mass*
    The mass of a body is the quantity of matter in the body and can be defined by Newton's laws as that quantity, which the bigger it is the smaller will be the acceleration produced by a given force. Mass thus represents the inertia or 'reluctance to accelerate'. It has the SI unit of kg.

4   *Force*
    Force is defined by Newton's laws as that which causes a body to accelerate and has the unit of the newton (N).

5   *Weight*

The weight of a body is the gravitational force acting on it and which has to be opposed if the body is not to fall. The weight of a body of mass m where the acceleration due to gravity is g is mg. Weight, as a force, has the SI unit of N.

Example

What is the weight of a block with a mass of 2 kg if the acceleration due to gravity is 9.8 m/s²?

Weight = mg = 2×9.8 = 19.6 N

6   *Pressure*

If a force F acts over an area A, the pressure p is F/A and has the SI unit of N/m², this being given the special name of pascal (Pa).

Problems

1   A thermometer has graduations at intervals of 0.5°C. What is the worst possible reading error?

2   An ammeter is quoted by the manufacturer as having an accuracy of ±2% f.s.d. on the 0 to 1 A scale. What will the error be in a reading of 0.80 A on that scale?

3   An instrument has a scale with graduations at intervals of 0.1 units. What is the worst possible reading error?

4   The accuracy of a digital meter, which gives a 1 digit display, is specified by the manufacturer as being ±1 digit. What will be the accuracy specified as a percentage of f.s.d.?

5   Determine the means for the following sets of results: (a) The times taken for 10 oscillations of a simple pendulum: 51, 49, 50, 49, 52, 50, 49, 53, 49, 52 s, (b) The diameter of a wire when measured at a number of points using a micrometer screw gauge: 2.11, 2.05, 2.15, 2.12, 2.16, 2.14, 2.16, 2.17, 2.13, 2.15 mm, (c) The volume of water passing through a tube per 100 s time interval when measured at a number of times: 52, 49, 54, 48, 49, 49, 53, 48, 50, 53 cm³.

6   Write the following data in scientific notation in the units indicated: (a) 20 mV in V, (b) 15 cm³ in m³, (c) 230 μA in A, (d) 20 dm³ in m³, (e) 15 pF in F, (f) 210 GPa in Pa, (g) 1 MV in V.

7   Write the following data in the units indicated: (a) 1.2×10^3 V in kV, (b) 2.0×10^5 Pa in MPa, (c) 0.20 m³ in dm³, (d) 2.4×10^{-12} F in pF, (e) 0.003 A in mA, (f) 12×10^9 Hz in GHz.

8   Round the following numbers to two significant figures: (a) 12.91, (b) 0.214, (c) 0.01391, (d) 191.9, (e) 0.0013499

9   The mass of a beaker was measured using a balance as 25.6 g. Water with a mass of 6.02 g was added to the beaker. What is the total mass to the appropriate number of significant figures?

10   What is (a) the density, (b) the relative density of a cube if it has a side of length 200 mm and a mass of 2000 g? The density of water is 1000 kg/m³.

11   What is the weight of a body of mass 20 g if the acceleration due to gravity is 9.8 m/s²?

2 Energy

2.1 Introduction

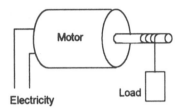

Figure 2.1 *Using electrical energy*

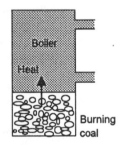

Figure 2.2 *Burning a fuel*

It is difficult to explain what is meant by the term *energy*; we can say that energy is involved in doing jobs, in making things move, in warming things up. For example:

1   We can use an electric motor to lift a load (Figure 2.1) and consider that we are supplying electrical energy to the motor to do the job of lifting the load.

2   We could use our own energy to lift a load and consider our energy comes from the chemical reactions in our body with the 'fuel', i.e. food, we have taken in.

3   We can heat a boiler to produce hot water by burning a fuel (Figure 2.2), e.g. coal, and consider the chemical reactions that are involved in the burning process supplying the energy to heat the water.

4   Alternatively, we could have used an electric immersion heater to heat water and consider that electrical energy was used to supply the energy to heat the water.

This chapter is an introduction to energy in its various forms, the idea that we can transform energy from one form to another and that in all the transformations the total amount of energy remains constant, i.e. the principle of the conservation of energy. Also introduced is the term *power*, it being the rate of transfer of energy. Examples of simple machines, in which both the input and output are mechanical, are considered.

2.2 Energy transformations

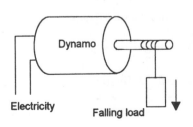

Figure 2.3 *Using 'falling load' energy*

Consider the situation described in Figure 2.1 of electrical energy being used to lift a load. When the load has been lifted we can consider there is energy associated with the lifted load. For example, we might use the lifted load to fall and drive a dynamo (Figure 2.3) and so reverse the process of Figure 2.1 and transform the energy associated with the lifted load into energy associated with motion and hence into electrical energy. The energy associated with a lifted load is termed *potential energy* and the energy associated with motion is termed *kinetic energy*. Thus, in Figure 2.1, we can consider there to be the energy transformations:

electrical energy → potential energy

and in Figure 2.2:

potential energy → kinetic energy → electrical energy

Figure 2.4 illustrates the sequence of energy transformations involved in using coal to heat a boiler to produce high-pressure steam which is then used to produce rotation of a turbine, the turbine then driving a generator to produce electricity which is then used to produce heat from an electric fire. The energy associated with the rotation of the turbine shaft is energy associated with motion and so is a form of kinetic energy. The energy transformations are thus:

fuel energy → heat energy → kinetic energy
    → electrical energy → heat energy from electric fire

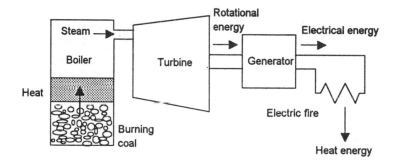

Figure 2.4 *Energy transformations*

2.2.1 Forms of energy

There are many forms that energy can take, the mechanism that is used for transferring from one form to another being work or heat.

1   *Work*
    Work is the transfer of energy that occurs when the point of application of a force moves through a distance (see Section 2.3).

2   *Heat*
    Heat is the transfer of energy that occurs between two systems when there is a temperature difference between them. Thus if we observe an object and find that its temperature is increasing, then a transfer of energy, as heat, is occurring into the object (see Chapter 3).

The following are forms that energy can take:

1   *Potential energy*
    This is energy associated with position, e.g. a lifted load. The term gravitational potential energy is often used for the potential energy involved in the lifting of a load against gravity. The term strain energy is often used for the potential energy associated with the

stretching of a spring or some other elastic material. For example, energy is stored in a stretched rubber band; when you release an unrestrained band this energy is transformed into kinetic energy.

2   *Kinetic energy*
    This is energy associated with motion, e.g. a rotating shaft or a falling load. An object can increase its kinetic energy by increasing its linear velocity or by increasing its rotational velocity.

3   *Chemical energy*
    This is the energy released as a result of chemical reactions, e.g. the burning of a fuel, the eating of food or the reactions involved when explosives go off.

4   *Electrical energy*
    This is the energy released when an electric current occurs.

5   *Magnetic energy*
    If we bring one magnet close to another, attraction or repulsion can occur and the other magnet is forced to move. Thus magnetic energy has been transformed into kinetic energy.

6   *Light energy*
    Passing an electric current through a lamp causes the filament to become hot and we talk of the electrical energy being converted into heat energy and light energy.

7   *Sound energy*
    Sound is a propagating pressure wave.

8   *Radio energy*
    Radio waves carry energy which can be transformed into electrical signals in radios and hence used to operate loudspeakers and produce sound as a result of the diaphragm of the speaker being caused to move back-and-forth.

9   *Nuclear energy*
    When nuclei split or are fused together, energy can be released.

2.2.2 Conservation of energy

Energy does not disappear when a useful job has been done, it just changes from one form to another. When you lift an object off the floor and on to a bench, energy is transferred from you to the object which gains potential energy. The energy you have is provided by food. To lift an object of mass 1 kg through a height of 1 m uses the energy you gain from, for example, about 2.5 milligrams of sugar (this is typically about four grains of sugar). When the object falls off the table and down to the floor it loses its potential energy but gains kinetic energy. When it is just about to hit the floor, all the potential energy that it has acquired in being lifted off the floor has been transformed into kinetic energy. When it hits the floor it stops moving and so all the kinetic energy vanishes. It might seem that energy has been lost. But this is not the case. The object and the floor show an increase in temperature. The kinetic energy is transferred via heat into a rise in temperature.

Energy is never lost, it is only transformed from one form to another or transferred from one object to another. This is the *principle of the conservation of energy*. In any process we never increase the total amount of energy, all we do is transform it from one form to another.

2.3 Work

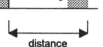

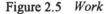
distance

Figure 2.5 *Work*

Work is said to be done when the energy transfer takes place as a result of a force pushing something through a distance (Figure 2.5), the amount of energy transferred W being the product of the force F and the displacement s of the point of application of the force in the direction of the force.

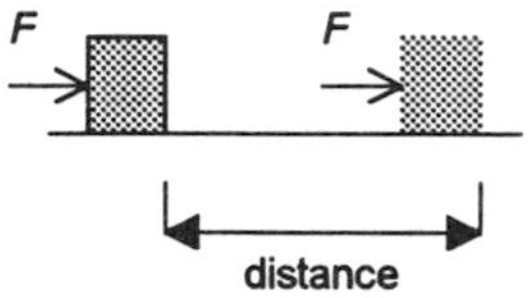

$$W = Fs$$

With force in newtons and distance in metres, the unit of work is the joule (J) with 1 J being 1 N m.

Example

What is the work done in using a hoist to lift a pile of bricks of mass 20 kg from the ground to the top of a building if the building has a height of 30 m? Take the acceleration due to gravity to be 9.8 m/s².

The work done is the force that has to be applied to lift the bricks, i.e. the weight mg, multiplied by the vertical distance through which the point of application of the force has to move the bricks. Thus:

work done $= mg \times h = 20 \times 9.8 \times 30 = 5880$ J

Example

The work done in moving an object through a distance of 20 m is 500 J. Assuming that the force acts in the direction of the motion and is constant, calculate the value of the force.

Using $W = Fs$ we have $500 = F \times 20$, hence $F = 500/20 = 25$ N.

Example

The locomotive of a train exerts a constant force of 120 kN on a train while pulling it at 40 km/h along a level track. What is the work done in 15 minutes?

In 15 minutes the train covers a distance of 10 km. Hence, work done $= 120 \times 1000 \times 10 \times 1000 = 1200\,000\,000$ J.

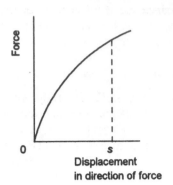

Figure 2.6 *Force–displacement graph*

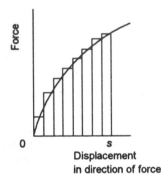

Figure 2.7 *Force–displacement graph*

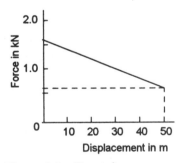

Figure 2.8 *Example*

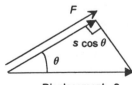

Figure 2.9 *An oblique force*

2.3.1 Work as area under force–distance graph

Consider the work done when the force moving an object is not constant but varying, e.g. in the manner shown in Figure 2.6. We can tackle such a problem by considering the displacement over some distance as being made up of a small number of displacements for each of which the force can be considered constant. Figure 2.7 illustrates this. For each small displacement the work done is the product of the force and the displacement and so is equal to the area of the strip. The total work done in giving a displacement from 0 to s is thus the sum of the areas of all the strips between 0 and s and so is equal to the area under the graph.

Example

A load is hauled along a track with a tractive effort F which varies with the displacement s in the direction of the force in the following manner:

F in kN	1.6	1.4	1.2	1.0	0.8	0.6
s in m	0	10	20	30	40	50

Determine the work done in moving the load from displacement 0 to 50 m.

Figure 2.8 shows the force–displacement graph. The work done is the area under the graph between displacements 0 and 50 m and hence is the area of a rectangle 600×50 J plus the area of the triangle $\frac{1}{2} \times (1600 - 600) \times 50$ J. The work done is thus 55 000 J.

2.3.2 Work due to an oblique force

Consider the work done by a force F when the resulting displacement s is at some angle θ to the force (Figure 2.9). The displacement in the direction of the force is $s \cos \theta$ and so the work done is:

work done $= F \times s \cos \theta$

Alternatively, we can consider the force can be resolved into two components, namely $F \cos \theta$ in the direction of the displacement and $F \sin \theta$ at right angles to it. There is no displacement in the direction of the $F \sin \theta$ component and so it does no work. Hence the work done by the oblique force is solely due to the $F \cos \theta$ component and so is, as before, $(F \cos \theta) \times s$.

Example

A barge is towed along a canal by a tow rope inclined at an angle of 20° to the direction of motion of the barge. Calculate the work done in moving the barge a distance of 100 m along the canal if the pull on the rope is 400 N.

We have an oblique force and so, since the displacement in the direction of the force is 100 cos 20°, the work done = 400 × 100 cos 20° = 37.6 kJ.

2.3.3 Work due to volume change of a fluid

In the case of fluids, i.e. liquids and gases, it is generally more convenient to consider the work done in terms of pressure and volume changes rather than forces and displacements. Consider a fluid at a pressure p trapped into a container by a piston of surface area A (Figure 2.10). Pressure is force/area and so the force acting on the piston is pA. The work done when the piston moves through a distance s is thus:

$$\text{work done } W = Fs = pAs$$

As is the change in volume of the trapped fluid. Thus:

$$\text{work done} = p \times \text{change in volume}$$

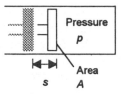

Figure 2.10 *Work done by pressure*

2.4 Potential energy

Consider an object of mass m being lifted from the floor through a vertical height h (Figure 2.11). If the object has a weight mg then the force that has to be applied to move the object is mg and the distance through which the point of application of the force is moved in the direction of the force is h. Thus the work done is mgh. This is the energy transferred to the body. Energy an object has by virtue of its position is called *potential energy*. Thus the object gains potential energy of;

$$\text{potential energy} = mgh$$

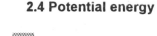

Figure 2.11 *Potential energy*

This form of potential energy is often called *gravitational potential energy* because it is the energy an object has by virtue of moving against a gravitational force. The unit of potential energy, indeed all forms of energy, is the joule (J).

Example

What is the potential energy of an object of mass 3.0 kg relative to the floor when it is lifted vertically from the floor to a height of 1.2 m above the floor? Take g at 9.8 m/s².

The potential energy relative to the floor is $mgh = 3.0 \times 9.8 \times 1.2 = 35.28$ J.

2.4.1 Strain energy

There are other forms of potential energy. Suppose we had the object attached to a spring and apply a force to the object which results in the spring being extended (Figure 2.12(a)). Work is done because the point of application of the force is moved through a distance. Thus the object gains potential energy as a result of the work that has been done. This

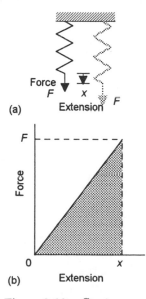

Figure 2.12 *Strain energy*

form of potential energy is termed *elastic potential energy or strain energy*.

For a spring, or a strip of material, being stretched, the force F is generally proportional to the extension x (Figure 2.12(b)) and so the average force is $\frac{1}{2}F$ and work done = $\frac{1}{2}Fx$. This is the energy stored in the spring as a result of it being extended; it is equal to the area under the force–extension graph from an extension of 0 to x. Thus:

strain energy = $\frac{1}{2}Fx$

Example

What is the energy stored in a spring when a force of 200 N is needed to stretch it by 20 mm?

The strain energy is $\frac{1}{2}Fx = \frac{1}{2} \times 200 \times 0.020 = 2$ J.

2.5 Kinetic energy

Consider an object of mass m which has been lifted from the floor through a vertical height h, so acquiring a potential energy of mgh. If the object now falls back down to the floor (Figure 2.13), it loses its potential energy of mgh and gains energy by virtue of its motion, this being termed *kinetic energy*. The potential energy has been transformed into kinetic energy. The kinetic energy (KE) gained is the potential energy lost. If the object starts from rest on the table and has a velocity v when it hits the floor, then the average velocity is $v/2$. The average velocity is the distance h covered over the time t taken. Hence, average velocity = $v/2$ = h/t and so $h = vt/2$. The average acceleration is the change in velocity divided by the time taken. Thus $g = v/t$. Hence:

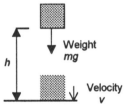

Figure 2.13   *Kinetic energy*

$$\text{KE gained} = mgh = m \times \frac{v}{t} \times \frac{vt}{2} = \tfrac{1}{2}mv^2$$

Example

Calculate the kinetic energy of an object of mass 5 kg moving at a velocity of 6 m/s.

$$\text{KE} = \tfrac{1}{2}mv^2 = \tfrac{1}{2} \times 5 \times 6^2 = 90 \text{ J}$$

2.6 Conservation of mechanical energy

In the absence of any dissipation of energy as heat, mechanical energy is conserved. For example, an object of height h above the ground has a potential energy mgh relative to the ground. When it falls, h decreases and so the potential energy decreases. But the velocity of the object increases from its initial zero value and so it gains kinetic energy. With the object just on the point of hitting the ground, the potential energy has become zero and the kinetic energy a maximum. At any point in the fall, the sum of the potential energy and the kinetic energy is a constant (Figure 2.14). Thus, when it is on the point of hitting the ground, the

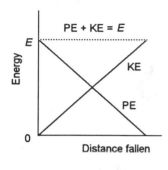

Figure 2.14   *PE + KE = E*

potential energy it has lost is equal to the gain it has made in kinetic energy.

Example

An object of mass 20 kg is allowed to fall freely from rest to the ground through a vertical height of 2.0 m. Calculate its potential and kinetic energies when the body (a) is 1.0 m above the ground, (b) hits the ground. Take g as 9.8 m/s².

(a) The potential energy at height 2.0 m is $mgh = 20 \times 9.8 \times 2.0 = 392$ J and initially there is no kinetic energy. The total mechanical energy initially is thus 392 J. After falling to height 1.0 m it will have lost half of this potential energy, i.e. 196 J, and will have a potential energy of 196 J. The potential energy lost will have been transformed into kinetic energy and so the kinetic energy gained will be 196 J.

(b) When the object hits the ground it loses all its potential energy. At the point of impact this will all have been transformed into kinetic energy and so the kinetic energy is 196 J.

Example

An object of mass 2 kg slides from rest down a smooth plane inclined at 30° to the horizontal. What will its velocity be when it has slid 2 m down the plane? Take g as 9.8 m/s².

The object is losing potential energy and gaining kinetic energy and since the plane is stated as being smooth we can assume that there are no frictional effects and so no energy required to overcome friction and be dissipated as heat. Thus, for the sum of the potential energy and kinetic energy to be a constant we have PE lost in sliding down the plane = KE gained.

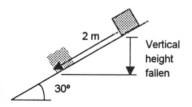

Figure 2.15   *Example*

The vertical distance through which the object has fallen (Figure 2.15) is 2 sin 30° and so the potential energy lost is $mgh = 2 \times 9.8 \times 2 \sin 30°$. The gain in kinetic energy is ½mv^2, where v is the velocity after the object has slid through 2 m. Thus we have:

$$2 \times 9.8 \times 2 \sin 30° = \tfrac{1}{2} \times 2 \times v^2$$

and so $v = 4.4$ m/s.

Example

A car of mass 850 kg stands on an incline of 5°. If the hand brake is released, what will be the velocity of the car after travelling 100 m down the incline if the resistances to motion total 60 N?

Energy is conserved and so we have loss in PE = gain in KE + energy to overcome resistive forces. In travelling 100 m the car 'falls' through a vertical height of 100 sin 5°. Hence the loss in potential energy is:

$$PE = mgh = 850 \times 9.8 \times 100 \sin 5° = 7.26 \times 10^4 \text{ J}$$

The work W done against friction during the motion is:

$$W = Fs = 60 \times 100 = 6000 \text{ J}$$

The kinetic energy gained by the car at the bottom of the incline is the potential energy given up minus the work done against friction and so is KE $= 72.6 \times 10^3 - 6.0 \times 10^3 = 66.6 \times 10^3$ J. But:

$$KE = \tfrac{1}{2}mv^2 = \tfrac{1}{2} \times 850v^2$$

Hence $v = 12.5$ m/s.

Example

A simple pendulum has a bob of mass 1 kg suspended by a light string of negligible mass and length 1 m (Figure 2.16). If the pendulum swings freely through an angle of ±30° from the vertical, find the speed at its lowest position if all resistances to motion can be ignored. Take g as 9.8 m/s².

At its heighest position the bob is stationary and has no kinetic energy but more potential energy than when it is swinging through its lowest position. Hence:

KE at lowest position = loss in PE in moving from highest to lowest position

$$\tfrac{1}{2} \times 1 \times v^2 = 1 \times 9.8 \times (1 - 1 \cos 30°)$$

and so $v = 1.6$ m/s.

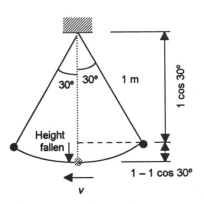

Figure 2.16 *Example*

2.7 Power

Power is the rate at which energy is transferred, i.e.

$$\text{power} = \frac{\text{energy transferred}}{\text{time taken}}$$

Hence, where the energy is transferred as a result of work, the power is the rate of doing work. When the energy is transferred as a result of heat, the power is the rate of heat transfer. When the unit of the energy transferred is the joule (J) and the time seconds (s), then power has the unit of J/s. This unit is given a special name and symbol, the watt (W). Thus 1 W is a rate of energy transfer of 1 J per second.

For an object on which work is done for a time t and results in a displacement s in that time, s/t is the average velocity v of the object over that time period and so:

$$\text{power} = \frac{W}{t} = \frac{Fs}{t} = F \times \frac{s}{t} = Fv$$

Example

In an experiment to measure his/her own power, a student of mass 60 kg raced up a flight of fifty steps, each step being 0.2 m high and found that it took 20 s. What is the power?

The gain in potential energy of the student in moving his/her mass through a vertical height of 50×0.2 m is $= mgh = 60 \times 9.8 \times 50 \times 0.2 = 5880$ J. Hence the power is $5880/20 = 294$ W.

Example

The locomotive of a train exerts a constant force of 120 kN on a train while pulling it at 40 km/h along a level track. What is the power?

Power $= Fv = 120 \times 10^3 \times 40 \times 10^3/3600 = 1.3 \times 10^6$ W $= 1.3$ MW.

Example

Calculate the power a car must exert if it is to maintain a constant velocity of 30 m/s when the resisting forces amount to 4.0 kN.

Power $= Fv = 4.0 \times 10^3 \times 30 = 120 \times 10^3$ W $= 120$ kW.

Example

What power will be required for a pump to extract water from a mine at 5 m³/s and pump it through a vertical height of 20 m. Water has a density of 1000 kg/m³. Take g as 9.8 m/s².

A volume of 5 m³ of water has a mass of 5000 kg and a weight of $5000g$ N. This weight of water has to be moved through a distance of 20 m in 1 s, i.e. an average velocity of 20 m/s. Hence:

Power $= Fv = 5000 \times 9.8 \times 20 = 980 \times 10^3$ W $= 980$ kW.

2.8 Machines

A *machine* can be defined as a mechanical device which enables an effort force to be magnified or reduced or applied in a more convenient line of action, or the displacement of the point of application of a force to be magnified or reduced. The term *effort* is used for the input force, the term *load* for the output force.

The *force ratio* or *mechanical advantage* MA is defined as:

$$MA = \frac{load}{effort}$$

The *movement ratio* or *velocity ratio* VR is defined as:

$$VR = \frac{distance\ moved\ by\ effort}{distance\ moved\ by\ load}$$

For an *ideal machine* where there are no frictional forces, mechanical energy is conserved and so the work done by the effort must equal the work done by the load:

$$effort \times distance\ moved\ by\ effort = load \times distance\ moved\ by\ load$$

Rearranging this gives:

$$\frac{load}{effort} = \frac{distance\ moved\ by\ effort}{distance\ moved\ by\ load}$$

$$MA = VR$$

2.8.1 Efficiency

For a non-ideal machine there are some energy losses due to friction. The *efficiency* of a machine may be defined as a fraction by:

$$efficiency = \frac{energy\ transferred\ from\ machine\ to\ the\ load}{energy\ transferred\ from\ effort\ to\ the\ machine}$$

i.e. the useful energy output/(energy input). Thus:

$$efficiency = \frac{load \times distance\ moved\ by\ load}{effort \times distance\ moved\ by\ effort} = \frac{MA}{VR}$$

If we consider the energy input and output per second, i.e. the powers, then we can also write:

$$efficiency = \frac{power\ output}{power\ input}$$

The above definitions give the efficiency as a fraction. It is, however, often written as a percentage; this just involves multiplying the fraction by 100.

Example

A machine is able to lift a mass of 50 kg through a vertical height of 5 m when it is supplied with an energy input of 4000 J. What is the efficiency of the machine?

The useful energy output from the machine results in a gain in potential energy $mgh = 50 \times 9.8 \times 5 = 2450$ J. Thus the efficiency, expressed as a percentage is $(2450/4000) \times 100 = 61.25\%$.

Example

The input power to a machine is 75 kW and the output power is 60 kW. What is its efficiency?

Efficiency $= (60/75) \times 100 = 80\%$.

2.8.2 Examples of machines

Levers are simple machines. There are three basic types of lever. One has the effort and load on opposite sides of the fulcrum and is a force amplifier (Figure 2.17(a)), e.g. scissors and pliers. Another has the effort and load on the same side of the fulcrum but with the effort at the greater distance and is a force amplifier (Figure 2.17(b)), e.g. a wheelbarrow. The third has the effort and load on the same side of the fulcrum but with the load at the greater distance and is a movement amplifier (Figure 2.17(c)), e.g. the human forearm. If moments are taken about the fulcrum then, whatever the type of lever:

effort × (effort to fulcrum distance)
= load × (load to fulcrum distance)

Thus:

$$MA = \frac{load}{effort} = \frac{effort \ to \ fulcrum \ distance}{load \ to \ fulcrum \ distance}$$

A single *pulley wheel* (Figure 2.18) is a machine which changes the line of action of the effort but not the size of the force, the load equalling the effort. Pulleys with more than one pulley wheel change not only the line of action but also the size of the force. For the two pulley system shown in Figure 2.19, at equilibrium we have equilibrium of each part of the system and so for the effort we have $E = T$, where T is the tension in the pulley rope. For the load we have, if we neglect the mass of the pulley, $L = 2T$ and so MA $= L/E = 2$. If the mass of the pulley m is not neglected then $L + mg = 2T$ and so MA $= (2T - mg)/T$. In general, when the mass of the pulleys is neglected, MA $= n$, where n is the number of pulley wheels.

An *inclined plane* is a machine. For a smooth inclined plane at an angle θ to the horizontal (Figure 2.20), the effort required to push a load L up the plane is the component of the load down the plane and so $E = L \sin \theta$. Thus MA $= 1/\sin \theta$. The wedge and the screw thread are examples of inclined planes. For a screw thread, the angle of the incline can be obtained by considering the unwinding of one turn of the thread. If p is the pitch of the thread, i.e. the amount by which the screw is raised by one rotation, and r the radius of the cylinder on which the thread is wound, then $p/2\pi r = \tan \theta$. since for small angles $\tan \theta$ is

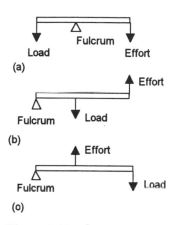

(a)

(b)

(c)

Figure 2.17 *Levers*

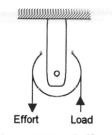

Effort   Load

Figure 2.18 *Pulley*

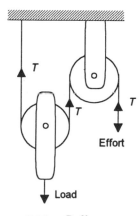

Figure 2.19 *Pulley system*

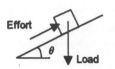

Figure 2.20 *Inclined plane*

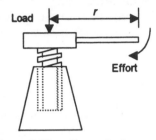

Figure 2.21 *Screw jack*

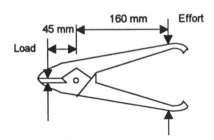

Figure 2.22 *Example*

approximately the same as sin θ, MA = $2\pi r/p$. The *screw jack* is a simple lifting machine (Figure 2.21) based on the screw. The load is moved vertically by the thread through a distance p when the effort is moved through one complete circle, i.e. $2\pi r$. Hence VR = $2\pi r/p$. For ideal conditions, MA = VR = $2\pi r/p$.

See the discussion in the Chapter 30 for a more detailed discussion of machines.

Example

What effort must be applied to the handle of a screw jack to lift a car when the load applied to the jack is 5 kN? The screw has a pitch of 12 mm and the jack handle has a radius of rotation of 400 mm.

MA = load/effort = $2\pi r/p$ and so:

$$\text{effort} = \frac{5 \times 10^3 \times 12 \times 10^{-3}}{2\pi \times 400 \times 10^{-3}} = 23.9 \text{ N}$$

Example

For the tin snips shown in Figure 2.22, what effort will need to be applied if the force required to cut the sheet in the snips is 1.4 kN?

MA = load/effort = (effort to fulcrum distance)/(load to fulcrum distance) and so:

$$\text{effort} = \frac{1.4 \times 10^3 \times 45 \times 10^{-3}}{160 \times 10^{-3}} = 394 \text{ N}$$

Activities

1 Determine the power that can be developed by a student by making him/her run up a flight of steps, so gaining potential energy, and measure time taken to gain this potential energy.
2 Determine the form of lever action and MA for (a) a pair of scissors, (b) a wheelbarrow.
3 Determine the MA and VR for a screw jack.

Problems

1 List the main energy transformations involved in: (a) a car being driven along a level road, (b) a ball rolling down a hill, (c) a firework rocket is ignited and soars upwards, (d) a stretched rubber band snapping, (e) a hydroelectric system where water from a high reservoir is to fall and drive a turbine and hence produce electricity.
2 Calculate the work done when a hydraulic hoist is used to lift a car of mass 1000 kg through a vertical height of 2 m.
3 Calculate the work done in pushing a broken-down car a distance of 20 m if a constant force of 300 N is required to keep it moving at a steady pace.

4 An object resting on a horizontal surface is moved 2.5 m along the plane under the action of a horizontal force of 20 N. Calculate the work done by the force.

5 A hoist raises 15 crates, each of mass 250 kg, a vertical distance of 3.0 m. Determine the work done by the hoist.

6 An object resting on a horizontal surface is moved 2.5 m along the plane under the action of a force of 20 N which is at an angle of 60° to the surface. Calculate the work done by the force.

7 A force of 200 N acts on a body. If the work done by the force is 30 kJ, through what distance and in what direction will the body move?

8 A spring is extended by 200 mm by a force which increases uniformly from zero to 500 N. Calculate the work done.

9 The following data gives the force F acting on a body in the direction of its motion when it has moved through a number of distances s from its initial position. Determine the work done when the body is moved from zero to 50 m.

F in N	100	180	225	240	200	150
s in m	0	10	20	30	40	50

10 A load with a mass of 2000 kg is hauled up an incline of 1 in 100. The frictional resistance opposing the motion up the plane is constant at 300 N. What energy is needed to get the load a distance of 4.0 m up the slope?

11 A pump delivers 2.5 m³ of water through a vertical height of 60 m. What is the work done? The density of water is 1000 kg/m³.

12 A block of mass 2 kg slides at a constant speed a distance of 0.8 m down a plane which is inclined at 30° to the horizontal. Determine the work done by the weight of the block.

13 Calculate the potential energy acquired by an object of mass 3.0 kg when it is lifted through a vertical distance of 1.4 m.

14 A car of mass 1200 kg starts from rest and reaches a speed of 20 m/s after travelling 250 m along a straight road. If the driving force is constant, what is its value?

15 If a drag racing car with a mass of 1000 kg can accelerate from rest to a speed of 120 m/s in 400 m what is (a) the work done by the driving force, (b) the value of the driving force if it is assumed to be constant?

16 Determine (a) the kinetic energy and (b) the velocity of an object of mass 10 kg when it has moved from rest under the action of a force of 20 N a distance of 4.0 m.

17 Determine the amount of kinetic energy lost when a car of mass 900 kg slows from 70 km/h to 50 km/h.

18 What is the constant force acting on a body of mass 8 kg if its velocity increases from 4 m/s to 6 m/s while it moves through a distance in the direction of the force of 5 m?

19 What distance will be needed to bring an object of mass 10 kg moving at a velocity of 5 m/s to rest if a constant force of 100 N is applied to the body in the opposite direction to its motion?

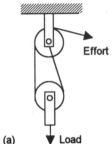

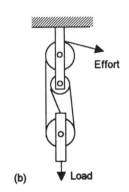

(a) Load

(b) Load

Effort

Effort

Figure 2.23 *Problem 30*

20 A car of mass 1200 kg travelling at 20 m/s brakes. The wheels lock and the car slides 30 m before coming to rest. What is the average frictional force bringing the car to rest?

21 Two masses of 2 kg and 4 kg are connected by a light string passing over a light, frictionless pulley. If the system is released from rest, determine the velocity of the 4 kg mass when it has descended a distance of 1.4 m.

22 A cutting tool operates against a constant resistive force of 2000 N. If the tool moves through a distance of 150 mm in 6 s, what is the power used?

23 A train moving along a level track has a maximum speed of 50 m/s. Determine the maximum power of the train engine if the total resistance to motion is 30 kN.

24 Determine the power required for a car to be driven along a straight road at a constant speed of 25 m/s if the resistance to motion is constant at 960 N.

25 A car is found to have a maximum speed of 140 km/h along a level road when the engine is developing a power of 50 kW. What is the resistance to motion?

26 The input power to a machine is 50 kW and the output power is 45 kW. What is its efficiency?

27 A hydroelectric power station has water falling from a reservoir a height of 100 m above the turbine. If the water flow is 0.25 m^3/s and the overall efficiency of the turbine plant is 60%, what will be the electrical power output? Take g as 9.8 m/s^2 and 1 m^3 of water to have a mass of 1000 kg.

28 A screw jack has a pitch of 6 mm and a handle giving a radius of turning of 600 mm. What effort has to be applied to the handle if the screw jack is to lift a load of 200 kg?

29 For a two-wheel pulley system, of the type shown in Figure 2.19, what effort is required to lift a load of 50 kg and what will be the tension in the pulley rope? The lower pulley has a mass of 2.0 kg.

30 Determine the mechanical advantage of the pulley systems shown in Figure 2.23 if the weight of the pulley can be neglected.

31 What is the force required to lift the handles of a loaded wheelbarrow, total mass 80 kg, if the distance of the centre of gravity, i.e. the distance at which the entire load can be considered to act, from the wheel axle is 300 mm and from the effort 1000 mm?

32 A machine lifts a load of 150 kg through a distance of 0.17 m when the effort of 200 N moves through a distance of 1.7 m. What is VR, MA and the efficiency?

33 A machine has an efficiency of 60%. If an effort of 200 N raises a load of 1000 N, what is the velocity ratio?

34 A force of 4 kN is used to jack up the rear axle of a car by 25 mm. If this requires the operator of the jack to exert a force of 250 N through a distance of 0.6 m, what is (a) the velocity ratio, (b) the mechanical advantage and (c) the efficiency of the jack?

3 Heat

3.1 Introduction

Heat is defined as the transfer of energy that occurs between two systems when there is a temperature difference between them. As with other forms of energy, the SI unit for heat is the joule (J). This chapter is a basic introduction to the effects of heat transfer, namely temperature changes, changes of state, expansion and pressure changes.

3.1.1 Temperature scales

Temperatures are expressed on the Celsius scale or the Kelvin scale. The *Celsius scale* has the melting point of ice as 0°C and the boiling point of water as 100°C. Temperatures on the *Kelvin scale* have the same size degree as the Celsius scale but the melting point of ice is 273.15 K and the boiling point of water is 373.15 K. Thus temperatures on the Kelvin scale equal temperatures on the Celsius scale plus 273.15. Temperatures on the Kelvin scale are usually denoted by the symbol T.

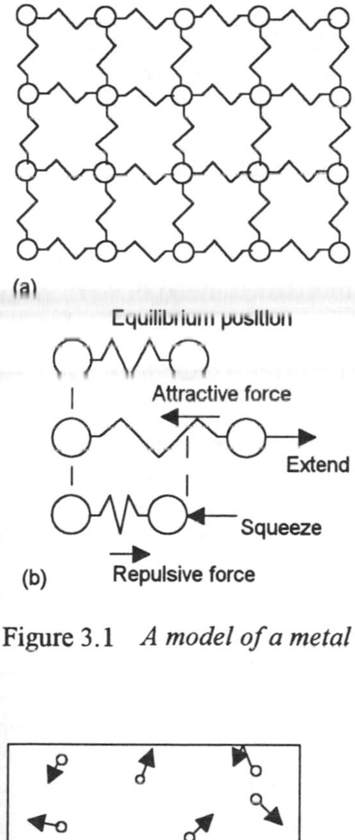

3.1.2 Basic structure of solids, liquids and gases

A simple model of a solid is that of closely packed spheres (Figure 3.1(a)), each sphere representing an atom. Each sphere is tethered to its neighbours by springs, these representing the inter atomic bonds. When forces are applied to stretch a material, then the springs are stretched and exert attractive forces pulling the material back to its original position; when forces are applied to compress the material then the springs are compressed and exert repulsive forces which push the atoms back towards their original positions. These forces thus keep a solid in a fixed shape. If a solid is heated, the heat causes the spheres to vibrate and the higher the temperature the greater the vibration. At a high enough temperature the vibration is sufficiently vigorous for the spheres to break out of their close-packed array and the solid turns into a liquid. The spheres are still, however, close enough to each other for there to be weak forces which are sufficient to hold the spheres within the confines of a drop of liquid. Further heating results in an increase in temperature and the spheres moving about faster within the confines of the liquid as a result of gaining kinetic energy. At a higher enough temperature they have so much energy that they break free and the result is a gas (Figure 3.2). In the gas the spheres can be considered to have moved so far apart they there are no forces between them. They thus move around within the confines of a container, bouncing off the walls of the container. It is this bouncing off the walls which gives rise to the pressure on the walls.

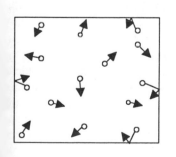

(a)

Equilibrium position

Attractive force

Extend

Squeeze

Repulsive force

(b)

Figure 3.1   *A model of a metal*

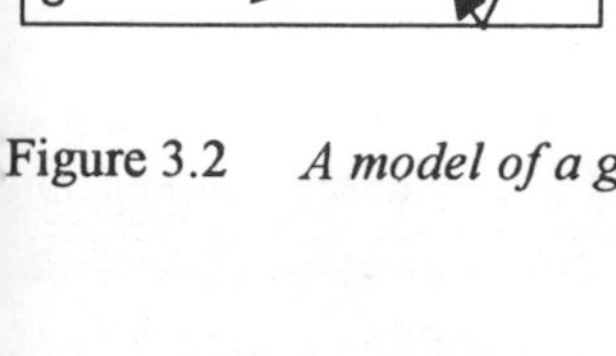

Figure 3.2   *A model of a gas*

3.2 Heat capacity

The *heat capacity C* of a body is the quantity of heat required to raise its temperature by 1 K and has the SI unit J/K. Thus the heat Q required to change the temperature by ΔT K is:

$$Q = C\Delta T$$

The *specific heat capacity c* is the heat required to raise the temperature of 1 kg of a body by 1 K and has the unit J kg^{-1} K^{-1}. Thus the heat Q required to change the temperature of m kg of a body by ΔT K is:

$$Q = mc\Delta T$$

Typical values of specific heat capacities are 4200 J kg^{-1} K^{-1} for water, 950 J kg^{-1} K^{-1} for aluminium, 500 J kg^{-1} K^{-1} for iron and 390 J kg^{-1} K^{-1} for copper.

Example

How much heat will an iron casting of mass 10 kg have to lose to drop in temperature from 200°C to 20°C? The specific heat capacity of the iron is 480 J kg^{-1} K^{-1}.
$$Q = mc\Delta T = 10 \times 480 \times (200 - 20) = 8.64 \times 10^5 \text{ J}$$

Example

What will be the rise in temperature of 0.5 kg of water in a container of capacity 50 J/K when 5 kJ of heat is transferred to the system? The specific latent heat of water is 4200 J kg^{-1} K^{-1}.

The heat transfer will increase the temperature of both the water and its container. Thus:

$$5000 = 0.5 \times 4200 \times \Delta T + 50 \times \Delta T$$

Hence $\Delta T = 2.3$ K.

3.2.1 Latent heat

Heat transfer to a substance which results in a change of temperature is said to be *sensible heat*, such a transfer increasing the energy of its atoms or molecules. In some circumstances, heat transfer to a body may result in no temperature change but a structural change such as from solid to liquid or liquid to vapour or vice versa. Such a change is said to be a change of *phase*. A *phase* is defined as a region in a material which has the same composition and structure throughout. Liquid water and ice have different structures and thus when ice melts there is a change of phase. The energy gained by the body in such a situation is used to change bonds between atoms or molecules. Heat which results in no temperature change is called *latent heat*.

The *specific latent heat L* of a material is defined as the amount of heat needed to change the phase of 1 kg of the material without any change in temperature and has the unit J/kg. Thus the heat transfer Q needed to change the phase of m kg of a material is:

$$Q = mL$$

The term *specific latent heat of fusion* is used when the change of phase is from solid to liquid and the term *specific latent heat of vaporisation* for the change from liquid to vapour. Typical values are: specific latent heat of fusion for water 335 kJ/kg, aluminium 387 kJ/kg, and iron 268 kJ/kg; specific latent heat of vaporisation for water 2257 kJ/kg, ethyl alcohol 857 kJ/kg.

Example

How much heat is required to change 1.2 kg of water at 20°C to steam at 100°C? The specific heat capacity of water in this temperature range is 4200 J kg^{-1} K^{-1} and the specific latent heat at 100°C for liquid to vapour is 2257 kJ/kg.

To raise the water from 20°C to 100°C: $Q = mc\Delta t = 1.2 \times 4200 \times (100 - 20) = 4.0 \times 10^5$ J. To change water from liquid to vapour at 100°C: $Q = mL = 1.2 \times 2257 \times 10^3 = 27.1 \times 10^5$ J. The total heat required is thus $4.0 \times 10^5 + 27.1 \times 10^5 = 31.1 \times 10^5$ J.

3.3 Expansion

Solids expand when their temperature is increased. The amount by which a length of solid expands depends on the change in temperature, the original length of the material and the material concerned. The *coefficient of linear expansion a* (or *linear expansivity*) is defined as:

$$a = \frac{\text{change in length}}{\text{original length} \times \text{change in temperature}}$$

The coefficient has the unit of /°C or /K. If L_θ is the length at temperature θ and L_0 the length at temperature 0°C, then:

$$a = \frac{L_\theta - L_0}{L_0\theta}$$

and so:

$$L_\theta = L_0(1 + a\theta)$$

Typical values of the coefficient are aluminium 0.000 023 /K, copper 0.000 017 /K, mild steel 0.000 011 /K and soda glass 0.000 009 /K.

Some practical implications of thermal expansion are:

1   Overhead telephone and electrical cables are hung so that they are slack in summer and so the contraction that occurs in winter when the temperature drops does not result in the wires breaking.

2 Steel bridges expand when the temperature rises and so the ends are often supported on rollers to allow them to expand and contract, freely.

3 In fitting a metal collar onto a shaft, the collar is often heated so that it expands and can be easily slid onto the shaft; when it cools it contracts and binds firmly to the shaft.

Example

A bar of copper has a length of 300 mm at 0°C. By how much will it expand when heated to 50°C? The coefficient of linear expansion for the copper is 0.000 017 /K.

Change in length = coefficient of linear expansion × original length × change in temperature = 0.000 017 × 300 × 50 = 0.255 mm.

Example

A copper telephone cable is to be fixed between two posts 40 m apart when the temperature is 25°C. How much slack should the engineers allow if the cable is not to become taut before the temperature reaches −10°C. The coefficient of linear expansion for the copper is 0.000 017 /K.

Slack = change in length = coefficient of linear expansion × original length × change in temperature = 0.000 017 × 40 × 35 = 0.0238 m.

3.3.1 Area and volume expansion of solids

Consider the expansion of the surface of a square sheet of material of side L_0 when the temperature increases by θ. Each side will expand to a length $L_\theta = L_0(1 + a\theta)$ and thus the area at θ is $A_\theta = L_\theta^2 = [L_0(1 + a\theta)]^2$. But the initial area $A_0 = L_0^2$ and so:

$$A_\theta = A_0(1 + 2a\theta + a^2\theta^2)$$

Since a is very small we can neglect the term involving a^2 and so:

$$A_\theta = A_0(1 + 2a\theta)$$

We thus have an area coefficient of expansion for a solid which is twice the linear coefficient.

Consider the expansion of the volume of a cube of material of side L_0 when the temperature increases by θ. Each side will expand to a length $L_\theta = L_0(1 + a\theta)$ and thus the volume at θ is $V_\theta = L_\theta^3 = [L_0(1 + a\theta)]^3$. But the initial volume $V_0 = L_0^3$ and so:

$$V_\theta = V_0(1 + 3a\theta + 3a^2\theta^2 + a^3\theta^3)$$

Since a is very small we can neglect the term involving a^2 and a^3 and so:

$$A_\theta = A_0(1 + 3a\theta)$$

We thus have a volume coefficient of expansion for a solid which is three times the linear coefficient.

Example

A block of metal has a volume of 5000 mm³ at 20°C. What will be the change in volume when the temperature rises to 100°C if the coefficient of linear expansion is 0.000 023 /K?

Change in volume = 3 × linear coefficient × original volume × change in temperature = 3 × 0.000 023 × 5000 × 100 = 34.5 mm³.

3.3.2 Expansion of liquids

When liquids expand as a result of an increase in temperature we can define a real volume coefficient expansion γ as:

$$\gamma = \frac{\text{change in volume}}{\text{original volume} \times \text{change in temperature}}$$

The coefficient has the unit of /°C or /K. If V_θ is the volume at temperature θ and V_0 the volume at temperature 0°C, then:

$$\gamma = \frac{V_\theta - V_0}{V_0\theta}$$

$$V_\theta = V_0(1 + \gamma\theta)$$

The term real is used for the coefficient when referring to the actual volume change occurring for the liquid. When the liquid is in a container, not only will the liquid expand but so will the container and the apparent volume of the liquid indicated by its level in the container is an underestimate of the true volume. We can thus define an apparent coefficient β as:

$$\beta = \frac{\text{apparent change in volume}}{\text{original volume} \times \text{change in temperature}}$$

The change in volume of the container will be $3a\theta V_0$, where a is the coefficient of linear expansion of the solid material used for the container. Hence the real change in volume of the liquid is the apparent change in volume plus $3a\theta V_0$ and so:

$$\gamma V_0\theta = \beta V_0\theta + 3a\theta V_0$$

$$\gamma = \beta + 3a$$

The real coefficient is thus the sum of the apparent coefficient of the liquid and the volume coefficient of the container material.

Example

A glass container will hold 1000 cm³ at 20°C. How much water will be needed to fill it when the temperature is 80°C? The real volume coefficient of expansion of water is 0.000 21 /K and the coefficient of linear expansion of the glass is 0.000 009 /K.

The apparent coefficient of expansion is 0.000 21 – 3 × 0.000 009 = 0.000 183 /K. Hence the apparent change in volume of the water = 0.000 183 × 1000 × 60 = 10.98 cm³ and so the volume required is 1010.98 cm³.

3.4 Gas laws

Gases are fluids, differing from liquids in that while liquids are practically incompressible and possess a definite volume, gases are readily compressible and can change their volume. The following are three laws which are found to be reasonably obeyed by the so-called permanent gases, e.g. oxygen and nitrogen, at temperatures in region of room temperature. A gas which is considered to obey these laws exactly is called an *ideal gas*.

1 *Boyle's law*

The volume V of a fixed mass of gas is inversely proportional to its pressure p if the temperature remains constant:

$$p \propto 1/V \text{ or } pV = \text{a constant}$$

Thus if the initial pressure is p_1 and the volume V_1 and we then change the pressure to p_2 to give volume V_2, without changing the temperature:

$$p_1 V_1 = p_2 V_2$$

Example

A gas occupies a volume of 0.10 m³ at a pressure of 1.5 MPa. What will be the gas pressure if the gas is allowed to expand to a volume of 0.15 m³ and the temperature remains constant?

Applying Boyle's law gives:

$$p_1 V_1 = p_2 V_2$$

$$1.5 \times 0.10 = p_2 \times 0.15$$

Hence $p_2 = 1.0$ MPa.

2 *Charles's law*

The volume V of a fixed mass of gas at constant pressure is proportional to the temperature T when the temperature is on the kelvin scale (K):

$$\frac{V}{T} = \text{a constant}$$

Thus if initially we have a volume V_1 at a temperature T_1 and we then, without changing the pressure, change the temperature to T_2 to give a volume V_2:

$$\frac{V_1}{T_1} = \frac{V_2}{T_2}$$

Example

A gas occupies a volume of 0.12 m³ at a temperature of 20°C. What will be its volume at 100°C if the gas is allowed to expand to maintain a constant pressure?

Using Charles's law and $V_1/T_1 = V_2/T_2$, with $T_1 = 20 + 273 = 293$ K and $T_2 = 100 + 273 = 373$ K:

$$\frac{0.12}{293} = \frac{V_2}{273}$$

Hence $V_2 = 0.11$ m³.

3   *Pressure law*
The pressure p for a fixed mass of gas at constant volume is proportional to the temperature T on the absolute or kelvin scale:

$$\frac{p}{T} = \text{a constant}$$

Thus if initially we have a pressure p_1 at a temperature T_1 and we then, without changing the volume, change the temperature to T_2 to give a pressure p_2:

$$\frac{p_1}{T_1} = \frac{p_2}{T_2}$$

Example

A fixed volume of gas is initially at a pressure of 140 kPa at 20°C. What will be its pressure if the temperature is increased to 100°C without the volume changing?

Using the pressure law $p_1/T_1 = p_2/T_2$ with $T_1 = 20 + 273 = 293$ K and $T_2 = 100 + 273 = 373$ K:

$$\frac{140}{293} = \frac{p_2}{373}$$

Hence $p_2 = 178$ kPa.

In addition to the above laws relating pressure, volume and temperature we have a law which applies to mixtures of gases, e.g. air:

4 *Dalton's law of partial pressures*

The total pressure of a mixture of gases occupying a given volume at a particular temperature is equal to the sum of the pressures of each gas when considered separately.

Example

An enclosed sample of air produces a pressure of 1.2 kPa. When water vapour is introduced into the air the total pressure rises to 1.3 kPa. What is the pressure due to just the water vapour at that temperature?

Using Dalton's law of partial pressures:

total pressure = pressure due to air along +
pressure to just the water vapour

Hence the water vapour pressure is 1.3 – 1.2 = 0.1 kPa.

A gas is said to be at *standard temperature and pressure* (STP) when it is at a temperature 0°C, i.e. 273 K, and the normal atmospheric pressure of 101.325 kPa.

3.4.1 Characteristic gas equation

Provided there is no change in the mass of a gas, we can combine Boyle's law, Charles's law and the pressure law to give the general equation:

$$\frac{p_1 V_1}{T_1} = \frac{p_2 V_2}{T_2}$$

$$\frac{pV}{T} = \text{a constant}$$

When 1 kg of gas is considered, the constant is called the *characteristic gas constant* and denoted by R. Its value depends on the gas concerned. For a mass m kg of gas, the constant has the value mR, hence:

$$\frac{pV}{T} = mR$$

The SI unit of R is thus $(\text{Pa} \times \text{m}^3)/(\text{K} \times \text{kg}) = (\text{N/m}^2 \times \text{m}^3)/(\text{K} \times \text{kg}) = (\text{N} \times \text{m})/(\text{K} \times \text{kg}) = \text{J}/(\text{kg} \times \text{K})$. Typical values for the characteristic gas constant R are oxygen 260 J kg^{-1} K^{-1}, hydrogen 4160 J kg^{-1} K^{-1} and air 287 J kg^{-1} K^{-1}.

Example

A gas has a volume of 0.10 m^3 at a pressure of 100 kPa and a temperature of 20°C. What will be the temperature of the gas when it is compressed to a volume of 0.04 m^3 by a pressure of 500 kPa?

For an ideal gas $p_1V_1/T_1 = p_2V_2/T_2$ and so:

$$\frac{100 \times 0.10}{293} = \frac{500 \times 0.04}{T_2}$$

Hence $T_2 = 586$ K $= 313°$C.

Example

What will be the volume of 2.0 kg of air at 20°C and a pressure of 1.1 MPa if the characteristic gas constant for air is 287 J kg⁻¹ K⁻¹?

For an ideal gas $pV/T = mR$ and thus:

$$V = \frac{mRT}{p} = \frac{2.0 \times 287 \times 293}{1.1 \times 10^6} = 0.15 \text{ m}$$

Example

A rigid gas container of internal volume 0.60 m³ contains a gas at 20°C and 250 kPa. If a further 1.5 kg of gas is pumped into the container, what will be the pressure when the temperature is back again at 20°C. Take R for air as 290 J kg⁻¹ K⁻¹.

For an ideal gas $pV/T = mR$ and thus initially we have:

$$m = \frac{pV}{RT} = \frac{250 \times 10^6 \times 0.60}{290 \times 293} = 1.8 \text{ kg}$$

Hence the new mass is $1.8 + 1.5 = 3.3$ kg and thus:

$$p = \frac{mRT}{V} = \frac{3.3 \times 290 \times 293}{0.60} = 467 \times 10^3 \text{ Pa} = 467 \text{ kPa}$$

3.4.2 Kinetic model of an ideal gas

A model of an ideal gas is of molecules, rather like little ball bearings, moving around in an enclosure and bouncing off each other and the container walls (Figure 3.3). The pressure exerted on the container walls is due to the walls being bombarded by the molecules. If we double the volume of a container then, assuming the mean velocity of the molecules does not change, it will take twice as long for molecules to cross from one side of the container to the other and so the number of collisions per second with the wall will be halved. Doubling the volume halves the pressure. We thus have an explanation of Boyle's law.

The molecules are considered to be minute and occupying such a small percentage of the total container volume that each molecule can effectively be considered to have the entire container volume in which to move. Each molecule is considered to move around in the container without its motion being influenced, other than at a point of collision, by the other molecules, i.e. there are no intermolecular forces other than at a point of collision. Thus if we have two gases in a container, the total pressure is due to the sum of the pressures due to each when considered

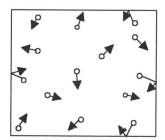

Figure 3.3 *Kinetic model*

to be alone occupying the container; hence we have an explanation for Dalton's law of partial pressures.

The internal energy of the gas, i.e. the mean kinetic energy of the molecules, is taken to be only a function of the temperature. Thus the higher the temperature the higher the mean kinetic energies of the molecules and the faster they move about in the container. For a given molecular mass, the higher the velocities of the molecules the greater the force they exert on the walls when they collide with it and so the greater the pressure. Thus increasing the temperature increases the mean kinetic energy of the molecules and the resulting increased velocity gives an increase in pressure in a fixed volume container.

Activities

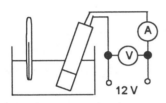

Figure 3.4   *Activity 1*

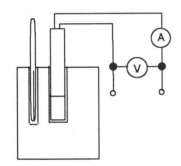

Figure 3.5   *Activity 2*

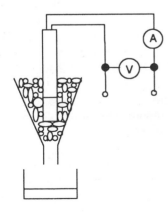

Figure 3.6   *Activity 3*

1 Determine the specific heat capacity of a liquid, e.g. water. A possible method is to use a small 12 V electric immersion heater (Figure 3.4). Thus 1 kg of water might be contained in a metal container and its temperature noted. Then, with the immersion heater wholly immersed in the water, the electricity is switched on for 5 minutes. During that time the heater is used to stir the water and the stirring is continued after the heater is switched off. The highest temperature reached is noted. The power *IV* delivered to the heater can be measured by an ammeter and voltmeter. The heat capacity of the container can reasonably be neglected when such a large mass of water is used.

2 Determine the specific heat capacity of a solid. One method that can be used involves the use of a metal block which has been drilled so that a 12 V electric immersion heater fits into one hole and a thermometer into another (Figure 3.5). The heater is switched on for 5 minutes. The highest temperature reached is noted. The power *IV* delivered to the heater can be measured by an ammeter and voltmeter. Neglect heat losses to the surroundings. To ensure that the thermometer makes good thermal contact with the block, a little oil should be placed in the hole.

3 Determine the specific latent heat of fusion of ice. One method that can be used involves the use of a 12 V electric immersion heater. The heater element is placed in a filter funnel surrounded with closely packed small pieces of ice (Figure 3.6). The heater is switched on for 3 minutes and the mass of water produced measured. The power *IV* delivered to the heater can be measured by an ammeter and voltmeter. Neglect heat gains from the surroundings.

4 Using the apparatus available in your laboratory, verify the gas laws.

5 Investigate the bicycle pump, determining how it functions and the pressures that can be obtained.

6 Using the apparatus available in your laboratory determine the linear coefficient of expansion of a metal.

Problems

1 What will be the temperature change of 3 kg of aluminium when supplied with 5 kJ of heat if the specific heat capacity of aluminium is 950 J kg^{-1} k^{-1}?

2 What will be the rise in temperature of 200 g of water in a container of capacity 40 J/K when 2 kJ of heat is transferred to the system? The specific latent heat of water is 4200 J kg^{-1} K^{-1}.

3 What heat is required to convert 5.0 kg of water at 20°C to steam at 100°C? The specific heat capacity of water in this temperature range is 4.19 kJ kg^{-1} K^{-1} and the specific latent heat at 100°C for liquid to vapour is 2257 kJ/kg.

4 What heat is required to change 10 kg of ice into liquid water at 0°C if the specific latent heat of fusion is 335 kJ/kg.

5 What heat is required to melt a 0.5 kg block of iron if initially it is at 20°C and the melting point of iron is 1200°C? The specific heat capacity of the iron is 0.50 kJ kg^{-1} K^{-1} and the specific latent heat of fusion is 270 kJ/kg.

6 The heating element in an electric kettle is rated as 2 kW. If the water in the kettle is at 100°C, how much water will be converted into steam in 1 minute? The specific latent heat of vaporisation of the water is 2257 kJ/kg.

7 A steel tape-measure is correct at 18°C. If it used to measure a distance of 100 m when the temperature is 8°C, what error will be made? The coefficient of linear expansion is 0.000 012 /K.

8 A steam pipe has a length of 20 m at 20°C. How much expansion will have to be allowed for when the steam increases the temperature of the pipe to 300°C? The coefficient of linear expansion is 0.000 012 /K.

9 A copper sphere has a diameter of 10 mm at 0°C. By how much will its volume increase when the temperature is raised to 100°C? The coefficient of linear expansion is 0.000 017 /K.

10 A plate has an area of 500 mm^2 at 18°C. By how much will this area increase when the temperature is raised to 100°C? The material has a coefficient of linear expansion of 0.000 017 /K.

11 A steel girder has a length of 8.0 m at 25°C. What will be its length at 0° if the coefficient of linear expansion of steel is 0.000 015 /K?

12 A mercury-in-glass thermometer has a distance of 300 mm between the 0°C and the 100°C marks. If the cross-sectional area of the tube is 0.15 mm^2, what will be the total volume of mercury in the thermometer at 0°C. The real volume coefficient of expansion of mercury is 0.000 18 /K and the coefficient of linear expansion of the glass is 0.000 009 /K.

13 Ethyl alcohol has a real coefficient of volume expansion of 0.0011 /K, by how much will a volume of 2000 mm^3 be reduced when the temperature falls from 20°C to 0°C?

14 A steel tank has a volume of 4.0 m^3 and is filled with heating oil at 15°C. How much of the oil will overflow if the temperature increases to 25°C? The steel has a coefficient of linear expansion of 0.000 011 /K and the heating oil a real coefficient of volume expansion of 0.001 20 /K.

15 Initially a gas has a volume of 0.14 m³ and a pressure of 300 kPa. What will be its volume when the pressure becomes 60 kPa if the temperature remains unchanged?

16 A gas with a volume of 2 m³ is compressed from a pressure of 100 kPa to a pressure of 500 kPa. If the temperature remains unchanged, what is the resulting volume?

17 A gas has a volume of 0.010 m³ at 18°C. What will be its volume at 85°C if the pressure acting on the gas remains unchanged?

18 A gas has a volume of 0.40 m³ at 10°C and is heated to a temperature of 120°C. What will be its volume if the pressure remains the same?

19 A gas has a volume of 0.100 m³ at a temperature of 25°C and a pressure of 140 kPa. What will be its volume when it is at a pressure of 700 kPa and a temperature of 60°C?

20 A gas cylinder contains 0.11 m³ of gas at an absolute pressure of 1000 kPa and a temperature of 15°C. What will be the volume of the gas at the atmospheric pressure of 101 kPa and a room temperature of 25°C?

21 A gas has a volume of 0.4 m³ at a pressure of 90 kPa and a temperature of 30°C. What will be its volume at s.t.p?

22 A gas with a volume of 0.40 m³ at s.t.p. is heated until it occupies a volume of 0.46 m³ at a pressure of 115 kPa. What will be its temperature?

23 A gas at a pressure of 250 kPa and temperature 20°C has a volume of 0.050 m³. What will be its volume at a pressure of 400 kPa and a temperature of 100°C?

24 What is the mass of a gas at a pressure of 500 kPa and a temperature of 50°C if it occupies a volume of 0.10 m³. The gas has a characteristic gas constant of 189 J kg⁻¹ K⁻¹.

25 A rigid gas container of internal volume 1.2 m³ holds 1.8 kg of a gas at 18°C. What is the pressure of the gas and to what value will it change if a further 0.8 kg of gas are added at the same temperature? The gas has a characteristic gas constant of 287 J kg⁻¹ K⁻¹.

26 What is the mass of a gas at a pressure of 350 kPa and a temperature of 35°C if it occupies a volume of 0.03 m³. The gas has a characteristic gas constant of 290 J kg⁻¹ K⁻¹.

27 A rigid container of internal volume 0.85 m³ contains a gas at a pressure of 275 kPa and temperature 15°C. What will be the pressure of the gas in the container if an additional 1.7 kg of the gas is pumped into the container at the same temperature? The gas has a characteristic gas constant of 290 J kg⁻¹ K⁻¹.

4 Light and Sound

4.1 Introduction

Light and sound are both examples of where energy is transmitted from one point to another by means of wave motion. Water waves are another example. With water waves, though waves move across the surface of the water and have a velocity, a piece of wood floating on the surface does not travel with the wave but bobs up and down as the waves pass it. There is thus a clear difference between the motion of the waves and the motion of objects affected by the waves.

In general we can describe the motion of a point in a medium through which a wave passes by a displacement–time graph (Figure 4.1). The displacement waveform oscillates from positive to negative values in a regular, periodic manner. One complete sequence of such an oscillation is called a *cycle*. The time T taken for one complete cycle is called the *periodic time* and the number of cycles occurring per second is called the *frequency f*. Thus $f = 1/T$. The unit of frequency is the hertz (Hz), 1 Hz being 1 cycle per second.

Now consider a snapshot of the disturbances in the medium through which a wave is travelling, i.e. a displacement–distance graph (Figure 4.2). The term *wavelength λ* is used for the distance between any two successive crests or any two successive troughs. If we consider a source beginning to emit waves of frequency *f*, then after a time *t* it will have emitted *ft* waves. The velocity of a wave *v* is the distance travelled by a particular crest or trough per second. Thus the first wave has moved a distance *ftλ* in a time *t*. Hence the velocity *v* is *ftλ*/*t*:

$$v = f\lambda$$

This chapter is about light and sound waves, their form and their basic properties.

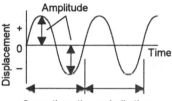

Figure 4.1   *Displacement–time graph*

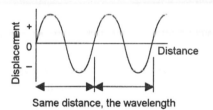

Figure 4.2   *Displacement–distance graph*

4.2 Light waves

Water waves are an example of a *transverse wave* in that the displacement is at right angles to the direction of propagation of the wave (Figure 4.3). Light is another example of a transverse wave motion. It is just one section of a group of waves termed *electromagnetic waves*; this including radio waves, infrared waves, ultraviolet waves and X-rays. All these waves have the same speed in a vacuum, namely 3×10^8 m/s, and all can be described in terms of 'displacements' which are oscillations of electric and magnetic fields. Figure 4.4 shows the range of electromagnetic waves, this being termed the *electromagnetic spectrum*.

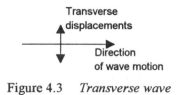

Figure 4.3   *Transverse wave*

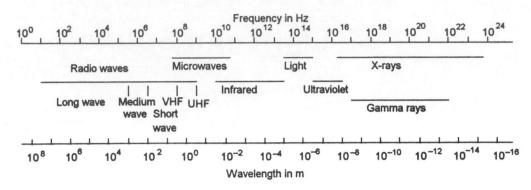

Figure 4.4 *The electromagnetic spectrum*

The group of waves known as light have a range of wavelengths from violet with wavelength 0.4 μm, through blue, to green, to yellow, to orange, to red with wavelength 0.7 μm.

Because of the very small wavelength of light waves when compared with the size of everyday objects, much of the behaviour of light waves can be explained in terms of light travelling in straight lines with straight line beams of light being referred to as rays of light.

4.2.1 Reflection of light

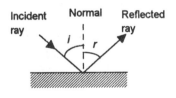

Figure 4.5 *Reflection*

The laws of reflections (Figure 4.5) can be stated as:

1 The incident ray, the normal to the surface at the point where the ray meets the surface, and the reflected ray all lie in the same plane.

2 The angle of reflection r, i.e. the angle between the reflected ray and the normal, equals the angle of incidence i, i.e. the angle between the incident ray and the normal.

With the reflection of light by a plane mirror, the eye sees an object by means of rays that have been bent by the mirror and as we think of light only travelling in straight lines, the object, or rather its image, appears to be behind the mirror (Figure 4.6). Such an image is called a *virtual image* because no light actually comes directly from the image, it only appears to. Because the angle of reflection equals the angle of reflection, angle ABD equals angle BDC and so we must have AB = BC. The image appears to be as far behind the mirror as the object is in front.

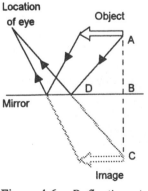

Figure 4.6 *Reflection at a plane mirror*

4.2.2 Refraction of light

When a ray of light passes from one medium into another, for example, air into glass, refraction occurs at the interface (Figure 4.7); the ray of light bends from its straight-line path in passing across the interface. This occurs because the velocity of the light changes on passing from one medium to another. The laws of refraction are:

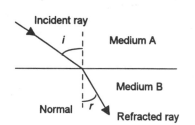

Figure 4.7   *Refraction*

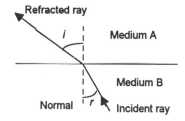

Figure 4.8   *Refraction*

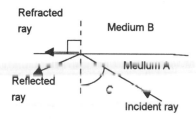

Figure 4.9   *Critical angle*

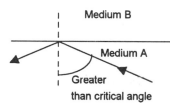

Figure 4.10   *Total internal reflection*

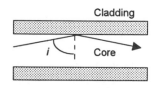

Figure 4.11   *Optical fibre*

1   The incident ray, the normal to the surface at the point where the ray meets the surface, and the refracted ray all lie in the same plane.

2   For a given pair of media, and light of a given frequency, i.e. colour, the sine of the angle of incidence divided by the sine of the angle of refraction is a constant. This is known as Snell's law and the ratio is called the refractive index.

The *refractive index* in going from medium A to medium B, $_An_B$, is given by:

$$\text{refractive index } _An_B = \frac{\sin i}{\sin r}$$

where i is the angle of incidence, i.e. the angle between the incident ray in medium A and the normal, and r the angle of refraction, i.e. the angle between the refracted ray in medium B and the normal.

If we reverse the path and have the ray of light passing from medium B into medium A (Figure 4.8). The same path is followed, but in the reverse direction, the ray now bending away from the normal. Thus we have, when using the same notation for the angles, a refractive index in this case of:

$$_Bn_A = \frac{\sin r}{\sin i} = \frac{1}{_An_B}$$

When we have this condition of the ray bending further away from the normal after refraction, then we can have a particular incident angle which results in the refracted ray of light bending through $90°$ and thus not being transmitted across the interface (Figure 4.9). The angle of incidence in such a case is termed the *critical angle C*. We then have:

$$_Bn_A = \frac{\sin C}{\sin 90°} = \sin C$$

For angles of incidence greater than the critical angle, the ray of light is totally reflected at the interface (Figure 4.10), there being no refracted ray.

As an illustration of the significance of the critical angle in the choice of optical materials, consider the material used for *fibre optics*. Light for which the angle of incidence is greater than the critical angle is transmitted along such a fibre by total internal reflection, none of such light being lost from the fibre by being refracted through the cladding (Figure 4.11).

Because of refraction, the apparent depth of an object immersed in, say, water is less when viewed from above the water surface than the real depth (Figure 4.12). We have $_an_w = \sin i/\sin r$ with $\sin i = AB/BI$ and $\sin r = AB/BO$. Thus:

$$_an_w = \frac{AB/BI}{AB/BO} = \frac{BO}{BI} = \frac{\text{real depth}}{\text{apparent depth}}$$

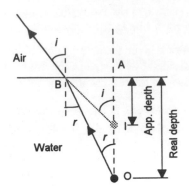

Figure 4.12 *Real and apparent depth*

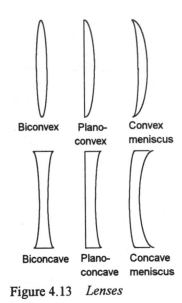

Figure 4.13 *Lenses*

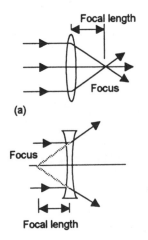

Figure 4.14 *The focus*

Example

Determine the critical angle for the glass/air interface if the glass has a refractive index of 1.5.

The refractive index of the glass is for light going from air to glass. Thus $1/1.5 = \sin C/\sin 90°$ and so $C = 41.8°$.

Example

If the refractive index for air to water is 1.3, what will be the apparent depth of an object 1.0 m below the surface of a still pond?

Refractive index = real depth/apparent depth and so apparent depth = 1.0/1.3 = 0.77 m.

4.2.3 Lenses

Thin lenses have either two spherical surfaces or one spherical surface and one plane surface; Figure 4.13 shows examples of such lenses. Light incident on the air to glass interface of the front of a lens suffers refraction, passes through the glass of the lens and is then refracted again at the glass to air interface before emerging into the air on the other side of the lens.

In the case of convex lenses, parallel light rays converge after passing through the lens to a point called the *focus* (Figure 4.14(a)). In the case of concave lenses, parallel light rays diverge after passing through the lens and appear to come from a point called the *focus* (Figure 4.14(b)). Because light can pass either way through a lens, each lens has two focal points, one on either side of the lens. Since a ray through the centre of a lens is undeviated, we can use these properties of rays to construct diagrams enabling us to determine the location and form of images produced. Figure 4.15 illustrates the use of such ray diagrams.

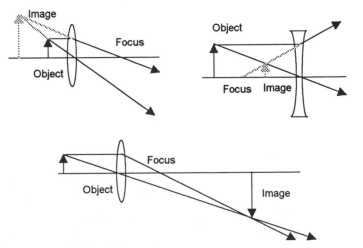

Figure 4.15 *Ray diagrams*

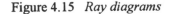

Lenses produce *real images* when the light actually passes through them or *virtual images* when the light only appears to come from the image and no light actually passes through them. The following are the basic characteristics of convex and concave lenses.

1 *Convex lenses*
   When the distance of an object from the lens is less than the focal length, the images produced are erect, magnified and virtual. When the distance of an object from the lens is between the focal length and twice the focal length, the images produced are inverted, magnified and real. When the distance of an object from the lens is greater than twice the focal length, the images are inverted, diminished and real.

2 *Concave lenses*
   For all distances of an object from the lens, the images are erect, diminished and virtual.

We can use the above ray properties, or by considering the refraction of the light at the surfaces, to derive an equation relating the object and image distances. Consider the rays in Figure 4.16 and an object of height O at a distance u from a lens and producing an image of height I a distance v from the lens, the focal length being f. For the two triangles with angle a we have $O/u = I/v$ and for the triangles with angle β we have $O/f = I/(v - f)$. Hence, $f/u = (v - f)/v$. We can rearrange this equation to give:

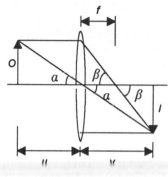

Figure 4.16 *Ray diagram*

$$\frac{1}{f} = \frac{1}{u} + \frac{1}{v}$$

The same equation can be used for convex lenses with virtual images and concave lens if, when numbers are substituted for object and image distances, positive numbers are used for distances to real objects and real images and negative numbers for distances to virtual objects and virtual images. Convex lenses have a positive focal length and concave lenses a negative focal length.

We can also use Figure 4.16 to derive a relationship for the magnification produced. For the two triangles with angle a we have $O/u = I/v$ and so:

$$\text{magnification} = \frac{I}{O} = \frac{v}{u}$$

Example

An object is 20 cm in front of a lens and produces a real image 10 cm on the other side of the lens. What is the focal length of the lens and the magnification produced?

The object distance and the image distances will both be positive. Thus:

$$\frac{1}{f} = \frac{1}{u} + \frac{1}{v} = \frac{1}{20} + \frac{1}{10} = \frac{3}{20}$$

and so f = +6.7 cm. The positive sign indicates that the lens is a convex lens.

$$\text{Magnification} = \frac{v}{u} = \frac{10}{20} = 0.5$$

The image is half the size of the object.

Example

An object is placed 20 cm in front of a concave lens of focal length 10 cm. What will be the image distance from the lens?

The object distance is positive and the focal length negative. Hence:

$$\frac{1}{-10} = \frac{1}{20} + \frac{1}{v}$$

$$\frac{1}{v} = -\frac{1}{10} - \frac{1}{20} = -\frac{3}{20}$$

and v = −6.7 cm. The image is thus virtual and so the same side as the object.

4.3 Sound waves

Sound is produced by a vibrating object, e.g. the string of a guitar or the cone of a loudspeaker. The vibration causes some medium, such as air, to assume an oscillatory motion as first it is pushed by the vibrating object one way and then pulled back in the reverse direction. We can simulate the behaviour of a medium subject to an oscillating disturbance by a row of objects, e.g. trolleys, linked by springs (Figure 4.17).

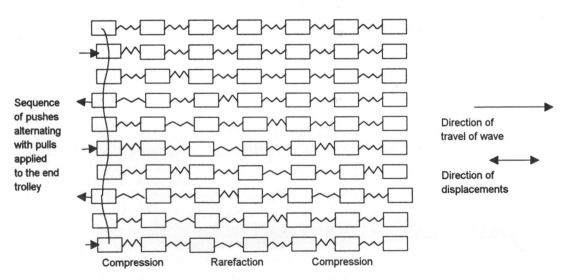

Sequence of pushes alternating with pulls applied to the end trolley

Direction of travel of wave

Direction of displacements

Compression   Rarefaction   Compression

Figure 4.17 *Transmission of a longitudinal wave*

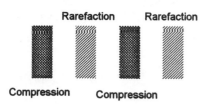

Figure 4.18 *Pressure wave*

When the end trolley is moved backwards and forwards in an oscillation, a disturbance is set up which travels along the row of trolleys. Note that each trolley oscillates back and forth with a disturbance which is in the same direction as that of the wave motion; this type of wave motion is called a *longitudinal wave*.

As can be seen from Figure 4.17, the appearance of the row of trolleys while the wave is travelling along it is of groups of trolleys closer together than their initial undisturbed positions alternating with groups of trolleys further apart than their initial positions. We can thus consider the wave motion as being transmitted through the medium of the trolleys as a series of compressions and rarefactions (Figure 4.18). In the case of a gas we would have alternate regions of pressure above the atmospheric pressure and regions below it. In the case of a solid we have alternate regions of atoms having being pushed closer together and regions where they are pulled further apart.

Sound is a longitudinal wave and requires, unlike electromagnetic waves, a material medium for its transmission. The frequencies of sound waves that are audible to humans are in the region 20 Hz to 20 000 Hz. The speed of sound in air at 0°C is 331 m/s and in water 1402 m/s, the speed depending on the temperature. At 18°C the speed in air rises to 342 m/s. The speed of longitudinal waves in solids is typically 5100 m/s in steel and 3970 m/s in copper.

Example

A vibrating loudspeaker cone produces sound waves with a frequency of 1 kHz. What will their wavelength in air be if the speed of sound in air is 340 m/s?

$v = f\lambda$ and so $\lambda = 340/1000 = 0.34$ m.

4.3.1 Amplitude, intensity, loudness and pitch

The *amplitude* of oscillation of air molecules for a sound frequency of 1 kHz is about 10^{-11} m for the faintest sound that can be heard and about 10^{-5} m for the loudest sound that can be tolerated.

The *intensity* of a wave is the rate at which the wave transmits energy through a particular area per second, i.e. the power per unit area. The threshold of hearing by humans for a frequency of 1 kHz is about 10^{-12} W/m^2 and a person talking is about 10^{-6} W/m^2.

The *loudness* of sound as perceived by a human depends on the amplitude and frequency of the sound, as well as the sensitivity of the hearer to the particular frequency being sounded. A sound described as loud by one person may not be described as loud by another. Figure 4.19 shows how typically the intensity to produce a particular perceived loudness depends on the frequency.

A *musical note* is heard with periodic sound waves that can be considered pleasant; noise is generally sound which is not periodic. Musicians use the term *pitch* to describe the frequency of a musical note as perceived by them; the pitch determines the position of a particular

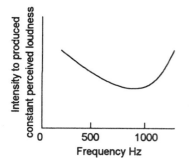

Figure 4.19 *Constant loudness*

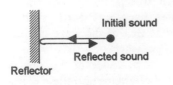

Figure 4.20 *Echo*

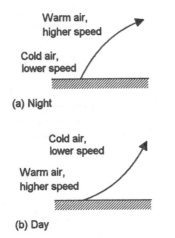

(a) Night

(b) Day

Figure 4.21 *Refraction*

note on a musical scale. The notes on the scale are labelled with the letters from A to G, which are then repeated for each octave. The standard pitch used for tuning the instruments in an orchestra is 440 Hz for note A, the pitch of the other notes then being defined in relationship to this frequency. A scientific standard pitch is based on the use of 256 Hz for middle C and this gives 288 Hz for D, 320 Hz for E, 342 Hz for F, 384 Hz for G, 426 Hz for A, 480 Hz for B and 512 Hz for top C.

4.3.2 Reflection and refraction of sound

As with other wave motions, sound can be reflected and refracted and obeys the same rules. Thus, for reflection the angle of incidence equals the angle of reflection and for refraction the ratio of the sine of the angle of incidence to the sine of the angle of refraction is a constant, this constant depending on the values of the speed of sound in the two media concerned.

Echoes are examples of reflection. A sound is heard and then some time later it is heard after travelling to a distant surface and being reflected back (Figure 4.20), the time difference between the initial sound and the reflected sound occurs because the sound takes a finite time to travel to the reflector and back again. Examples of refraction occur when sound passes from one medium to another where its speed is different. Because the speed of sound in air depends on the temperature, when there is a difference in temperature between the air near the ground and the air higher up there is a difference in speed and so refraction occurs (Figure 4.21).

Example

A gun is fired and then 2 s later an echo is heard. If the speed of sound is 340 m/s, how far away from the gun is the reflecting surface?

In 2 s the sound travels 640 m and so the reflector is 340 m away.

Activities

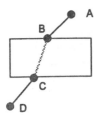

Figure 4.22 *Activity 1*

1 Determine the refractive index of a rectangular glass block by plotting a ray of light through it and measuring the angle of incidence and the angle of refraction. This can be done by placing two pins A and B in line on the far side of the block, as in Figure 4.22, and then placing two pins C and D in the same line as perceived when the A and B are viewed through the block.
2 Determine the focal length of a convex lens by throwing the image of a distant object onto a white screen. The distance from lens to screen is the focal length.
3 Use a convex lens with a real object to throw a real image on a screen. Measure the object distance and image distances and hence determine the focal length.
4 Use a translucent scale, illuminated from behind, as the object for a convex lens and throw a real image of it onto a screen to which a

scale, e.g. a piece of graph paper, has been attached. Record the size of the image for a number of different object distances. Plot a graph of the magnification against the image distance and hence, since $m = (v/f) - 1$, obtain the focal length from the slope of the graph.

Problems

1 Two plane mirrors are parallel and face each other. If they are 20 cm apart and a small object is placed 5 cm from one of them, find the distance from the object of the image produced after four reflections.

2 If the refractive index for light passing from air into glass is 1.6, what will be the critical angle for light passing from glass to air?

3 An object is 20 cm below the surface of water of refractive index 1.33. What will be the apparent depth of the object?

4 Light passes from air into a glass block. If the angle of incidence is 30°, what is the angle of refraction? The glass has a refractive index of 1.5.

5 A ray of light in passing from air to ice is found to have an angle of incidence of 45° and an angle of refraction of 30°. What is the refractive index for light passing from air to ice and what would be the angle of refraction for light at an angle of incidence of 60°?

6 If the refractive index for light passing from air into glass is 1.5, what is the refractive index for light passing from the glass into air?

7 Determine the image distance and say whether the image is real or virtual for (a) a real object 20 cm from a convex lens of focal length 10 cm, (b) a real object 15 cm from a concave lens of focal length 5 cm, (c) a real object 60 cm from a convex lens of focal length 24 cm, (d) a real object 12 cm from a convex lens of focal length 24 cm, (e) a real object 22 cm from a concave lens of focal length 16 cm, (f) a real object 3 cm from a convex lens of focal length 6 cm, (g) a real object 6 cm from a concave lens of focal length 12 cm.

8 Determine the size of the image and state whether it is real or virtual when an object of height 2 cm is placed (a) 30 cm from a convex lens of focal length 5 cm, (b) 48 cm from a concave lens of focal length 24 cm, (c) 10 cm from a convex lens of focal length 15 cm.

9 A small illuminated object is 30 cm from a convex lens of focal length 20 cm and produces an image on a screen. What is the distance between the object and the screen?

10 How far is the slide from the lens in a slide projector if the image is produced on a screen 4.0 m from the lens which is of focal length 15 cm? What is the magnification?

11 A slide projector is to be used 10 m from a screen that is 1.4 m wide. What focal length lens should be used if a 35 mm wide slide is just to fill the screen?

12 What is the wavelength in air of a sound of frequency 256 Hz if the speed of sound is 340 m/s?

13 At night, the land loses heat more rapidly than the sea and thus the air temperatures over the land will be lower than that over the sea. What type of path would you expect for a sound produced over the land and directed towards the sea?

5 Statics

5.1 Introduction

Statics is concerned with the equilibrium of bodies under the action of forces. Thus, this chapter deals with situations where forces are in equilibrium and the determination of resultant forces and their moments.

5.1.1 Scalar and vector quantities

Scalar quantities can be fully defined by just a number; mass is an example of a scalar quantity. To specify a scalar quantity all we need to do is give a single number to represent its size. Quantities for which both the size and direction have to be specified are termed *vectors*. For example, force is a vector and if we want to know the effect of, say, a 100 N force then we need to know in which direction the force is acting. To specify a vector quantity we need to indicate both size and direction. To represent vector quantities on a diagram we use arrows. The length of the arrow is chosen according to some scale to represent the size of the vector and the direction of the arrow, with reference to some reference direction, the direction of the vector, e.g. Figure 5.1 to represent a force of 300 N acting in a north-east direction from A to B.

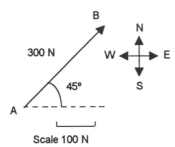

Figure 5.1   *A vector*

In order to clearly indicate in texts when we are referring to a vector quantity, rather than a scalar or just the size of a vector quantity, it is common practice to use a bold letter such as **a**, or when hand-written by underlining the symbol a. When we are referring to the vector acting in directly the opposite direction to **a**, we would use −**a**, with the minus sign being used to indicate that it is in the opposite direction to **a**. If we want to just refer to the size of **a** we write *a*.

5.1.2 Internal and external forces

The term *external forces* is used for the forces applied to an object from outside (by some other object). The term *internal forces* is used for the forces induced in the object to counteract the externally applied forces (Figure 5.2). To illustrate this, consider a strip of rubber being stretched. External forces are applied to the rubber to stretch it. The stretching pulls the atoms in the rubber further apart and they exert forces inside the material to resist this; these are the internal forces. The existence of the internal forces is apparent when the strip of rubber is released; it contracts from its extended length under the action of the internal forces. The internal force produced as a result of an applied external force is a force produced in reaction to their application and so is often called the *reactive force* or *reaction*.

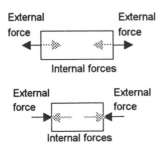

Figure 5.2   *External and internal forces*

5.2 Forces in equilibrium

If when two or more forces act on an object there is no resultant force (the object remains at rest or moving with a constant velocity), then the forces are said to be in *equilibrium*. This means that the sum of the force vectors is zero. If we have a mechanical system involving a number of connected bodies, e.g. a truss, then at equilibrium for the system we must have equilibrium for each point in the system. Thus if we draw a diagram representing the forces acting at some particular point in the system, such a diagram being called a *free-body*, then we can apply the conditions for equilibrium to those forces.

5.2.1 Two forces in equilibrium

For two forces to be in equilibrium (Figure 5.3) they have to:

1   Be equal in size.

2   Have lines of action which pass through the same point (such forces are said to be *concurrent*).

3   Act in exactly opposite directions.

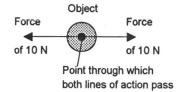

If any one of the above conditions is not met, there will be a resultant force in some direction and so the forces will not be in equilibrium.

Figure 5.3   *Two forces in equilibrium*

5.2.2 Three forces in equilibrium

For three forces to be in equilibrium (Figure 5.4) they have to:

1   All lie in the same plane (such forces are said to be *coplanar*).

2   Have lines of action which pass through the same point, i.e. they are concurrent.

3   Give no resultant force. If the three forces are represented on a diagram in magnitude and direction by the vectors F_1, F_2 and F_3, then their arrow-headed lines when taken in the order of the forces must form a triangle if the three forces are to be in equilibrium. This is known as the *triangle of forces*. If the forces are not in equilibrium, a closed triangle would not be formed.

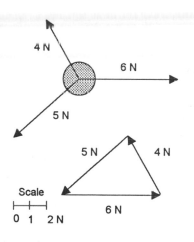

Figure 5.4   *Three forces in equilibrium*

Drawing the triangle of forces involves the following steps:

1   Select a suitable scale to represent the sizes of the forces.

2   Draw an arrow-headed line to represent one of the forces.

3   Take the forces in the sequence they occur when going, say, clockwise. Draw the arrow-headed line for the next force and draw it so that its line starts with its tail end from the arrowed end of the first force.

4   Then draw the arrow-headed line for the third force, starting with its tail end from the arrow end of the second force.

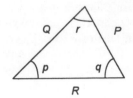

Figure 5.5 *Sine rule*

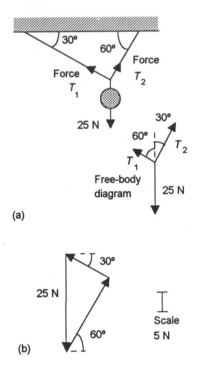

(a)

(b)

Figure 5.6 *Example*

5 If the forces are in equilibrium, the arrow end of the third force will coincide with the tail end of the first arrow-headed line to give a closed triangle.

Triangle of forces problems can be solved graphically or by calculation; possibly using the *sine rule* which can be stated as:

$$\frac{P}{\sin p} = \frac{Q}{\sin q} = \frac{R}{\sin r}$$

where P is the length of the side of the triangle opposite the angle p, Q the length of the side opposite the angle q and R the length of the side opposite the angle r (Figure 5.5).

Example

Figure 5.6(a) shows an object supported by two wires and the free-body diagram for the forces acting on the object. If the weight of the object is 25 N, what are the forces in the two wires when the object is in equilibrium?

When the object is in equilibrium, the three forces must complete a triangle of force. For the weight we know the magnitude and direction. For the forces in the two wires we only know the directions. Figure 5.6(b) shows the triangle of forces produced by utilising this information. We can obtain the forces in the wires from a scale diagram of the triangle of forces. Thus we might use a scale of, say, 5 N to 1 cm. The 25 N force is then represented by a vertical line of length 5 cm. The T_2 force will then be a line drawn from the arrow end of the 25 N force at an angle to the horizontal of 60°. We do not know its length but we do know that the T_1 line must join the tail end of the 25 N force at an angle of 30° below the horizontal. When we draw this line we can find the point of intersection of the T_1 and T_2 lines and hence their sizes.

Alternatively we can use the sine rule. The angle between the two wires is, from Figure 5.6(a), 90°. Thus

$$\frac{T_1}{\sin (90° - 60°)} = \frac{25}{\sin 90°}$$

and $T_1 = 12.5$ N. For the other string:

$$\frac{T_2}{\sin (90° - 30°)} = \frac{25}{\sin 90°}$$

and $T_2 = 21.7$ N.

5.3 Resultant forces

A single force which is used to replace a number of forces and has the same effect is called a *resultant force*. What we are doing in finding a resultant force is adding a number of vectors to find their sum.

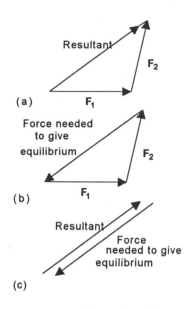

(a)

(b)

(c)

Figure 5.7 *Triangle rule*

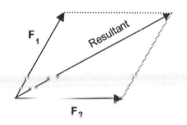

Figure 5.8 *Parallelogram rule*

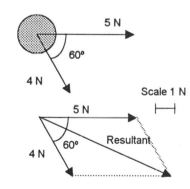

Figure 5.9 *Example*

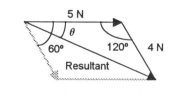

Figure 5.10 *Example*

We can use the triangle rule to add two forces; this is because if we replace the two forces by a single force it must be equal in size and in the opposite direction to the force needed to give equilibrium. For determining the resultant, the triangle rule can be stated as: to add two forces F_1 and F_2 we place the tail of the arrow representing one vector at the head of the arrow representing the other and then the line that forms the third side of the triangle represents the vector which is the resultant of F_1 and F_2 (Figure 5.7(a)). Note that the directions of F_1 and F_2 go in one sense round the triangle and the resultant, goes in the opposite direction. Figure 5.7(b) shows the triangle rule with F_1 and F_2 in equilibrium with a third force and Figure 5.7(c) compares the equilibrium force with the resultant force.

An alternative and equivalent rule to the triangle rule for determining the sum of two vectors is the *parallelogram rule*. This can be stated as: if we place the tails of the arrows representing the two vectors F_1 and F_2 together and complete a parallelogram, then the diagonal of that parallelogram drawn from the junction of the two tails represents the sum of the vectors F_1 and F_2. Figure 5.8 illustrates this. The triangle rule is just the triangle formed between the diagonal and two adjacent sides of the parallelogram.

The procedure for drawing the parallelogram is as follows:

1 Select a suitable scale for drawing lines to represent the forces.

2 Draw an arrowed line to represent the first force.

3 From the start of the first line, i.e. its tail end, draw an arrowed line to represent the second force.

4 Complete the parallelogram by drawing lines parallel to these force lines.

5 The resultant is the line drawn as the diagonal from the start point, the direction of the resultant being outwards from the start point.

Example

An object is acted on by two forces, of magnitudes 5 N and 4 N, at an angle of 60° to each other. What is the resultant force on the object?

Figure 5.9 shows how these two vectors can be represented by the sides of a parallelogram. The resultant can be determined from a scale drawing by measurements of its length and angle or by calculation. From a scale drawing, the diagonal represents a force of 7.8 N acting at an angle of about 26° to the 5 N force.

By calculation, we have a triangle with adjacent sides of 5 and 4 and an angle of $180° - 60° = 120°$ between them (Figure 5.10) and need to determine the length of the other side. The cosine rule can be used; the square of a side is equal to the sum of the squares of the other two sides minus twice the product of those sides times the cosine of the angle between them. Thus:

$$(\text{resultant})^2 = 5^2 + 4^2 - 2 \times 5 \times 4 \cos 120° = 25 + 16 + 20$$

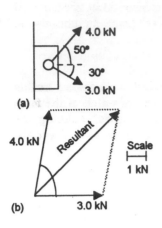

Hence the resultant is 7.8 N.

We can determine the angle θ between the resultant and the 5 N force by the use of the sine rule. Thus:

$$\frac{4}{\sin \theta} = \frac{7.8}{\sin 120°}$$

Hence, $\sin \theta = 0.444$ and $\theta = 26.4°$.

Example

Determine the size and direction of the resultant of the two forces shown acting on the bracket in Figure 5.11(a).

Arrowed lines are drawn to scale, with an angle of 80° between them, to represent the two forces. The parallelogram is then completed and the diagonal drawn (Figure 5.11(b)). The diagonal represents a force of 5.4 kN at 47° to the 3.0 kN force.

Figure 5.11 *Example*

5.4 Resolving forces

A single force can be replaced by two forces at right angles to each other. This is known as *resolving* a force into its *components*. It is done by using the parallelogram of forces in reverse, i.e. starting with the diagonal of the parallelogram and finding the two forces which would 'fit' the sides of such a parallelogram (Figure 5.12). If we have a force F, then its components are given by (horizontal component)/F = cos θ and (vertical component)/F = sin θ as:

horizontal component = $F \cos \theta$

vertical component = $F \sin \theta$

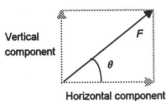

Figure 5.12 *Resolving a force*

Example

Determine the horizontal and vertical components of a force of 5 N at an angle of 40° to the horizontal.

Using the parallelogram rule we can draw the horizontal and vertical components as the sides of the parallelogram, as shown in Figure 5.13. Then, the horizontal component = 5 cos 40° = 3.8 N and the vertical component = 5 sin 40° = 3.2 N. We could therefore replace the 5 N force by the two components of 3.8 N and 3.2 N and the effect would be precisely the same.

Figure 5.13 *Example*

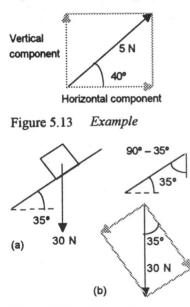

Figure 5.14 *Example*

Example

An object of weight 30 N rests on an incline which is at 35° to the horizontal (Figure 5.14(a)). What are the components of the weight acting at right angles to the incline and along the incline?

Figure 5.14(b) shows how the parallelogram rule can be used to determine the components. Thus the component down incline = 30 sin 35° = 17.2 N and the component at right angles to incline = 30 cos 35° = 24.6 N.

5.5 Moment of a force

The *moment of a force* about an axis is the product of the force F and its perpendicular distance r from the axis to the line of action of the force (Figure 5.15). An alternative, but equivalent, way of defining the moment of a force about an axis is as the product of the force F and the radius r of its potential rotation about the axis. Thus:

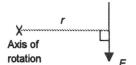

Axis of rotation

$$\text{moment} = Fr$$

Figure 5.15 *Moment of a force*

The SI unit of the moment is the newton metre (N m).

If an object fails to rotate about some axis then the turning effect of one force must be balanced by the opposite direction turning effect of another force. Thus for the situation shown in Figure 5.16 of a beam balanced on a pivot, the clockwise moment of force F_1 about the pivot axis must be balanced by the anticlockwise moment of force F_2 about the same axis, i.e. $F_1 r_1 = F_2 r_2$. Thus, when there is no rotation, *the algebraic sum of the clockwise moments about an axis must equal the algebraic sum of the anticlockwise moments about the same axis*. This is known as the *principle of moments*.

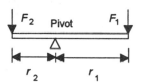

Figure 5.16 *Balanced beam*

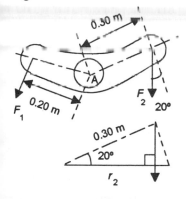

Example

Calculate the moments about the axis through A in Figure 5.17 of forces F_1 and F_2, when F_1 is 200 N and F_2 is 400 N.

The perpendicular distance of the line of action of force F_1 from the pivot axis A is $r_1 = 0.20$ m; hence the moment is $F_1 r_1 = 200 \times 0.2 = 40$ N m anticlockwise.

The perpendicular distance of the line of action of force F_2 from the pivot axis A is $r_2 = 0.30 \cos 20°$ m; hence the moment is $F_2 r_2 = 400 \times 0.30 \cos 20° = 113$ N m clockwise.

Figure 5.17 *Example*

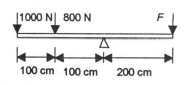

100 cm   100 cm   200 cm

Figure 5.18 *Example*

Example

Calculate the force required at the right-hand end of the pivoted beam in Figure 5.18 if the beam is to balance and not rotate.

Taking moments about the pivot axis: anticlockwise moments = $1000 \times 200 + 800 \times 100 = 280\ 000$ N cm. This must equal the clockwise moment about the same axis and thus $200F = 280\ 000$ and so $F = 1400$ N.

Example

If a force of 1000 N is applied to the car brake pedal shown in Figure 5.19, what will be the resulting force F in the brake cable?

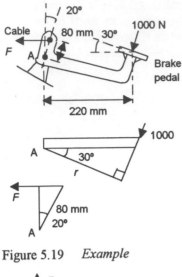

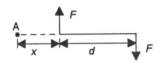

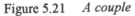

Figure 5.19   *Example*

Figure 5.20   *A couple*

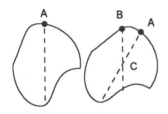

Figure 5.21   *A couple*

Taking moments about the axis through A. The perpendicular distance between the axis through A and the line of action of the 1000 N force is 0.220 cos 30° and so:

$$\text{moment} = 1000 \times 0.220 \cos 30° = 191 \text{ N m clockwise}$$

The perpendicular distance between the axis through A and the line of action of force F is 0.080 cos 20° and so:

$$\text{moment of } F = F \times 0.080 \cos 20° \text{ anticlockwise}$$

As these moments balance we have $F \times 0.080 \cos 20° = 191$ and so $F = 2541$ N.

5.5.1 Couples

A *couple* is two coplanar parallel forces of the same size with their lines of action separated by some distance and acting in opposite directions (Figure 5.20). The *moment of a couple* about any axis is the algebraic sum of the moments due to each of the forces. Thus, taking moments about the axis through A in Figure 5.21 gives clockwise moment = $F(d + x)$ and anticlockwise moment = Fx. Hence:

$$\text{moment of couple} = F(d + x) - Fx = Fd$$

Hence, *the moment of a couple is the product of the force size and the perpendicular distance between the forces.*

5.6 Centre of gravity

The weight of an object is made up of the weights of each particle, each atom, of the object and so we have a multitude of forces which do not act at a single point. However, it is possible to have the same effect by replacing all the forces of an object by a single weight force acting at a particular point; this point is termed the *centre of gravity*.

If an object is suspended from some point A on its surface (Figure 5.22), equilibrium will occur when the moments of its constituent particles about the axis through A balance. When this occurs we can think of all the individual weight forces being replaced by a single force with a line of action vertically passing through A. We can thus consider the centre of gravity to lie somewhere on the vertical line through A. If the object is now suspended from some other point B, when equilibrium occurs the centre of gravity will lie on the vertical line through B. The intersection of these lines gives the location C of the centre of gravity.

Consider an object to be made up of a large number of small elements: segment 1: weight w_1 a distance x_1 from some axis, segment 2: weight w_2 a distance x_2 from the same axis, segment 3: weight w_3 a distance x_3 from the same axis, and so on for further segments. The total moment of the segments about the axis will be the sum of all the moment terms, i.e. $w_1x_1 + w_2x_2 + w_3x_3 + \ldots$ If a single force W, i.e. the total weight, acting through the centre of gravity is to be used to replace the forces due to each

Figure 5.22   *Centre of gravity*

segment then we must have a moment about the axis of Wx, where x is the distance of the centre of gravity from the axis:

Wx = sum of all the moment terms due to the segments

$$x = \frac{w_1 x_1 + w_2 x_2 + w_3 x_3 + \dots}{W}$$

For symmetrical homogeneous objects, the centre of gravity is the geometrical centre. Thus, for a sphere the centre of gravity is at the centre of the sphere. For a cube, the centre of gravity is at the centre of the cube. For a rectangular cross-section homogeneous beam, the centre of gravity is halfway along its length and in the centre of the cross-section. The centre of gravity of a triangular section lies at the intersection of lines drawn from each apex and bisecting the opposite sides. This locates the centre of gravity as being two-thirds of the way down any one of these median lines from an apex.

For composite objects, the centre of gravity can be determined by considering the weight to be made up of a number of smaller objects, each with its weight acting through its own centre of gravity. This is illustrated in the examples that follow.

Example

A homogeneous square cross-section beam of side 100 mm has a length of 2.0 m; at what point will be its centre of gravity?

Because the beam is homogeneous and of constant cross-section, the centre of gravity will lie halfway along the beam, i.e. 1.0 m from each end, and in the centre of the section, i.e. 50 mm from face.

Example

Determine the position of the centre of gravity of the object shown in Figure 5.23, the two parts are different sized, homogeneous, rectangular objects of constant thickness sheet.

The centre of gravity of each piece is at its centre. Thus, one piece has its centre of gravity a distance along its centre line from X of 50 mm and the other a distance from X of 230 mm. If the first piece has a weight 6 N and the second piece a weight of 4 N, then taking moments about X gives for the moment total $(6 \times 50) + (4 \times 230) = 1220$ N mm. If the total weight of $6 + 4 = 10$ N is considered to act at the centre of gravity a distance x from X then $10\ x = 1220$ and so $x = 122$ mm.

Example

Determine the position of the centre of gravity for the homogeneous, constant thickness, sheet shown in Figure 5.24.

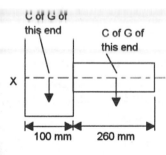

Figure 5.23   *Example*

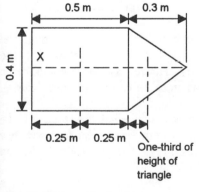

Figure 5.24   *Example*

The centre of gravity of the rectangular part will lie at its centre and that of the triangular part at two-thirds of the distance from an apex along a line bisecting the opposite side. Thus, for the rectangular part the centre of gravity is 0.25 m from point X and for the triangular part $0.5 + 0.1 = 0.6$ m. The weight of each part will be proportional to its area and so, if the weight per unit area is w, the weight of the rectangular part $= 0.4 \times 0.5w = 0.2w$ and the weight of the triangular part $= \frac{1}{2} \times 0.4 \times 0.3w = 0.06w$. Hence, taking moments about X:

$$\text{moment} = 0.2w \times 0.25 + 0.06w \times 0.6 = 0.086w$$

The total weight of the shape is $0.2w + 0.06w = 0.26w$. Hence the centre of gravity is a distance x from X, where:

$$x = \frac{0.086w}{0.26w} = 0.33 \text{ m}$$

The centre of gravity will lie this distance along the axis from X and half the thickness of the sheet in from the sheet surface.

5.7 Static equilibrium

An object is said to be in *static equilibrium* when there is no movement or tendency to movement in any direction. This requires that, for coplanar but not necessarily concurrent forces:

1   There must be no resultant force in any direction, i.e. the total of the upward components of forces must equal the downward components and the total of the rightward components of forces must equal the total of leftward components.

2   The sum of the anticlockwise moments about any axis must equal the sum of the clockwise moments about the same axis.

Example

Figure 5.25 shows a beam resting on supports at each end. If the beam carries loads at the positions shown, what are the reactive forces at the supports? The weight of the beam may be neglected.

Taking moments about the left-hand end of the beam gives the clockwise moments as $2 \times 1 + 3 \times 2.5 = 9.5$ kN m and the anticlockwise moments as $3.5R_2$. Hence, as the anticlockwise moments and clockwise moments must be equal, we have $3.5R_2 = 9.5$ and so $R_2 = 2.7$ kN.

The total of the upward forces must equal the total of the downward forces; the reactive forces must be vertical forces if they are to balance the downward forces, i.e. $R_1 + R_2 = 2 + 3 = 5$ kN, and so $R_1 = 2.3$ kN.

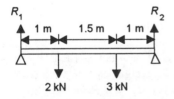

Figure 5.25   *Example*

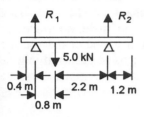

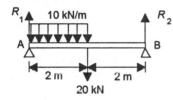

Figure 5.26   *Example*

Example

Determine the reactive forces at the supports for the beam shown in Figure 5.26. The beam is homogeneous, of constant cross-section, and has a weight of 0.5 kN.

The weight of the beam can be considered to act at its midpoint, i.e. 2.3 m from one end. Taking moments about the left-hand pivot gives clockwise moments = $5.0 \times 0.8 + 1.9 \times 0.5 = 4.95$ kN m and anticlockwise moments as $3.0 \times R_2$. Hence, since equilibrium is assumed, the anticlockwise moments equal the clockwise moments and so $3.0R_2 = 4.95$. Hence $R_2 = 1.65$ kN and, since $R_1 + R_2 = 5.5$ kN, then $R_1 = 3.85$ kN.

Example

Figure 5.27 shows a uniform beam of length 4 m which is subject to a uniformly distributed load of 10 kN/m over half its length and has a weight of 20 kN. It is supported at each end. Determine the reactions at the supports.

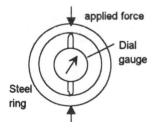

Figure 5.27   *Example*

The weight of the beam can be considered to act at its midpoint. The uniformly distributed load can be considered to act at the midpoint of the section of the beam over which it is acting and have a total force of $10 \times 2 = 20$ kN. Hence, taking moments about A gives $20 \times 1 + 20 \times 2 = 4R_2$ and so $R_2 = 15$ kN. Equating the upwards and downwards directed forces gives $R_1 + R_2 = 20 + 20$ and so R_1 25 kN.

5.8 Measurement of force

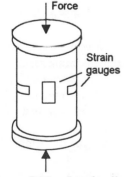

Figure 5.28   *Proving ring*

Methods that are commonly used for the measurement of forces are:

1   *Elastic element methods* which depend on the force causing some element to stretch, or become compressed, and so change in length. This change then becomes a measure of the force.

   The simplest example of such a method is the *spring balance* the extension of the spring being proportional to the force. Direct reading spring balances are not, however, capable of high accuracy since the extensions produced are relatively small.

   A more accurate form is the *proving ring*. This is a steel ring which becomes distorted when forces are applied across a diameter, the amount of distortion being proportional to the force. Figure 5.28 shows a ring where a dial gauge is used to monitor the displacement; another form has strain gauges attached to the ring. Proving rings are capable of high accuracy and are typically used for forces in the range 2 kN to 2000 kN.

   The proving ring is just one form of force measurement system in which forces are used to produce displacement which is monitored; the term *load cell* is commonly used for such systems and can take other forms such as columns, tubes or cantilevers. Strain gauges are commonly used to monitor the strain produced as a result of the forces applied to the cell member (Figure 5.29). Strain gauges are

Figure 5.29   *Load cell*

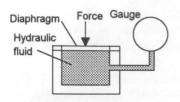

Figure 5.30   *Hydraulic force measurement system*

essentially lengths of wire coils which, when stretched or compressed produce electrical resistance changes. The resistance change is proportional to the strain experienced by the strain gauge and hence the amount by which it has been increased or reduced in length as a result of the forces. Such load cells are generally used for forces between about 500 N and 6000 kN with an accuracy of about ±0.2% of the full-scale reading.

2   *Hydraulic pressure methods* use the change in pressure of hydraulic fluid that is produced by the application of a force as a measure of the force. A chamber containing hydraulic fluid (Figure 5.30) is connected to a pressure gauge, possibly a Bourdon tube pressure gauge. The chamber has a diaphragm to which the force is applied. The force causes the diaphragm to move and produces a change in pressure in the fluid which then shows up on the pressure gauge. Such methods tend to be used for forces up to 5 MN and typically have an accuracy of the order of ±1%.

Activities

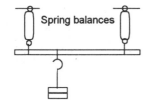

Figure 5.31   *Activity 2*

1   Determine the centre of gravity of an irregular shaped sheet of card or metal by suspending it from points on its rim. The centre of gravity will lie on the vertical line through the point of suspension and thus if two points of suspension are used, the intersection of their lines gives the location of the centre of gravity.

2   Using the arrangement shown in Figure 5.31, investigate how the forces at the supports of the beam depend on the value and position of the load and account for the results obtained.

3   Clamp a retort stand near the edge of the bench and butt one end of a ruler up against it, the other end of the rule being attached by string to a spring balance and the upper end of the retort stand (Figure 5.32). Suspend weights from the free end of the rule and determine the reading on the balance when the rule is horizontal. Explain the results you obtain.

Problems

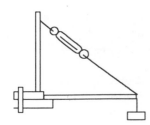

Figure 5.32   *Activity 3*

1   Determine the resultant forces acting on objects subject to the following forces acting at a point on the object:
(a) A force of 3 N acting horizontally and a force of 4 N acting at the same point on the object at an angle of 60° to the horizontal,
(b) A force of 3 N in a westerly direction and a force of 6 N in a northerly direction,
(c) Forces of 4 N and 5 N with an angle of 65° between them,
(d) Forces of 3 N and 8 N with an angle of 50° between them,
(e) Forces of 5 N and 7 N with an angle of 30° between them,
(f) Forces of 9 N and 10 N with an angle of 40° between them,
(g) Forces of 10 N and 12 N with an angle of 105° between them.

2   In a plane structure a particular point is acted on by forces of 1.2 kN and 2.0 kN in the plane, the angle between the forces being 15°. What is the resultant force?

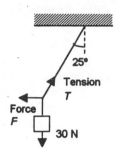

Figure 5.33   *Problem 4*

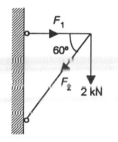

Figure 5.34   *Problem 5*

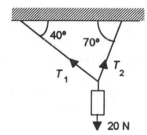

Figure 5.35   *Problem 6*

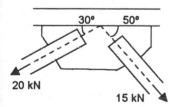

Figure 5.36   *Problem 10*

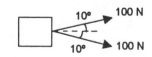

Figure 5.37   *Problem 11*

3 An object of weight 30 N is suspended from a horizontal beam by two chains. The two chains are attached to the same point on the object and are at 30° and 40° to the vertical. Determine the tensions in the two chains.

4 An object of weight 30 N is suspended by a string from the ceiling. What horizontal force F must be applied to the object if the string is to become deflected and make an angle of 25° with the vertical (Figure 5.33)?

5 An object of weight 20 N is supported from the ceiling by two cables inclined at 40° and 70° to the ceiling (Figure 5.34). Determine the tensions in the cables.

6 Determine the forces F_1 and F_2 in the jib shown in Figure 5.35.

7 An object of weight 5 N hangs on a vertical string. At what angle to the vertical will the string be when the object is held aside by a force of 3 N in a direction which is 20° above the horizontal?

8 Determine the resultant force acting on an object when two forces of 20 kN and 40 kN are applied to the same point on the object and the angle between the lines of action of the forces is 90°.

9 Three girders in the same plane meet at a point. If there are tensile forces of 70 kN, 80 kN and 90 kN in the girders, what angles will the girders have to be at if there is equilibrium?

10 What is the resultant force acting on the gusset plate shown in Figure 5.36 as a result of the forces shown?

11 An object is being pulled along the floor by two people, as illustrated in Figure 5.37. What will be the resultant force acting on the object?

12 Determine the horizontal and vertical components of the following forces: (a) 10 N at 40° to the horizontal, (b) 15 kN at 70° to the horizontal, (c) 12 N at 20° to the horizontal, (d) 30 kN at 80° to the horizontal.

13 An object of weight 30 N rests on an incline which is at 20° to the horizontal. What are the components of the weight at right angles to the incline and parallel to the incline?

14 A cable exerts a force of 15 kN on a bracket. If the cable is at an angle of 35° to the horizontal, what are the horizontal and vertical components of the force?

15 An object of weight 10 N rests on an incline which is at 30° to the horizontal. What are the components of this weight in a direction at right angles to the incline and parallel to the incline?

16 Masses of 2 kg and 4 kg are attached to opposite ends of a uniform horizontal 3.0 m long beam. At what point along the beam should a single support be placed for the beam to be in equilibrium? What will be the reactive force at the support? Neglect the mass of the beam.

17 A uniform horizontal beam 10 m long is supported at its ends. Loads of 2 kN and 3 kN are placed 2.0 m and 3.0 m, respectively, from one end. What are the reactive forces at the ends of the beam? Neglect the mass of the beam.

18 A uniform horizontal beam of length 4.5 m is supported at its ends. Masses of 5 kg and 20 kg are placed 1.0 m and 2.5 m, respectively,

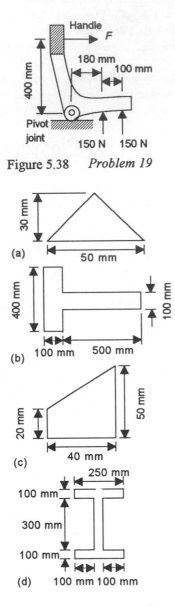

Figure 5.38   *Problem 19*

Figure 5.39   *Problem 22*

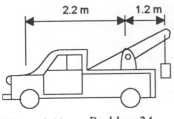

Figure 5.40   *Problem 24*

from one end. What are the reactive forces at the supports? Neglect the mass of the beam.

19 Figure 5.38 shows a control lever. Determine the size of the force F required to just maintain the lever in the position shown.

20 A homogeneous rod has a length of 200 mm and a uniform cross-section of 20 mm × 30 mm. At what point is the centre of gravity?

21 A rod has a length of 1.0 m. The first 0.5 m of the rod has a cross-section of 30 mm × 30 mm and the second 0.5 m a cross-section of 10 mm × 10 mm. If the material is of constant density, where will be the location of the centre of gravity?

22 The shapes in Figure 5.39 were cut from constant thickness sheet of uniform density. Determine the positions of the centre of gravity.

23 A loaded bus is checked at a weigh bridge and the front axle is found to be carrying a load of 3200 kg, the rear axle 4300 kg. How far is the centre of gravity of the bus from the front axle if the distance between the two axles is 8.0 m?

24 What is the position of the centre of gravity of the 3000 kg tow truck shown in Figure 5.40 if the maximum load it can lift before the front wheels come off the ground is 4000 kg?

25 A uniform, homogeneous, beam of length 2.4 m rests on two supports. If the supports are positioned 0.2 m from one end and 0.4 m from the other, what will be the reactions in the supports if the beam has a mass of 0.5 kg?

26 A beam AB of weight 50 N and length 2.4 m has its centre of gravity 0.8 m from end A. If it is suspended horizontally by two vertical ropes, one attached at end A and the other at a point which is 2.0 m from A, what are the tensions in the ropes?

27 A non-uniform beam of length 6.0 m and mass 25 kg is pivoted about its midpoint. When a mass of 35 kg is put on one end and a 30 kg mass at the other end, the beam assumes a horizontal position. What is the position of the centre of gravity?

28 A uniform beam of length 1.2 m has a weight of 15 N and is supported at 0.4 m from the left-hand end and at the right-hand end. What will be the reactions at the supports if it carries a point load of 20 N at the left-hand end and a uniformly distributed load of 75 N/m over a length of 0.4 m from the right-hand end?

29 A uniform beam of length 1.2 m has a weight of 15 N and is supported at its ends. What are the reactions at the supports if it carries a uniformly distributed load of 75 N/m 0.2 m from the left-hand end to the mid-span position?

30 A uniform beam of length 6 m has a weight of 180 kN and is supported at its ends. If it carries a uniformly distributed load of 60 kN/m over a length of 3 m from the left-hand end to the mid-span point, what are the reactions at the supports?

6 Stress and strain

6.1 Introduction

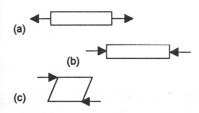

Figure 6.1 *(a) Tension, (b) compression, (c) shear*

When a material is subject to external forces, then internal forces are set up in the material which oppose the external forces. The material can be considered to be rather like a spring. A spring, when stretched by external forces, sets up internal opposing forces which are readily apparent when the spring is released and they force it to contract. A material subject to external forces which stretch it is said to be in *tension* (Figure 6.1(a)). A material subject to forces which squeeze it is said to be in *compression* (Figure 6.1(b)). If a material is subject to forces which cause it to twist or one face slide relative to an opposite face then it is said to be in *shear* (Figure 6.1(c)). This chapter is a consideration of the action of tensile and compressive forces on materials.

6.2 Direct stress and strain

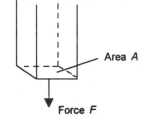

Figure 6.2 *Stress*

In discussing the application of forces to materials an important aspect is often not so much the size of the force as the force applied per unit cross-sectional area. Thus, for example, if we stretch a strip of material by a force F applied over its cross-sectional area A (Figure 6.2), then the force applied per unit area is F/A. The term *stress*, symbol σ, is used for the force per unit area.

$$\text{stress} = \frac{\text{force}}{\text{area}}$$

Stress has the units of pascal (Pa), with 1 Pa being a force of 1 newton per square metre, i.e. 1 Pa = 1 N/m². Multiples of the pascal are generally used, e.g. the megapascal (MPa) which is 10^6 Pa and the gigapascal (GPa) which is 10^9 Pa. Because the area over which the forces are applied is more generally mm² rather than m², it is useful to recognise that 1 GPa = 1 GN/m² = 1 kN/mm² and 1 MPa = 1 MN/m² = 1 N/mm². The area used in calculations of the stress is generally the original area that existed before the application of the forces. The stress is thus sometimes referred to as the *engineering stress*, the term *true stress* being used for the force divided by the actual area existing in the stressed state.

The stress is said to be *direct stress* when the area being stressed is at right angles to the line of action of the external forces, as when the material is in tension or compression. Shear stresses are not direct stresses since the forces being applied are in the same plane as the area being stressed.

When a material is subject to tensile or compressive forces, it changes in length (Figure 6.3). The term *strain*, symbol ε, is used for:

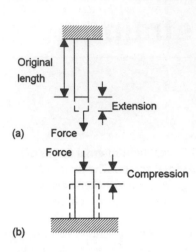

Original length

Extension

(a)   Force

Force

Compression

(b)

Figure 6.3   *(a) Tensile strain, (b) compressive strain*

$$\text{strain} = \frac{\text{change in length}}{\text{original length}}$$

Since strain is a ratio of two lengths it has no units; note that both lengths must be in the same units of length. Thus we might, for example, have a strain of 0.01. This would indicate that the change in length is 0.01 × the original length. However, strain is frequently expressed as a percentage:

$$\text{Strain as a \%} = \frac{\text{change in length}}{\text{original length}} \times 100\%$$

Thus the strain of 0.01 as a percentage is 1%, i.e. this is when the change in length is 1% of the original length.

Example

A bar of material with a cross-sectional area of 50 mm² is subject to tensile forces of 100 N. What is the tensile stress?

Tensile stress = force/area = $100/(50 \times 10^{-6})$ = 2×10^6 Pa = 2 MPa.

Example

A pipe has an outside diameter of 50 mm and an inside diameter of 45 mm and is acted on by a tensile force of 50 kN. What is the stress acting on the pipe?

The cross-sectional area of the pipe is $\frac{1}{4}\pi(D^2 - d^2)$, where D is the external diameter and d the internal diameter. Thus, the cross-sectional area = $\frac{1}{4}\pi(50^2 - 45^2)$ = 373 mm². Hence:

$$\text{Stress} = \frac{\text{force}}{\text{area}} = \frac{50 \times 10^3}{373 \times 10^{-6}} = 134 \times 10^6 \text{ Pa} = 134 \text{ MP}$$

Example

A strip of material has a length of 50 mm. When it is subject to tensile forces it increases in length by 0.020 mm. What is the strain?

$$\text{Strain} = \frac{\text{change in length}}{\text{original length}} = \frac{0.020}{50} = 0.000\,04 \text{ or } 0.04$$

Example

A tensile test piece has a gauge length of 50 mm. This increases by 0.030 mm when subject to tensile forces. What is the strain?

$$\text{Strain} = \frac{\text{change in length}}{\text{original length}} = \frac{0.030}{50} = 0.000\,06 \text{ or } 0.06$$

6.3 Stress–strain graphs

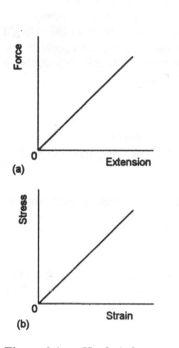

(a)

(b)

Figure 6.4   *Hooke's law*

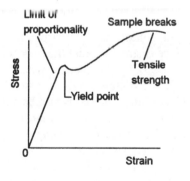

Figure 6.5   *Stress–strain graph for mild steel*

If gradually increasing tensile forces are applied to, say, a strip of mild steel then initially when the forces are released the material springs back to its original shape. The material is said to be *elastic*. If measurements are made of the extension at different forces and a graph plotted, then the extension is found to be proportional to the force and the material is said to obey *Hooke's law*. Figure 6.4(a) shows a graph when Hooke's law is obeyed. Such a graph applies to only one particular length and cross-sectional area of a particular material. We can make the graph more general so that it can be applied to other lengths and cross-sectional areas of the same material by dividing the extension by the original length to give the strain and the force by the cross-sectional area to give the stress (Figure 6.4(b)). Then we have, for a material that obeys Hooke's law, the stress proportional to the strain:

stress ∝ strain

Figure 6.5 shows the type of stress–strain graph which would be given by a sample of mild steel. Initially the graph is a straight line and the material obeys Hooke's law. The point at which the straight line behaviour is not followed is called the *limit of proportionality*.

With low stresses the material springs back completely to its original shape when the stresses are removed, the material being said to be *elastic*. At higher forces this does not occur and the material is then said to show some *plastic* behaviour. The term plastic is used for that part of the behaviour which results in permanent deformation. The stress at which the material starts to behave in a non-elastic manner is called the *elastic limit*. This point often coincides with the point on a stress–strain graph at which the graph stops being a straight line, i.e. the *limit of proportionality*.

The *strength* of a material is the ability of it to resist the application of forces without breaking. The term *tensile strength* is used for the maximum value of the tensile stress that a material can withstand without breaking, i.e.

$$\text{tensile strength} = \frac{\text{maximum tensile forces}}{\text{original cross-sectional area}}$$

The *compressive strength* is the maximum compressive stress the material can withstand without becoming crushed. The unit of strength is that of stress and so is the pascal (Pa), with 1 Pa being 1 N/m^2. Strengths are often millions of pascals and so MPa is often used, 1 MPa being 10^6 Pa or 1 000 000 Pa. Typically, carbon and low alloy steels have tensile strengths of 250 to 1300 MPa, copper alloys 80 to 1000 MPa and aluminium alloys 100 to 600 MPa.

With some materials, e.g. mild steel, there is a noticeable dip in the stress–strain graph at some stress beyond the elastic limit and the strain increases without any increase in load. The material is said to have yielded and the point at which this occurs is the *yield point*. For some materials, such as mild steel, there are two yield points termed the upper

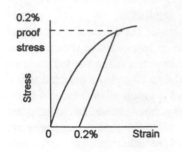

Figure 6.6   *0.2% proof stress*

yield point and the lower yield point. A carbon steel typically might have a tensile strength of 600 MPa and a yield stress of 300 MPa.

Some materials, such as aluminium alloys (Figure 6.6), do not show a noticeable yield point and it is usual here to specify *proof stress*. The 0.2% proof stress is obtained by drawing a line parallel to the straight line part of the graph but starting at a strain of 0.2%. The point where this line cuts the stress–strain graph is termed the 0.2% yield stress. A similar line can be drawn for the 0.1% proof stress.

Example

A material has a yield stress of 200 MPa. What tensile forces will be needed to cause yielding with a bar of the material with a cross-sectional area of 100 mm²?

Since stress = force/area, then the yield force = yield stress × area = $200 \times 10^6 \times 100 \times 10^{-6} = 20\ 000$ N.

Example

Calculate the maximum tensile force a steel bar of cross-section 20 mm × 10 mm can withstand if the tensile strength of the material is 400 MPa.

Tensile strength = maximum stress = maximum force/area and so the maximum force = tensile strength × area = $400 \times 10^6 \times 0.020 \times 0.010 = 80\ 000$ N = 80 kN.

6.3.1 Stiffness

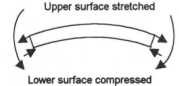

Figure 6.7   *Bending*

The *stiffness* of a material is the ability of a material to resist bending. When a strip of material is bent, one surface is stretched and the opposite face is compressed, as illustrated in Figure 6.7. The more a material bends the greater is the amount by which the stretched surface extends and the compressed surface contracts. Thus a stiff material would be one that gave a small change in length when subject to tensile or compressive forces. This means a small strain when subject to tensile or compressive stress and so a small value of strain/stress, or conversely a large value of stress/strain. For most materials a graph of stress against strain gives initially a straight line relationship, as illustrated in Figure 6.8. Thus a large value of stress/strain means a steep slope of the stress–strain graph. The quantity stress/strain when we are concerned with the straight line part of the stress–strain graph is called the *modulus of elasticity* (or sometimes *Young's modulus*).

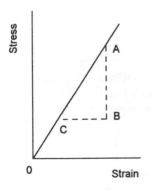

Figure 6.8   *Modulus of elasticity = AB/BC*

$$\text{Modulus of elasticity} = \frac{\text{stress}}{\text{strain}}$$

The units of the modulus are the same as those of stress, since strain has no units. Engineering materials frequently have a modulus of the order of 1 000 000 000 Pa, i.e. 10^9 Pa. This is generally expressed as GPa, with 1 GPa = 10^9 Pa. A stiff material has a high modulus of elasticity.

For most engineering materials the modulus of elasticity is the same in tension as in compression. Typical values are: steels 200 to 210 GPa, aluminium alloys 70 to 80 GPa, copper alloys 100 to 160 GPa.

Example

For a material with a tensile modulus of elasticity of 200 GPa, what strain will be produced by a stress of 4 MPa? Assume that the limit of proportionality is not exceeded.

Since the modulus of elasticity is stress/strain then:

$$\text{strain} = \frac{\text{stress}}{\text{modulus}} = \frac{4 \times 10^6}{200 \times 10^9} = 0.000\,02$$

Example

Calculate the strain that will be produced by a tensile stress of 10 MPa stretching a bar of aluminium alloy with a tensile modulus of 70 GPa. Assume that the limit of proportionality is not exceeded.

Since the modulus of elasticity is stress/strain then:

$$\text{strain} = \frac{\text{stress}}{\text{modulus}} = \frac{10 \times 10^6}{70 \times 10^9} = 0.000\,14$$

Example

A tie bar has two holes a distance of 1.0 m apart. By how much does this distance increase when a tensile load of 20 kN is applied to the tie bar? The tie bar is a rectangular section 40 mm × 10 mm and the material of which the bar is made has a tensile modulus of 210 GPa. Assume that the limit of proportionality is not exceeded.

$$\text{Stress} = \frac{\text{force}}{\text{area}} = \frac{20 \times 10^3}{0.040 \times 0.010} = 50 \times 10^6 \text{ Pa}$$

If the extension is e then the strain is change in length/original length = $e/4.0$. Since the modulus of elasticity is stress/strain then:

$$210 \times 10^9 = \frac{50 \times 10^6}{(e/4.0)}$$

Hence $e = 0.95$ mm.

Example

A machine is mounted on a rubber pad. The pad has to carry a load of 6 kN and have a maximum compression of 2 mm under this load. The maximum stress that is allowed for the rubber is 0.25 MPa. What is the size of the pad that would be appropriate for these maximum conditions. The modulus of elasticity for the rubber can be taken as being constant at 5 MPa.

Since compressive stress = force/area then the area required is

$$\text{area} = \frac{\text{force}}{\text{stress}} = \frac{6 \times 10^3}{0.25 \times 10^6} = 24 \times 10^{-3} \text{ m}^2$$

Since modulus of elasticity = stress/strain then:

$$\text{strain} = \frac{\text{stress}}{\text{modulus}} = \frac{0.25 \times 10^6}{5 \times 10^6} = 0.05$$

Strain = change in length/original length and so:

$$\text{length} = \frac{\text{change in length}}{\text{strain}} = \frac{0.002}{0.05} = 0.040 \text{ m} = 40 \text{ m}$$

The pad would thus require an area of 24 000 mm² and a thickness of 40 mm.

6.3.2 Ductility/brittleness

Glass is a *brittle material* and if you drop a glass it breaks; however it is possible to stick all the pieces together again and restore the glass to its original shape. If a car is involved in a collision, the bodywork of mild steel is less likely to shatter like the glass but more likely to dent and show permanent deformation (the term *permanent deformation* is used for changes in dimensions which are not removed when the forces applied to the material are taken away). Materials which develop significant permanent deformation before they break are called *ductile*. Ductile materials permit manufacturing methods which involve bending them to the required shapes or using a press to squash the material into the required shape. Brittle materials cannot be shaped in this way.

The stress–strain graph for cast iron (Figure 6.9) shows that very little plastic deformation occurs, no sooner has the stress risen to the yield point then failure occurs. Thus the length of a piece of cast iron after breaking is not much different from the initial length. Figure 6.10 shows the stress–strain graph for mild steel and this shows a considerable amount of plastic strain before breaking; it is ductile.

A measure of the ductility of a material is obtained by determining the length of a test piece of the material, then stretching it until it breaks and then, by putting the pieces together, measuring the final length of the test piece (Figure 6.11). A brittle material will show little change in length from that of the original test piece, but a ductile material will indicate a significant increase in length. The *percentage elongation* of a test piece after breaking is thus used as a measure of ductility:

$$\% \text{ elongation} = \frac{\text{final length} - \text{initial length}}{\text{initial length}} \times 100\%$$

A reasonably ductile material, such as mild steel, will have a percentage elongation of about 20%, a brittle material such as a cast iron less than 1%.

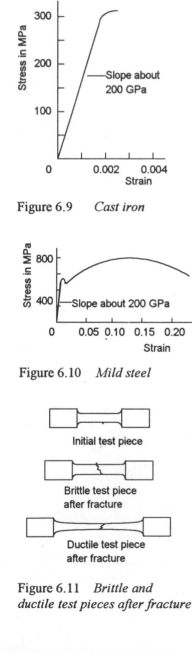

Figure 6.9   *Cast iron*

Figure 6.10   *Mild steel*

Figure 6.11   *Brittle and ductile test pieces after fracture*

Example

A 200 mm length of a material has a percentage elongation of 10%, by how much longer will a strip of the material be when it is broken?

Change in length = percentage elongation × original length/100 = $10 \times 200/100 = 20$ mm.

6.4 Factor of safety

When designing a structure a *factor of safety* has to be considered to make certain that working stresses keep within safe limits. For a brittle material, the factor of safety is defined in terms of the tensile strength:

$$\text{factor of safety} = \frac{\text{tensile strength}}{\text{maximum working stress}}$$

For ductile materials, the factor of safety is more usually defined in terms of the yield or proof stress:

$$\text{factor of safety} = \frac{\text{yield strength}}{\text{maximum working stress}}$$

For dead loads, a factor of safety of 4 or more is often used.

Example

Calculate the minimum diameter a mild steel bolt should have to withstand a load of 600 kN if the steel has a yield stress of 200 MPa and the factor of safety is to be 4.

Since factor of safety = (yield stress)/(maximum working stress) then the maximum working stress = (yield stress)/factor of safety = 200/4 = 50 MPa. Since stress = force/area, then the minimum area = force/stress = $(600 \times 10^3)/(50 \times 10^6) = 12 \times 10^{-3}$ m². If d is the minimum bolt diameter, then area = $\frac{1}{4}\pi d^2 = 12 \times 10^{-3}$ m² and so $d = 0.124$ m = 124 mm.

6.5 Poisson's ratio

When a piece of material is stretched, there is a transverse contraction of the material (Figure 6.12). The ratio of the transverse strain to the longitudinal strain is called *Poisson's ratio*:

$$\text{Poisson's ratio} = -\frac{\text{transverse strain}}{\text{longitudinal strain}}$$

The minus sign is because when one strain is tensile and giving an increase in length the other is compressive and giving a reduction in length. since it is a ratio, there are no units. For most engineering metals, Poisson's ratio is about 0.3.

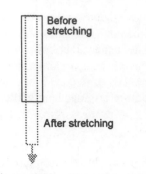

Figure 6.12   *Poisson's ratio*

Example

A bar of mild steel of length 100 mm is extended by 0.01 mm. By how much will the width of the bar contract if initially the bar has a width of 10 mm? Poisson's ratio = 0.31.

Longitudinal strain = extension/(original length) = 0.01/100 = 0.0001. Since Poisson's ratio = –(transverse strain)/(longitudinal strain) then transverse strain = – (Poisson's ratio) × (longitudinal strain) = –0.31 × 0.0001 = –0.000 031. Since, transverse strain = (change in width)/(original width), change in width = (transverse strain) × (original width) = –0.000 031 × 10 = 0.000 31 mm

6.6 Strain energy

Consider a length of material being stretched by tensile forces. Suppose we apply a force F to a strip of material and produce an extension x, as in Figure 6.13. The area under the graph is $\frac{1}{2}Fx$. But this is the product of the average force and the extension produced and so is the work done in stretching the material. If the volume of the material is AL then the work done per unit volume is $\frac{1}{2}Fx/AL$ and, since F/A is the stress and x/L is the strain:

work done per unit volume = ½ stress × strain

This is the energy stored in the material as a result of the stretching and is termed the *strain energy*. The energy is in joules (J) when stress is in Pa and volume in m³.

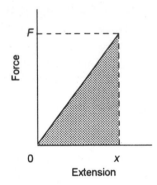

Figure 6.13   *Area under graph*

Example

A crane has a steel cable of length 10 m and cross-sectional area 1200 mm². What will be the strain energy stored in the cable when an object of mass 3000 kg is lifted by it? The modulus of elasticity of the steel is 210 GPa.

The strain energy per unit volume = ½ stress × strain and since stress/strain = modulus of elasticity E, if the cross-sectional area is A and the length L:

$$\text{strain energy} = \tfrac{1}{2}\frac{F}{A} \times \frac{F/A}{E} \times AL = \frac{F^2L}{2AE}$$

$$= \frac{(3000 \times 9.8)^2 \times 10}{2 \times 1200 \times 10^{-6} \times 210 \times 10^9} = 17.2 \text{ J}$$

Activities   1   Carry out the following simple experiments to obtain information about the tensile properties of materials when commercially made tensile testing equipment is not available. *Safety note*: when doing experiments involving the stretching of wires or other materials, the specimen may fly up into your face when it breaks. When a taut wire

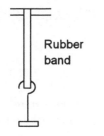

Rubber band

Figure 6.14 *Tensile test for rubber*

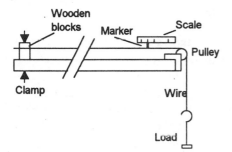

Figure 6.15 *Tensile test for a metal wire*

snaps, a lot of stored elastic energy is suddenly released. *Safety spectacles should be worn.*

Obtain a force–extension graph for rubber by hanging a rubber band (e.g. 74 mm by 3 mm by 1 mm band) over a clamp or other fixture, adding masses to a hanger suspended from it and measuring the extension with a ruler (Figure 6.14). In a similar way, obtain a force–extension graph for a nylon fishing line, the fishing line being tied to form a loop (e.g. about 75 cm long).

Determine the force–extension graph for a metal wire. Figure 6.15 shows the arrangement that can be used with, for example, iron wire with a diameter of about 0.2 mm, copper wire about 0.3 mm diameter or steel wire about 0.08 mm diameter, all having lengths of about 2.0 m.

Measure the initial diameter d of the wire using a micrometer screw gauge and the length L of the wire from the clamped end to the marker (a strip of paper attached by Sellotape) using a rule, a small load being used to give an initially taut wire. Add masses to the hanger and note the change in length e from the initial position. Hence plot a graph of force (F) against extension (e). Since $E =$ stress/strain $= (F/\frac{1}{4}\pi d^2)/(e/L)$ then $F = (E\pi d^2/4L)e$, determine the modulus of elasticity E from the gradient of the initial straight-line part of the graph.

Problems

1   What is the tensile stress acting on a test piece if a tensile force of 1.0 kN is applied to a cross-sectional area of 50 mm²?

2   A pipe has an external diameter of 35 mm and an internal diameter of 30 mm. What is the tensile stress acting on the pipe if it is subject to a tensile axial load of 800 N?

3   A tensile force acting on a rod of length 300 mm causes it to extend by 2 mm in the direction of the force. What is the strain?

4   A steel rod of length 100 mm has a constant diameter of 10 mm and when subject to an axial tensile load of 10 kN increases in length by 0.06 mm. What is (a) the stress acting on the rod, (b) the strain produced, (c) the modulus of elasticity?

5   A round tensile test piece has a gauge length of 100 mm and a diameter of 11.28 mm. If the material is an aluminium alloy with a modulus of elasticity of 70 GPa, what extension of the gauge length might be expected when the tensile load applied is 200 N?

6   A steel rod has a length of 1.0 m and a constant diameter of 20 mm. What will be its extension when subject to an axial tensile load of 60 kN if the modulus of elasticity of the material is 200 GPa?

7   A steel rod with a constant diameter of 25 mm and a length of 500 mm is subject to an axial tensile load of 50 kN. If this results in an extension of 0.25 mm, what is the modulus of elasticity?

8   An aluminium alloy has a tensile strength of 200 MPa. What force is needed to break a bar with a cross-sectional area of 250 mm²?

9   The following results were obtained from a tensile test of a steel. The test piece had a diameter of 10 mm and a gauge length of

50 mm. Plot the stress–strain graph and determine (a) the tensile strength, (b) the yield stress, (c) the tensile modulus.

Load/kN 0 5   10   15   20   25   30   32.5   35.8
Ext./mm 0 0.016 0.033 0.049 0.065 0.081 0.097 0.106 0.250

10 The following data were obtained from a tensile test on a stainless steel test piece. Determine (a) the limit of proportionality stress, (b) the tensile modulus.

Stress/MPa   0   90 170 255 345 495 605
Strain/$\times 10^{-4}$ 0   5   10   15   20   30   40

Stress/MPa   700 760 805 845 880 895
Strain/$\times 10^{-4}$ 50 60 70 80 90 100

11 Deduce the modulus of elasticity and the 0.1% proof stress for the material that gave the following tensile test results. The test piece had a diameter of 5.64 mm and a gauge length of 25 mm.

Load/kN 0 0.5   1.0   1.5   2.0   2.5   3.0   3.5
Ext./mm 0 0.0063 0.0125 0.0188 0.0250 0.0340 0.0431 0.0751

12 A flat tensile test piece of steel has a gauge length of 100.0 mm. After fracture, the gauge length was 131.1 mm. What is the percentage elongation?

13 A mild steel member in a truss has a uniform cross-section of 200 mm × 100 mm. What would be the maximum permissible tensile force for the member if the material has a yield stress of 200 MPa and a factor of safety of 4 is used?

14 A mild steel rod is required to withstand a maximum tensile load of 250 kN with a factor of safety of 4. What will be a suitable diameter for the rod if the material has a tensile strength of 540 MPa?

15 A flat tensile test piece has a gauge length of 100 mm. After fracture the gauge length was 112 mm. What is the percentage elongation?

16 Determine the thickness and diameter of a rubber pad of circular cross-section that can be used to support a machine if it is to carry a load of 4 kN and to compress by no more than 4 mm under this load. The stress in the rubber is not to exceed 250 kPa. The elastic modulus of the rubber is 2 MPa.

17 A steel pin of length 100 mm and cross-sectional area 500 mm² is subject to an axial load of 10 kN. What is the extension of the pin and the strain energy stored in it. The steel has an elastic modulus of 210 GPa.

18 A steel bar with a rectangular cross-section 50 mm × 30 mm and length 0.6 m is subject to an axial tensile load of 200 kN. If the steel has a tensile modulus of 210 GPa, what is the strain energy stored in the bar?

7 Linear motion

7.1 Introduction

This chapter is a review of the basic terms used in describing linear motion with a derivation of the equations used in tackling problems involving such motion. Also considered are the graphs of distance–time and velocity–time and the data that can be extracted from such graphs. The vector nature of velocity is considered and its resolution into components, this enabling problems to be tackled which involve projectiles. Chapter 8 extends this consideration of motion to that of angular motion and torque.

7.1.1 Basic terms

The following are basic terms used in the description of linear motion:

1 *Distance* is the distance along the path of an object, whatever the form of the path. Thus, if we say the distance covered in the motion of a car was, say, 3 km then the 3 km could have been covered along a straight road and the car be 3 km away from its start point. Another possibility is, however, that the 3 km was round a circular track and the car at the end of its 3 km might have been back where it started.

2 *Displacement* is the distance in a straight line between the start and end points of some motion. Thus a displacement of 3 km would mean that at the end of the motion that an object was 3 km away from the start point.

3 *Speed* is the rate at which distance is covered. Thus a car might be stated as having a speed of 50 km/h.

4 *Average speed* is the distance covered in a time interval divided by the time taken:

$$\text{average speed} = \frac{\text{distance covered}}{\text{time taken}}$$

A car which covers 80 km in 1 hour will have an average speed of 80 km/h over that time. During the hour it may, however, have gone faster than 80 km/h for part of the time and slower than that for some other part.

5 A *constant* or *uniform speed* occurs when equal distances are covered in equal intervals of time, however small we consider the time interval. Thus a car with an average speed of 60 km/h for 1 hour will be covering distance at the rate of 60 km/h in the first

minute, the second minute, over the first quarter of an hour, over the second half hour, indeed over any time interval in that hour.

6   *Velocity* is the rate at which displacement along a straight line changes with time. Thus an object having a velocity of 5 m/s means that the object moves along a straight line path at the rate of 5 m/s.

7   *Average velocity* is the displacement along a straight line occurring in a time interval divided by that time:

$$\text{average velocity} = \frac{\text{displacement occurring}}{\text{time taken}}$$

Thus an object having a displacement of 3 m along a straight line in a time of 2 s will have an average velocity in the direction of the straight line of 1.5 m/s over that time. During the 2 s there may be times when the object is moving faster or slower than 1.5 m/s.

8   A *constant* or *uniform velocity* occurs when equal displacements occur in the same straight line direction in equal intervals of time, however small the time interval. Thus an object with a constant velocity of 5 m/s in a particular direction for a time of 30 s will cover 5 m in the specified direction in each second of its motion.

9   *Acceleration* is the rate of change of velocity with time. The term *retardation* is often used to describe a negative acceleration, i.e. when the object is slowing down rather than increasing in velocity.

10   *Average acceleration* is the change of velocity occurring over a time interval divided by the time:

$$\text{average acceleration} = \frac{\text{change of velocity}}{\text{time taken}}$$

Thus if the velocity changes from 2 m/s to 5 m/s in 10 s then the average acceleration over that time is $(5 - 2)/10 = 0.3$ m/s². If the velocity changes from 5 m/s to 2 m/s in 10 s then the average acceleration over that time is $(2 - 5)/10 = -0.3$ m/s², i.e. it is a retardation.

11   A *constant* or *uniform acceleration* occurs when the velocity changes by equal amounts in equal intervals of time, however small the time interval. Thus an object with a constant acceleration of 5 m/s² in a particular direction for a time of 30 s will change its velocity by 5 m/s in the specified direction in each second of its motion.

7.2 Straight line motion   The equations that are derived in the following discussion all relate to uniformly accelerated motion in a straight line. If u is the initial velocity, i.e. at time $t = 0$, and v the velocity after some time t, then the change in velocity in the time interval t is $(v - u)$. Hence the acceleration a is $(v - u)/t$. Rearranging this gives:

$$v = u + at \qquad\qquad \text{[Equation 1]}$$

If the object, in its straight line motion, covers a distance s in a time t, then the average velocity in that time interval is s/t. With an initial velocity of u and a final velocity of v at the end of the time interval, the average velocity is $(u + v)/2$. Hence $s/t = (u + v)/2$ and so:

$$s = \left(\frac{u+v}{2}\right)t$$

Substituting for v by using the equation $v = u + at$ gives:

$$s = \left(\frac{u + (u + at)}{2}\right)t = \left(\frac{2u + at}{2}\right)t = ut + \tfrac{1}{2}at^2 \qquad \text{[Equation 2]}$$

Consider the equation $v = u + at$. Squaring both sides of this equation gives:

$$v^2 = (u + at)^2 = u^2 + 2uat + a^2t^2 = u^2 + 2a(ut + \tfrac{1}{2}at^2)$$

Hence, substituting for the bracketed term using equation 2:

$$v^2 = u^2 + 2as \qquad\qquad \text{[Equation 3]}$$

The equations [1], [2] and [3] are referred to as the equations for straight-line motion. The following examples illustrate their use in solving engineering problems.

Example

An object moves in a straight line with a uniform acceleration. If it starts from rest and takes 12 s to cover 100 m, what is the acceleration? If it continues with the same acceleration, how long will it take to cover the next 100 m and what will be its velocity after the 200 m?

For the first 100 m, we have $u = 0$, $s = 100$ m, $t = 12$ s and are required to obtain a. Using $s = ut + \frac{1}{2}at^2$:

$$100 = 0 + \tfrac{1}{2}a \times 12^2 = 0 + 72a$$

Hence $a = 100/72 = 1.4$ m/s^2.

To determine the time to cover the next 100 m we can use $s = ut + \frac{1}{2}at^2$ for the 200 m and subtract the time taken for the first 100 m. We have $u = 0$, $s = 200$ m and $a = 1.4$ m/s^2, hence:

$$200 = 0 + \tfrac{1}{2} \times 1.4 \times t^2$$

and so $t = 16.9$ s. Thus the time taken to cover the 100 m is $16.9 - 12 = 4.9$ s.

The velocity v after the 200 m can be determined using $v^2 = u^2 + 2as$ with $u = 0$, $a = 1.4$ m/s^2 and $s = 200$ m.

$$v^2 = 0 + 2 \times 1.4 \times 200$$

and $v = 23.7$ m/s.

Example

A car travelling at 25 m/s brakes and slows down with a uniform retardation of 1.2 m/s^2. How long will it take to come to rest?

A retardation is a negative acceleration in that the final velocity is less than the starting velocity. Thus we have $u = 25$ m/s, a final velocity of $v = 0$, retardation $a = -1.2$ m/s^2 and are required to obtain t. Using $v = u + at$, then:

$$0 = 25 + (-1.2)t = 25 - 1.2t$$

Thus $1.2t = 25$ and so $t = 25/1.2 = 20.8$ s.

Example

A car is moving with a velocity of 10 m/s. It then accelerates at 0.2 m/s^2 for 100 m. What will be the time taken for the 100 m to be covered?

We have $u = 10$ m/s, $a = 0.2$ m/s^2 and $s = 100$ m and are required to obtain t. Using $s = ut + \frac{1}{2}at^2$:

$$100 = 10t + \frac{1}{2} \times 0.2t^2$$

We can write this equation as:

$$0.1t^2 + 10t - 100 = 0$$

This is a quadratic equation. Using the formula for the solution of quadratic equations:

$$t = \frac{-b \pm \sqrt{b^2 - 4ac}}{2a}$$

then:

$$t = \frac{-10 \pm \sqrt{100 - 4 \times 0.1(-100)}}{2 \times 0.1} = \frac{-10 \pm \sqrt{140}}{0.2}$$

Hence $t = -50 - 59.2 = -109.2$ s or $t = -50 + 59.2 = 9.2$ s. Since the negative time has no significance, the answer is 9.2 s. The answer can be checked by substituting it back in the original equation.

Example

A stationary car A is passed by car B moving with a uniform velocity of 15 m/s. Two seconds later, car A starts moving with a constant acceleration of 1 m/s² in the same direction. How long will it take for car A to draw level with car B?

Both the cars will have travelled the same distance s when they have drawn level. We will measure time t from when car B initially passes car A. For car B, since $u = 15$ m/s and $a = 0$, using $s = ut + \frac{1}{2}at^2$ gives:

$$s = 15t + 0$$

For car A, since $u = 0$ and $a = 1$ m/s², we can write, with $(t - 2)$ for the time since car A is in motion for a time 2 s less than t:

$$s = 0 + \frac{1}{2} \times 1 \times (t - 2)^2$$

These two simultaneous equations can be solved by substituting for s in the second equation. Thus:

$$15t = 0.5(t - 2)^2 = 0.5(t^2 - 4t + 4) = 0.5t^2 - 2t + 2$$

Hence we obtain the quadratic equation:

$$0.5t^2 - 17t + 2 = 0$$

Using the formula for the solution of a quadratic equation:

$$t = \frac{-b \pm \sqrt{b^2 - 4ac}}{2a} = \frac{17 \pm \sqrt{17^2 - 4 \times 0.5 \times 2}}{2 \times 0.5}$$

Hence $t = 17 \pm \sqrt{285}$ and $t = 17 + 16.9 = 33.9$ s or $t = 17 - 16.9 = 0.1$ s. With the data given, this second answer is not feasible and so the time is 33.9 s after car B has passed car A.

7.3 Velocity and acceleration as vectors

Velocity and acceleration are vector quantities (see Chapter 5). A *vector* quantity is one for which both its magnitude and direction have to be stated for its effects to be determined; they have to be added by methods which take account of their directions, e.g. the parallelogram method.

Example

A projectile is thrown vertically upwards with a velocity of 10 m/s. If there is a horizontal wind blowing at 5 m/s, what will be the velocity with which the projectile starts out?

Figure 7.1 shows the vectors representing the two velocities and their resultant, i.e. sum, determined from the parallelogram of

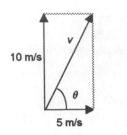

Figure 7.1 *Example*

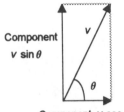

Figure 7.2 *Resolving a velocity into components*

vectors. We can use a scale drawing to obtain the resultant or, because the angle between the two velocities is 90° we can use the Pythagoras theorem to give $v^2 = 10^2 + 5^2$. Hence $v = 11.2$ m/s. This velocity will be at an angle θ to the horizontal, where $\tan \theta = 10/5$ and so $\theta = 63.4°$.

7.3.1 Resolution into components

A single vector can be resolved into two components at right angles to each other by using the parallelogram of vector method of summing vectors in reverse. For example, a velocity v at an angle θ to the horizontal can be resolved into a horizontal component of $v \cos \theta$ and a vertical component of $v \sin \theta$ (Figure 7.2).

Example

A projectile is fired from a gun with a velocity of 200 m/s at 30° to the horizontal. What is (a) the horizontal velocity component, (b) the vertical velocity component?

(a) Horizontal component = $v \cos \theta = 200 \cos 30° = 173$ m/s.
(b) Vertical component = $v \sin \theta = 200 \sin 30° = 100$ m/s.

7.4 Motion under the action of gravity

All freely falling objects in a vacuum fall with the same uniform acceleration directed towards the surface of the earth as a result of a gravitational force acting between the object and the earth. This acceleration is termed the *acceleration due to gravity g*. For most practical purposes, the acceleration due to gravity at the surface of the earth is taken as being 9.81 m/s². The equations for motion of a falling object are those for motion in a straight line with the acceleration as g.

When an object is thrown vertically upwards it suffers an acceleration directed towards the surface of the earth. An acceleration directed in the opposite direction to which an object is moving is a retardation, i.e. a negative acceleration since it results in a final velocity less than the initial velocity. The result is that the object slows down. The object slows down until its velocity upwards eventually becomes zero, it then having attained its greatest height above the ground. Then the object reverses the direction of its motion and falls back towards the earth, accelerating as it does; we have then the acceleration of $+g$. Figure 7.5 illustrates these points.

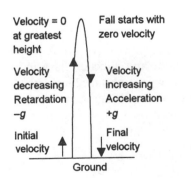

Figure 7.5 *Vertical motion under gravity*

Example

A stone is dropped down a vertical shaft and reaches the bottom in 5 s. How deep is the shaft? Take g as 9.8 m/s².

We have $u = 0$, $t = 5$ s and $a = g = 9.8$ m/s². We have to determine the distance fallen, h. Thus, using $s = ut + \frac{1}{2}at^2$, we have $h = 0 + \frac{1}{2} \times 9.8 \times 5^2 = 122.5$ m

Example

An object is thrown vertically upwards with a velocity of 8.0 m/s. What will be the greatest height reached and the time taken to reach that height? Take g as 9.8 m/s^2.

We have $u = 8.0$ m/s, $a = -g = -9.8$ m/s^2 and, at the greatest height H, $v = 0$. We have initially to determine H. The velocities and displacement are measured in an upward direction and so the acceleration due to gravity is negative. Thus, using $v^2 = u^2 + 2as$, $0 = 8.0^2 + 2(-9.8)H$ and so $H = 64/19.6 = 3.27$ m. To determine t we can use $v = u + at$ to give $0 = 8.0 + (-9.8)t$. Thus $9.8t = 8.0$, and so $t = 8.0/9.8 = 0.82$ s.

7.4.1 Motion down an inclined plane

For free fall the acceleration is g downwards. However, for objects moving down a smooth inclined plane, as in Figure 7.6, vertical motion is not possible. The result is an acceleration down the plane which is due to the resolved component of g in that direction:

acceleration down plane due to gravity = $g \sin \theta$

Example

A smooth inclined chute is used to send boxed goods down from the store to delivery trucks. If the ramp has a maximum vertical height of 4.0 m and is inclined at 30° to the horizontal, what will be the velocity of the boxes at the bottom of the chute and the time taken?

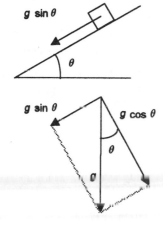

Figure 7.6 *Motion down a smooth plane under gravity*

With $u = 0$, $a = g \sin \theta$, $s = 4/\sin \theta$ (i.e. the length of the slope):

$$v^2 = u^2 + 2as = 0 + 2 \times 9.8 \sin 30° \times (4/\sin 30°)$$

and so $v = 8.9$ m/s. Using $v = u + at$ gives $8.9 = 0 + (9.8 \sin 30°) \times t$ and hence $t = 1.8$ s.

7.5 Graphs of motion

This section is a discussion of how graphs can be used to describe the motion of an object.

7.5.1 Distance–time graphs

If the distance moved by an object in a straight line, from some reference point on the line, is measured for different times then a distance–time graph can be plotted. For the distance–time graph shown in Figure 7.7, if the distance changes from s_1 to s_2 when the time changes from t_1 to t_2, the average velocity is $(s_2 - s_1)/(t_2 - t_1)$. But this is just the gradient of the graph. Thus:

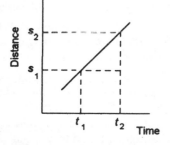

Figure 7.7 *Distance–time graph*

velocity = gradient of distance–time graph

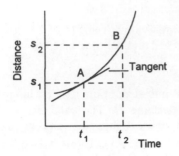

Figure 7.8 *Distance–time graph*

With a straight-line graph, the gradient is constant and the distance changes by equal amounts in equal intervals of time, however small a time a time interval we consider. There is thus a uniform velocity.

When the graph is not a straight line, as in Figure 7.8, the gradient and hence the velocity is changing. $(s_2 - s_1)$ is the distance travelled in a time of $(t_2 - t_1)$ and the average velocity over that time is $(s_2 - s_1)/(t_2 - t_1)$. The smaller we make the times between A and B then the more the average is taken over a smaller time interval and so the more it approximates to the instantaneous velocity. An infinitesimal small time interval means we have the tangent to the curve. Thus if we want the velocity at an instant of time, say A, then we have to determine the gradient of the tangent to the graph at A. Thus:

instantaneous velocity = gradient of tangent to the distance–time graph at that instant

Example

A car has a constant velocity of 4 m/s for the time from 0 to 3 s and then zero velocity from 3 s to 5 s. Sketch the distance–time graph.

From 0 to 3 s the gradient of the distance–time graph is 4 m/s. From 3 s to 5 s the velocity is zero and so the gradient is zero. The graph is thus as shown in Figure 7.9.

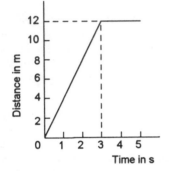

Figure 7.9 *Example*

Example

Figure 7.10 shows an example of a distance–time graph. Describe how the velocity is changing in the motion from A to F.

From the start position at A to B, the distance increases at a constant rate with time. The gradient is constant over that time and so the velocity is constant and equal to $(4 - 0)/(2 - 0) = 2$ m/s.

From B to C there is no change in the distance from the reference point and so the velocity is zero, i.e. the object has stopped moving. The gradient of the line between B and C is zero.

From C to E the distance increases with time but in a non-uniform manner. The gradient changes with time. Thus the velocity is not constant during that time.

The velocity at an instant of time is the rate at which the distance is changing at that time and so is the gradient of the graph at that time. Thus to determine the velocity at point D we draw a tangent to the graph curve at that instant. So at point D my estimate of the velocity is about $(6.0 - 3.2)/(5 - 3) = 1.4$ m/s.

At point E the distance–time graph shows a maximum. The gradient changes from being positive prior to E to negative after E. At point E the gradient is momentarily zero. Thus the velocity changes from being positive prior to E to zero at E and then negative after E. At E the velocity is zero.

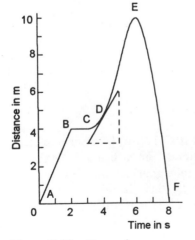

Figure 7.10 *Example*

From E to F the gradient is negative and so the velocity is negative. A negative velocity means that the object is going in the opposite direction and so is moving back to its starting point, i.e. the distance is becoming smaller rather than increasing. In this case, the object is back at its starting point after a time of 8 s.

7.5.2 Velocity–time graphs

If the velocity of an object is measured at different times then a velocity–time graph can be drawn. Acceleration is the rate at which the velocity changes. Thus, for the graph shown in Figure 7.11, the velocity changes from v_1 to v_2 when the time changes from t_1 to t_2. Thus the acceleration over that time interval is $(v_2 - v_1)/(t_2 - t_1)$. But this is the gradient of the graph. Thus:

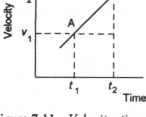

Figure 7.11 *Velocity–time graph*

acceleration = gradient of the velocity–time graph

With a straight line graph, the gradient is the same for all points and so we have a uniform acceleration. When the graph is not straight line, as in Figure 7.12, the acceleration is no longer uniform. $(v_2 - v_1)$ is the change in velocity in a time of $(t_2 - t_1)$ and thus $(v_2 - v_1)/(t_2 - t_1)$ represents the average acceleration over that time. The smaller we make the times between A and B then the more the average is taken over a smaller time interval and more closely approximates to the instantaneous acceleration. An infinitesimal small time interval means we have the tangent to the curve. Thus if we want the acceleration at an instant of time then we have to determine the gradient of the tangent to the graph at that time, i.e.

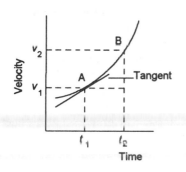

Figure 7.12 *Velocity–time graph*

instantaneous acceleration = gradient of tangent to the
velocity–time graph at that instant

The distance travelled by an object in a particular time interval is average velocity over a time interval × the time interval. Thus if the velocity changes from v_1 at time t_1 to v_2 at time t_2, as in Figure 7.13, then the distance travelled between t_1 and t_2 is represented by the product of the average velocity and the time interval. But this is equal to the area under the graph line between t_1 and t_2. Thus:

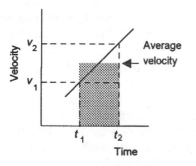

Figure 7.13 *Distance travelled*

distance travelled between t_1 and t_2 = area under the graph between
these times

Example

Figure 7.14 shows the velocity–time graph for a train travelling between two stations. What is (a) the distance between the stations and (b) the initial acceleration?

(a) The distance travelled is the area under the graph. This is the sum of the areas of the triangle ABE, the rectangle BCEF and the

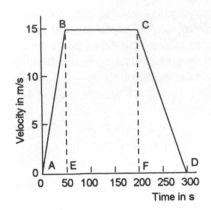

Figure 7.14   *Example*

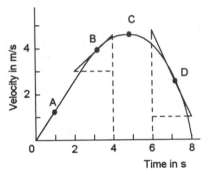

Figure 7.15   *Example*

triangle CDF. Hence, distance travelled = ½ × 15 × 50 + 15 × 50 + ½ × 15 × 100 = 3375 m.

(b) The acceleration is the gradient of the graph and so acceleration = 15/50 = 0.3 m/s².

Example

For the velocity–time graph shown in Figure 7.15, determine the acceleration at points A, B and C and estimate the distance covered in the 8 s.

Initially the graph is a straight line of constant gradient. The gradient, and hence the acceleration, at A is 2/2 = 1 m/s².

At B the graph is no longer a straight line. The tangent to the curve at B has a gradient, and hence an acceleration, of about (4.5 – 3.0)/(4 – 2) = 0.75 m/s².

At C the velocity is not changing with time. Thus there is a uniform velocity with zero acceleration.

At D the velocity is decreasing in a non-uniform manner with time. The tangent to the curve at that point has a gradient of about (0.5 – 4.8)/(8 – 6) = –1.15 m/s².

The distance covered in the 8 s is the area under the graph. Using the mid-ordinate rule for the estimation of the area with the time interval divided into four strips gives the distance as 1.2 × 2 + 3.8 × 2 + 4.6 × 2 + 2.6 × 2 = 24.4 m. Alternatively we could have estimated the area by counting the number of graph squares under the graph and multiplying the result by the area of one graph square.

Activities

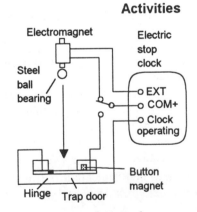

Figure 7.16   *Activity 1*

1 Determine the acceleration due to gravity by an experiment involving the free fall of some object. One possible method is to use an electric stop clock to measure the time t taken for a steel ball bearing to fall over a distance s of about 1 m (Figure 7.16). With the two-way switch in the 'up' position, the ball bearing is held by an electromagnet. When the switch is changed to the 'down' position, the magnet releases the ball and simultaneously starts the clock. When the ball reaches the end of its fall, it strikes a 'trap-door' switch which opens and stops the clock. The equation $s = \frac{1}{2}gt^2$ can then be used to compute g.

2 Analyse the motion of a car as it accelerates from rest and is changed through the gears from measurements made of the speedometer reading at different times during the motion. From your results obtain velocity–time and speed–time graphs.

Problems

1 A train is moving with a velocity of 10 m/s. It then accelerates at a uniform rate of 2.5 m/s² for 8 s. What is the velocity after the 8 s?

2 An object starts from rest and moves with a uniform acceleration of 2 m/s² for 20 s. What is the distance moved by the object?

3 A cyclist is moving with a velocity of 1 m/s. He/she then accelerates at 0.4 m/s^2 for 100 m. What will be the time taken for the 100 m?

4 A car accelerates from 7.5 m/s to 22.5 m/s at 2 m/s^2. What is (a) the time taken, (b) the distance travelled during this acceleration?

5 A conveyor belt is used to move an object along a production line with an acceleration of 1.5 m/s^2. What is the velocity after it has moved 3 m from rest?

6 A train starts from rest and moves with a uniform acceleration so that it takes 300 s to cover 9000 m. What is the acceleration?

7 A car is initially at rest. It then accelerates at 2 m/s^2 for 6 s. What will be the velocity after that time?

8 A car accelerates at a uniform rate from 15 m/s to 35 m/s in 20 s. How far does the car travel in this time?

9 An object has an initial velocity of 10 m/s and is accelerated at 3 m/s^2 for a distance of 800 m. What is the time taken to cover this distance?

10 A stone is thrown vertically upwards with an initial velocity of 15 m/s. What will be the distance from the point of projection and the velocity after 1 s?

11 A stone is thrown vertically upwards with an initial velocity of 5 m/s. What will be the time taken for it to reach the greatest height?

12 A stone falls off the edge of a cliff. With what velocity will it hit the beach below if the height of the cliff above the beach is 50 m?

13 What is the velocity attained by an object falling from rest through a distance of 4.9 m?

14 An object is thrown vertically upwards with an initial velocity of 8 m/s. How long will it take for the object to return to the same point from which it was thrown?

15 A fountain projects water vertically to a height of 5 m. What is the velocity with which the water must be leaving the fountain nozzle?

16 A stone is thrown vertically upwards with an initial velocity of 9 m/s. What is the greatest height reached by the stone and the time taken?

17 An object is thrown vertically upwards with a velocity of 20 m/s from an initial height h above the ground. It takes 5 s from the time of being projected upwards before the object hits the ground. Determine h.

18 An object is thrown vertically upwards with a velocity of 30 m/s and, from the same point at the same time, another object is thrown vertically downwards with a velocity of 30 m/s. How far apart will the objects be after 3s?

19 An object moving in a straight line gives the following distance–time data. Plot the distance–time graph and hence determine the velocities at times of (a) 2 s, (b) 3 s, (c) 4 s.

distance in mm	0	10	36	78	136	210
time in s	0	1	2	3	4	5

20 An object moving in a straight line gives the following distance–time data. Plot the distance–time graph and hence determine the velocities at times of (a) 5 s, (b) 8 s, (c) 11 s.

distance in m	0	1	15	31	49	75	110
time in s	0	1	4	6	8	10	12

21 A stone is thrown vertically upwards. The following are the velocities at different times. Plot a velocity–time graph and hence determine (a) the acceleration at the time $t = 1$ s, (b) the acceleration at $t = 2.5$ s, (c) the distance travelled to the maximum height where the velocity is zero.

velocity in m/s	15	10	5	0	−5	−10	−15
time in s	0	0.5	1.0	1.5	2.0	2.5	3.0

22 The velocity of an object varies with time. The following are values of the velocities at a number of times. Plot a velocity–time graph and hence estimate the acceleration at a time of 2 s and the total distance travelled in the 5 s.

velocity in m/s	0	1.2	2.4	3.6	3.6	3.6
time in s	0	1	2	3	4	5

23 The velocity of an object changes with time. The following are values of the velocities at a number of times. Plot the velocity–time graph and hence determine the acceleration at a time of 2 s and the total distance travelled between 1 s and 4 s.

velocity in m/s	25	24	21	16	9	0
time in s	0	1	2	3	4	5

24 State whether the distance–time graphs and the velocity–time graphs will be straight line with a non-zero gradient or zero gradient or curved in the following cases: (a) the velocity is constant and not zero, (b) the acceleration is constant and not zero, (c) the distance covered in each second of the motion is the same, (d) the distance covered in each successive second doubles.

25 Car A, travelling with a uniform velocity of 25 m/s, overtakes a stationary car B. Two seconds later, car B starts and accelerates at 6 m/s². How far will B have to travel before it catches up A?

8 Angular motion

This chapter is concerned with describing angular motion, deriving and using the equations for such motion and relating linear motion of points on the circumference of rotating objects with their angular motion. The term torque is introduced and the equation derived for the work done and power developed when an object is rotated by a torque.

8.1.1 Basic terms

The following are basic terms used to describe angular motion.

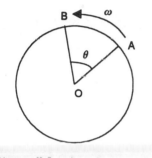

Figure 8.1 *Angular motion*

1 *Angular displacement*
    The angular displacement is the angle swept out by the rotation and is measured in radians. Thus, in Figure 8.1, the radial line rotates through an angular displacement of θ in moving from OA to OB. One complete rotation through 360° is an angular displacement of 2π rad; one quarter of a revolution is 90° or $\pi/2$ rad. As 2π rad = 360°, then 1 rad = 360°/2π or about 57°.

2 *Angular velocity*
    Angular velocity ω is the rate at which angular displacement occurs, the unit being rad/s.

3 *Average angular velocity*
    The average angular velocity over some time interval is the change in angular displacement during that time divided by the time. Thus, in Figure 8.1, if the angular displacement θ takes a time t then the average angular velocity over that time interval ω is θ/t.
    If a body is rotating at f revolutions per second then the angular displacement in 1 s is $2\pi f$ rad and so it has an average angular velocity given by:

$$\omega = 2\pi f$$

4 *Constant angular velocity*
    A constant or uniform angular velocity occurs when equal angular displacements occur in equal intervals of time, however small we consider the time interval.

5 *Angular acceleration*
    Angular acceleration is the rate at which angular velocity is changing, the unit being rad/s².

6   *Average angular acceleration*
The average angular acceleration over some time interval is the change in angular velocity during that time divided by the time.

7   *Constant angular acceleration*
A constant or uniform angular acceleration occurs when the angular velocity changes by equal amount in equal intervals of time, however small we consider the time interval.

Example

What is the angular displacement if a body makes 5 revolution?

Since 1 revolution is an angular displacement of 2π rad then 5 revolutions is $5 \times 2\pi = 31.4$ rad.

Example

Express the angular velocity of 6 rad/s in terms of the number of revolutions made per second.

Using $\omega = 2\pi f$ then $f = \omega/2\pi = 6/2\pi = 0.95$ rev/s.

Example

A body rotates at 2 rev/s. What is its angular velocity in rad/s?

Using $\omega = 2\pi f$ then $\omega = 2\pi \times 2 = 12.7$ rad/s.

8.2 Equations of motion

For a body rotating with a constant angular acceleration a, when the angular velocity changes from ω_0 to ω in a time t, then $a = (\omega - \omega_0)/t$ and hence we can write:

$$\omega = \omega_0 + at \qquad \text{[Equation 1]}$$

This should be compared with $v = u + at$ for linear motion.

The average angular velocity during this time is $\frac{1}{2}(\omega + \omega_0)$. If the angular displacement during the time is θ then the average angular velocity is θ/t and so $(\theta/t) = \frac{1}{2}(\omega + \omega_0)$. Substituting for ω using equation [1] gives:

$$\frac{\theta}{t} = \frac{\omega_0 + at + \omega_0}{2}$$

$$\theta = \omega_0 t + \frac{1}{2}at^2 \qquad \text{[Equation 2]}$$

This should be compared with $s = ut + \frac{1}{2}at^2$ for linear motion.

Squaring equation [1] gives:

$$\omega^2 = (\omega_0 + at)^2 = \omega_0^2 + 2a\omega_0 + a^2t^2 = \omega_0^2 + 2a(\omega_0 + \tfrac{1}{2}at^2)$$

Hence, using equation [2]:

$$\omega^2 = \omega_0^2 + 2a\theta \qquad \text{[Equation 3]}$$

This should be compared with $v^2 = u^2 + 2as$ for linear motion.

Example

An object which was rotating with an angular velocity of 4 rad/s is uniformly accelerated at 2 rad/s. What will be the angular velocity after 3 s?

$$\omega = \omega_0 + at = 4 + 2 \times 3 = 10 \text{ rad/s}.$$

Example

The blades of a fan are uniformly accelerated and increase in frequency of rotation from 500 to 700 rev/s in 3.0 s. What is the angular acceleration?

Since $\omega = 2\pi f$, the equation $\omega = \omega_0 + at$ gives:

$$2\pi \times 700 = 2\pi \times 500 + a \times 3.0$$

Hence $a = 419 \text{ rad/s}^2$.

Example

A flywheel, starting from rest, is uniformly accelerated from rest and rotates through 5 revolutions in 8 s. What is the angular acceleration?

The angular displacement in 8 s is $2\pi \times 5$ rad. Hence, using the equation $\theta = \omega_0 t + \frac{1}{2}at^2$:

$$2\pi \times 5 = 0 + \frac{1}{2}a \times 8^2$$

Hence the angular acceleration is 0.98 rad/s^2.

Example

A wheel starts from rest and accelerates uniformly with an angular acceleration of 4 rad/s^2. What will be its angular velocity after 4 s and the total angle rotates in that time?

$$\omega = \omega_0 + at = 0 + 4 \times 4 = 16 \text{ rad/s}$$

$$\theta = \omega_0 t + \frac{1}{2}at^2 = 0 + \frac{1}{2} \times 4 \times 4^2 = 32 \text{ rad}.$$

8.3 Relationship between linear and angular motion

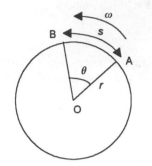

Figure 8.2 *Angular motion*

Consider the radial arm of radius r in Figure 8.2 rotating from OA to OB. When the radial arm rotates through angle θ from OA to OB, the distance moved by the end of the radial arm round the circumference is AB. One complete revolution is a rotation through 2π rad and point A moves completely round the circumference, i.e. a distance of $2\pi r$. Thus a rotation through an angle of 1 rad has A moving a circumferential distance of r and so a rotation through an angle θ has point A moving through a circumferential distance of $r\theta$. Hence, if we denote this circumferential distance by s then:

$$s = r\theta$$

If the point is moving with constant angular velocity ω then in time t the angle rotated will be ωt and so $s = r\omega t$. But s/t is the linear speed v of point A round the circumference. Hence:

$$v = r\omega$$

Now consider the radial arm rotating with a constant angular acceleration. If the point A had an initial linear velocity u then its angular velocity ω_0 would be given by $u = r\omega_0$. If it accelerates with a uniform angular acceleration a to an angular velocity ω in a time t then $a = (\omega - \omega_0)/t$. If the point A now has a linear velocity v then $v = r\omega$. Hence:

$$a = \frac{(v/r) - (u/r)}{t}$$

The linear acceleration a of the point A is $(v - u)/t$ and so:

$$ra = \frac{v - u}{r} = a$$

Thus:

$$a = ra$$

Example

What is the peripheral velocity of a point on the rim of a wheel when it is rotating at 3 rev/s and has a radius of 200 mm?

$$v = r\omega = r \times 2\pi f = 0.200 \times 2\pi \times 3 = 3.8 \text{ m/s.}$$

Example

The linear speed of a belt passing round a pulley wheel of radius 150 mm is 20 m/s. If there is no slippage of the belt on the wheel, how many revolutions per second are made by the wheel?

$$\omega = v/r = 20/0/150 = 133 \text{ rad/s, hence } f = \omega/2\pi = 133/2\pi = 21 \text{ rev/s.}$$

Example

The wheels of a car have a diameter of 700 mm. If they increase their rate of rotation from 50 rev/min to 1100 rev/min in 40 s, what is the angular acceleration of the wheels and the linear acceleration of a point on the tyre tread?

Using $\omega = \omega_0 + at$, then:

$$2\pi \times \frac{1100}{60} = 2\pi \times \frac{50}{60} + a \times 40$$

Hence the angular acceleration is 2.75 rad/s^2. Using $a = ra$, then $a = 0.350 \times 2.75 = 0.96$ m/s.

8.4 Torque

The turning effect of a force F about a point O is the moment of the force about that point and is the product of the force and the perpendicular distance r of O from the line of action of the force (Figure 8.3):

moment of F about O $= Fr$

If the force F is applied to the surface of a shaft of radius r (Figure 8.4), then the turning moment of the force about the centre of the shaft O is Fr. A reactive force R is set up which is equal in magnitude and opposite in direction to F, i.e. $R = -F$, and can be considered to act at O. This pair of oppositely directed but equal in magnitude forces which are not in the same straight line is called a *couple*. The turning moment of F about O is Fr in an anticlockwise direction and that of the reactive force R about A is $Rr = -Fr$ and this still gives an anticlockwise rotational moment. Indeed if we take moments about any point between OA we obtain the same result, an anticlockwise moment of Fr.

The turning moment of a couple is called the *torque* T. Thus:

$$T = Fr$$

With the force in N and r in m, the torque is in units of N m.

Figure 8.3 *Moment*

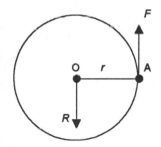

Figure 8.4 *Couple*

Example

Determine the torque acting on the teeth of a gear wheel of effective radius 150 mm, if a tangential force of 200 N is applied to the wheel.

$$T = Fr = 200 \times 0.150 = 30 \text{ N}.$$

8.4.1 Torque and work done

When a force is used to rotate an object then work is done since the point of application of the force moves through some distance. Thus if a force F causes a rotation from A to B in Figure 8.5, and angle θ is swept out, then the work = force × distance = $Fs = Fr\theta$. The torque $T = Fr$, hence:

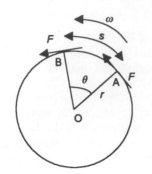

Figure 8.5 *Angular motion*

work = $T\theta$

If the rotation is f revolutions per second, then in 1 s there are f revolutions and, since each revolution is 2π radians, the angle rotated in 1 s is $2\pi f$. Hence the work done per second = $T \times 2\pi f$. The work done per second is the power and so, since the angular velocity $\omega = 2\pi f$:

power = $T\omega$

Example

A constant torque of 50 N m is used to keep a flywheel rotating at a constant 5 rev/s. What is the power input to the flywheel?

Since power = $T\omega$ and $\omega = 2\pi f$, then power = $T \times 2\pi f = 50 \times 2\pi \times 5$ = 1571 W = 1.571 kW.

Activities

1 A stroboscope flashing light emits pulses of light at regular intervals of time and if a rotating object is viewed by means of such light it can be made to appear to still if the flashing light is at such a frequency that every time the object is viewed it has rotated to the same position. The minimum frequency the light has to have to achieve this is the same frequency as that of the rotation of the object. Use such a flashing light to determine the frequency of rotation of some object such as a hand power tool.

2 The output power of a motor can be determined by a form of mechanical brake; Figure 8.6 shows one such form. A belt or rope is wrapped round a pulley driven by the motor. One end of the belt is attached by means of a spring balance to a horizontal support and the other end to a weight. With the arrangement shown in the figure, when the pulley is rotated in a clockwise direction the reading of the balance is taken and the value of the weight noted. The torque due to brake friction applied to the pulley is $r(F_1 - F_2)$ where r is the the the radius of the pulley, F_1 is equal to the weight and F_2 is equal to the reading on the spring balance. If the motor has a rotational frequency of f then the output power is $r(F_1 - F_2) \times 2\pi f$. The output power of the motor is converted by the brake into heat. Use this, or a similar method, to determine the output power of a motor.

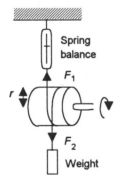

Figure 8.6 *Activity 2*

Problems

1 Express (a) an angular rotation of 100 revolutions as an angular displacement in radians, (b) an angular displacement of 600 radians into the number of revolutions, (c) an angular rotation of 15 rev/s into an angular velocity in rad/s, (d) an angular speed of 8 rad/s into the number of revolutions made per second, (e) an angular rotation of 90 rev/min into an angular velocity in rad/s.

2 What is the angular velocity in rad/s of a pulley which completes one revolution every ten seconds?

3   What is the angular velocity in rad/s of a turntable which rotates at 33.3 rev/min?

4   A flywheel rotating at 3.5 rev/s is accelerated uniformly for 4 s until it is rotating at 9 rev/s. Determine the angular acceleration and the number of revolutions made by the flywheel in the 4 s.

5   A flywheel rotating at 20 rev/min is accelerated uniformly for 10 s until it is rotating at 40 rev/min. Determine the angular acceleration and the number of revolutions made by the flywheel in the 10 s.

6   A flywheel rotating at 210 rev/min is uniformly accelerated to 250 rev/min in 5 s. Determine the angular acceleration and the number of revolutions made by the flywheel in that time.

7   A flywheel rotating at 0.5 rev/s is uniformly accelerated to 1.0 rev/s in 10 s. Determine the angular acceleration and the number of revolutions made by the flywheel in that time.

8   A grinding wheel is rotating at 50 rev/s when the power is switched off. It takes 250 s to come to rest. What is the average angular retardation?

9   A wheel of diameter 350 mm rotates with an angular velocity of 6 rad/s. What is the speed of a point on its circumference?

10 A flywheel of diameter 360 mm increases its angular speed uniformly from 10.5 to 11.5 rev/s in 11 s. Determine (a) the angular acceleration of the wheel, (b) the linear acceleration of a point on the wheel rim.

11 What is the peripheral speed of the tread on a car tyre of diameter 700 mm if the wheel rotates about its axle with an angular velocity of 6 rad/s?

12 A car has wheels of diameter 550 mm and is travelling along a straight road with a constant speed of 20 m/s. What is the angular velocity of the wheel?

13 A pulley attached to a shaft has a radius of 50 mm and rotates at 24 rev/s. What is the linear speed of the pulley belt wrapped round the wheel if no slippage occurs?

14 A bicycle has wheels of diameter 620 mm and is being pedalled along a road at 6.2 m/s. What is the angular velocity of the wheels?

15 A cord is wrapped around a wheel of diameter 400 mm which is initially at rest (Figure 8.7). When the cord is pulled, a tangential acceleration of 4 m/s² is applied to the wheel. What is the angular acceleration of the wheel?

16 Determine the torque acting on a wheel of radius 20 mm if a tangential force of 1 kN is applied to the wheel rim.

17 A 100 mm diameter bar is being turned in a lathe. If the force on the cutting tool, which is tangential to the surface of the bar, is 1.2 kN, what is the torque being applied?

18 At what speed must an electric motor be running if it develops a torque of 6 kN m and a power of 200 kW?

19 What is the power developed by a motor running at 30 rev/s and developing a torque of 4 kN m?

Figure 8.7   *Problem 15*

9 Dynamics

9.1 Introduction

Dynamics is the study of objects in motion. This chapter thus follows on from Chapter 7, where the terms and equations used in describing linear motion were introduced. Here we now consider the forces responsible for motion and deal with the forces, linear momentum and energy involved with linear motion; the torques, angular momentum and energy involved with angular motion, motion in a circular path and the motion of oscillating bodies are dealt with in later chapters.

9.2 Newton's laws

The fundamental laws involved with the study of dynamics are *Newton's laws of motion*. These can be stated as:

Law 1
> *A body will continue in a state of rest or uniform motion in a straight line unless it is compelled to change that state by an externally applied force.*

> Thus if an object is at rest then, unless a resultant force is applied to it there will be no motion. If an object is moving with a constant velocity then it will keep on in this motion until some externally applied force causes it to change its direction of motion and/or the magnitude of its velocity.

Law 2
> *The rate of change of momentum of a body is proportional to the applied external force and takes place in the direction of action of that force.*

> *Momentum* is defined as being the product of mass m and velocity v of a body. It is a vector quantity with the basic unit of kg m/s. The second law can thus be written as:

force $\propto$ rate of change of momentum (mv)

The unit of force, the newton (N), is defined so that when the mass is in kg and the velocity in m/s, the force is in N and so:

force = rate of change of momentum (mv)

If mass m is constant, the above expression becomes:

force = $m \times$ rate of change of $v = ma$

where the rate of change of velocity is the acceleration a. This law thus enables us to calculate the force needed to change the momentum of a body and to accelerate it.

Law 3
If one body exerts a force on a second body then the second body exerts an equal and opposite force on the first, i.e. to every action there is an opposite and equal reaction.

Thus if a gun fires bullets in one direction, i.e. exerts a force on them to propel them in that direction, the gun will experience a recoil force which is equal in size but in the opposite direction to the force propelling the bullets out of the gun.

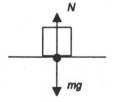

Figure 9.1   *The reaction force*

The *mass* of a body is a measure of the quantity of matter it contains. As the equation $F = ma$ indicates, it can also be considered to be the quantity which determines the measure of the resistance of a body to be accelerated by a force; the more mass a body has the greater the force needed to give it a particular acceleration and so the greater *inertia* it has. The *weight* of a body at rest at the earth's surface is equal to the gravitational force the earth exerts on it. If we allow a body to freely fall it will accelerate at the acceleration due to gravity g under the action of the gravitational force. Thus, using force = ma:

weight = mass × acceleration due to gravity

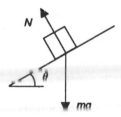

Figure 9.2   *The reaction force*

When an object of mass m is at rest on a horizontal surface there can be no resultant force acting on it (Newton's first law). Thus the weight mg of the object, which acts at right angles to the horizontal surface, must be balanced by some reaction force N which is at right angles, i.e. normal, to the surface and acting on the mass (Figure 9.1). If the object is resting on an inclined plane the reaction force is normal to the surface and must be equal to the component of the weight which is at right angles to the surface (Figure 9.2). Thus $N = mg \cos \theta$.

Example

A caravan of mass 1100 kg is towed by a car with an acceleration of 0.15 m/s². If the resistance to motion is 150 N, what is the force exerted by the car through the tow bar?

The force acting on the caravan to give it this acceleration is $F = ma$ = 1100 × 0.15 = 165 N. Thus the total force exerted by the tow bar is 165 + 150 = 315 N

Example

A block of mass 4.0 kg rests on a smooth inclined plane which is at 30° to the horizontal. If the block is connected to a 5.0 kg block by a light string in the way shown in Figure 9.3, what is the acceleration of the blocks and the tension in the string?

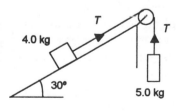

Figure 9.3   *Example*

The resultant force acting on the 5.0 kg block is $(5.0g - T)$. Hence, using $F = ma$:

$$5.0g - T = 5.0a$$

where a is the acceleration. The 4.0 kg block has a weight of $4.0g$ acting vertically downwards. The component of this force which is acting parallel to the slope is $4.0g \sin 30°$. Hence the force exerted on the 4.0 kg block by the string must be $(T - mg \sin 30°)$ and so:

$$T - mg \sin 30° = 4.0a$$

Adding the above two equations gives:

$$5.0g - 4.0g \sin 30° = 5.0a + 4.0a$$

Hence $a = 3.3$ m/s². Substituting this value in one of the equations gives $T = 32.7$ N.

Example

Calculate the average recoil force experienced by a machine gun firing 120 shots per minute, each bullet having a mass of 2 g and a muzzle velocity of 300 m/s.

The momentum of one bullet is $mv = 0.002 \times 300 = 0.6$ kg m/s. Thus the momentum given to 120 bullets is $120 \times 0.6 = 72$ kg m/s. This is the momentum given to the bullets in 60 s. The force acting on the bullets to give them this change of momentum must be:

force = rate of change of momentum = 72/60 = 1.2 N

The force experienced by the gun is opposite and equal to the force on the bullets (Newton's third law); hence, the force experienced by the gun = 1.2 N

Example

A hammer of mass 2.0 kg moving with a velocity of 8.0 m/s hits a nail and comes to rest in 0.1 s. What is the force acting on the nail?

The change in momentum of the hammer is $2.0 \times 8.0 = 16$ kg m/s. The rate of change of momentum of the hammer is thus 16/0.1 = 160 N. The force responsible for bringing the hammer to rest must by opposite and equal to the force acting on the nail (Newton's third law). Thus the force acting on the nail is 160 N.

Example

A jet of water with a diameter of 300 mm and moving with a velocity of 8 m/s strikes the stationary vane of a water wheel along a line at right angles to the vane. Calculate the force exerted by the jet on the vane if the water does not rebound from the vane. The density of water is 1000 kg/m³.

If v is the velocity of the water, all the water within a distance vt of the vane will hit it in a time t. If A is the cross-sectional area of the jet and ρ its density, then:

mass of water hitting the vane in time $t = \rho v t A$

momentum of water in time $t = \rho v t A v$

Because the water does not rebound, all this momentum is lost by the water in time t. Thus:

rate of change of momentum $= \rho v t A v / t = \rho v^2 A$

force acting on the water $= \rho v^2 A$

The force F acting on the vane is opposite and equal to the force acting on the water (Newton's third law). Hence:

$$F = \rho v^2 A = 1000 \times 8^2 \times \tfrac{1}{4}\pi \times 0.3^2 = 4.5 \times 10^3 \text{ N.}$$

9.2.1 Conservation of momentum

Consider an object of mass m_1 and velocity u_1 and another object of mass m_2 and velocity u_2, as in Figure 9.4(a). If they collide (Figure 9.4(b)), then on impact we must have, according to Newton's third law, the force on m_1 exerted by m_2 as opposite and equal to force on m_2 exerted by m_1. Thus the rate of change of momentum of m_1 must be opposite and equal to the rate of change of momentum of m_2. Thus, if the duration of the contact on impact is t:

$$m_1 \frac{(v_1 - u_1)}{t} = -m_2 \frac{(v_2 - u_2)}{t}$$

Hence, rearranging this equation gives:

$$m_1 u_1 + m_2 u_2 = m_1 v_1 + m_2 v_2$$

The *sum of the momentum before the collision equals the sum of the momentum after it*. This is known as the *conservation of momentum*.

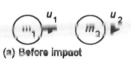

(a) Before impact

(b) At impact

Opposite and equal forces

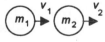

(c) After impact

Figure 9.4 *A collision*

Example

A pile driver of mass 120 kg falls vertically from rest through a height of 2.0 m onto a pile of mass 100 kg. If no rebound occurs, what will be the velocity of the pile driver and pile immediately after the impact?

For the pile driver, if v is the velocity on impact, the initial velocity $u = 0$, the acceleration is that of free fall g, and the distance fallen s = 2.0 m. Then $v^2 = u^2 + 2as = 0 + 2 \times 9.8 \times 2.0$ and so the velocity at impact $v = 6.3$ m/s. By the conservation of momentum:

$$120 \times 6.3 = (120 + 100)V$$

where V is the combined velocity of pile and driver after impact. Hence $V = 3.4$ m/s.

Example

A truck with a mass of 40 kg and moving with a velocity of 4.0 m/s collides with a truck of mass 60 kg moving in the same straight line but in the opposite direction with a velocity of 2.0 m/s. When the trucks collide they lock together. What will be their velocity immediately after the collision?

By the conservation of momentum and taking account of the direction of the velocities which gives the 60 kg truck a negative momentum before the collision because it is moving in the opposite direction to the 40 kg truck:

$$40 \times 4.0 - 60 \times 2.0 = (40 + 60)V$$

Thus $V = 0.40$ m/s and, since it is positive, it is in the same direction as the initial velocity of the 40 kg truck.

9.2.2 Impulse

The term *impulse* is used for the product of a force F and the time t for which it acts. If the velocity of a body of mass m changes from u to v in a time t then the change of momentum is $mv - mu$ and the rate of change of momentum is $(mv - mu)/t$. But the rate of change of momentum is equal to the force F responsible for the change (Newton's second law). Thus $F = (mv - mu)/t$ and so:

impulse = $mv - mu$ = change in momentum

Example

If a body is at rest on a smooth horizontal surface and a horizontal force of 2 N acts on the body for 6 s, what will be (a) the impulse given to the body, (b) the change in momentum?

(a) Impulse = force × time = $2 \times 6 = 12$ N s.
(b) Impulse = change in momentum and so the resulting change in momentum = 12 N s (kg (m/s)/s).

9.3 Friction   The term *frictional force* is used to describe the force that arises when two bodies are in contact with one another and that opposes the motion of one body relative to the other. Think of a crate resting on the floor. If you try to slide the crate over the floor you need to exert a big enough force to overcome the frictional force that exists between the crate and the floor before any motion can occur. Also, when you have got the crate moving, you have to keep on applying a force to overcome friction in order to keep it moving at a steady pace. The term *static friction* is used when the bodies are at rest and the frictional force is opposing attempted motion; the term *kinetic friction* is used when the bodies are moving with respect to each other and the frictional force is opposing motion with constant velocity.

If two objects are in contact and at rest, a force applied to one object which does not cause motion must be balanced by an opposing and equal size frictional force so that the resultant force is zero (Newton's first law). Increasing the applied force results in a force being reached when motion just starts. The value of the frictional force that has to be overcome before motion starts is called the *limiting frictional force*. When motion occurs there must be a resultant force acting on the object accelerating it (Newton's second law) and thus the applied force must have become greater than the frictional force.

The following are the basic laws of friction:

Law 1

>The frictional force is always in such a direction as to oppose relative motion and is always tangential to the surfaces in contact.

Law 2

>The frictional force is independent of the areas of the surfaces in contact.

Law 3

>The frictional force depends on the surfaces in contact and its limiting value is directly proportional to the normal reaction between the surfaces.

The *coefficient of static friction* μ_s is the ratio of the limiting frictional force F to the normal reaction N:

$$\mu_s = \frac{F}{N}$$

The *coefficient of kinetic friction* μ_k is the ratio of the kinetic frictional force F to the normal reaction N:

$$\mu_k = \frac{F}{N}$$

Table 9.1 gives some typical values for these coefficients.

Table 9.1 *Coefficients of friction*

Surfaces	Static coefficient	Kinetic coefficient
Steel on steel	0.7	0.6
Copper on steel	0.5	0.4
Glass on glass	0.9	0.4

Example

A block of wood of mass 2 kg rests on a horizontal surface, the coefficient of static friction being 0.6. What horizontal force is needed to start the block in horizontal motion? Take *g* to be 9.8 m/s².

The limiting frictional force, i.e. the force to start the object is motion, is $F = \mu N = 0.6 \times 2 \times 9.8 = 11.76$ N.

Example

What acceleration is produced when a horizontal force of 50 N acts on a block of mass 2.5 kg resting on a horizontal surface where the coefficient of kinetic friction is 0.5? Take *g* to be 9.8 m/s².

The normal force $N = 2.5g$ and so the frictional force F is:

$$F = \mu N = 0.5 \times 2.5 \times 9.8 = 12 \text{ N}$$

The resultant horizontal force acting on the object is thus 50 − 12 = 38 N and so the resulting acceleration *a* is:

$$a = \frac{\text{resultant force}}{m} = \frac{38}{2.5} = 15 \text{ m/s}^2.$$

9.3.1 The angle of static friction

Consider a block of material resting on a plane, the angle of inclination of which is slowly increased. At some angle θ the block just starts to slide down the plane. Figure 9.5 shows the situation when the block is about to slide. The normal reaction force N is equal to the component of the weight of the block which is at right angles to the surfaces of contact and is thus $mg \cos \theta$. The block is just on the point of sliding and so the limiting frictional force F must be equal to the component of the weight of the block down the plane, i.e. $F = mg \sin \theta$. Hence:

$$\mu_s = \frac{F}{N} = \frac{mg \sin \theta}{mg \cos \theta} = \tan \theta$$

This angle is called the *angle of static friction*.

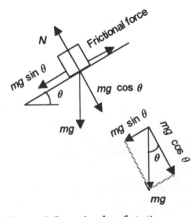

Figure 9.5 *Angle of static friction*

Example

On a production line, cans slide down a chute. If the angle of the chute to the horizontal is 20°, its length is 3.0 m and the coefficient of kinetic friction between the cans and the chute surface is 0.2, what will be the velocity of the cans at the bottom of the slide if they start at the top with a velocity of 1.0 m/s? Take g as 9.8 m/s².

If m is the mass of a can then the normal reaction force on a can is $mg \cos \theta$ (Figure 9.6). Hence the frictional force is $\mu_k mg \cos \theta$. The component of the weight of a can down the plane is $mg \sin h$. Hence the resultant force acting down the plane $= mg \sin \theta - \mu_k mg \cos \theta$. This resultant force will cause an acceleration a where:

$$mg \sin \theta - \mu_k mg \cos \theta = ma$$

Hence $a = 9.8 \sin 20° - 0.2 \times 9.8 \cos 20° = 1.5$ m/s². Thus, with $u = 1.0$ m/s, $a = 1.5$ m/s², $s = 3.0$ m, we have $v^2 = u^2 + 2as = 1.0^2 + 2 \times 1.5 \times 3.0$ and so $v = 3.2$ m/s.

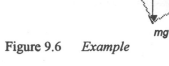

Figure 9.6　　Example

9.4 Linear motion and energy

The work done by the forces acting on a body is equal to the increase in its mechanical energy, i.e. its kinetic and potential energies. In applying this principle to the work done by a resistive force, it should be noted that the work done by a resistance is a negative quantity.

Example

A car of mass 1000 kg travelling at 20 m/s collides head on into a stationary car which has a mass of 800 kg. If the two vehicles lock together after impact, what will be their velocity and the loss in kinetic energy?

Using the principle of the conservation of momentum $1000 \times 20 = (1000 + 800)v$ and so $v = 11.1$ m/s. The loss in kinetic energy $= \frac{1}{2} \times 1000 \times 20^2 - \frac{1}{2}(1000 + 800) \times 11.1^2 = 8.9 \times 10^4$ J.

Example

A car of mass 1000 kg is driven up an incline of length 750 m and inclination 1 in 25 (Figure 9.7). Determine the driving force required from the engine if the speed at the foot of the incline is 25 m/s and at the top is 20 m/s and resistive forces can be neglected.

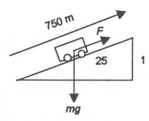

Figure 9.7　　Example

The work done by the driving force F is $F \times 750$ J. There is a change in both potential energy and kinetic energy as a result of the work done. The gain in potential energy is $1000 \times 9.81 \times 750 \sin \theta$, where θ is the angle of elevation of the incline. Since $\sin \theta$ is given as 1/25 then the gain in potential energy is 294.3 kJ. The initial kinetic energy at the foot of the incline is $\frac{1}{2} \times 1000 \times 25^2 = 312.5$ kJ and at the top of the incline is $\frac{1}{2} \times 1000 \times 20^2 = 200$ kJ. Hence there is a loss in kinetic energy of 112.5 kJ in going up the slope. The work done must equal the total change in energy and so:

$$F \times 750 = 294.3 \times 10^3 - 112.5 \times 10^3$$

and the driving force is 242.4 N.

Example

A car of mass 850 kg stands on an incline of 5° to the horizontal. If the hand brake is released, what will be the velocity of the car after travelling 100 m if the resistance to motion total 60 N?

The car 'falls' through a vertical height of 100 sin 5° and so the loss in potential energy is $mgh = 850 \times 9.8 \times 100$ sin 5° = 72.6×10^3 J. The work done against the resistances during the motion is $Fs = 60 \times 100 = 6000$ J. The kinetic energy gained by the car at the bottom of the incline is equal to the potential energy given up minus the work done against the resistances and so is $72.6 \times 10^3 - 6.0 \times 10^3 = 66.6 \times 10^3$ J. Thus $\frac{1}{2}mv^2 = 66.6 \times 10^3$ and so $v = 12.5$ m/s.

Activities

1 Determine the coefficient of static friction between two surfaces, e.g. a wood block on a wooden surface. One method that can be used involves finding the force needed to start the block in motion along a horizontal surface. For the arrangement shown in Figure 9.8, weights are added to the hanger until the block just starts to move. This can be repeated with weights placed on the block to give a number of different values of normal reaction force. Hence a graph can be plotted of the limiting frictional force against the normal reaction force and the coefficient is then the slope of the graph. An alternative method that can be used to determine the coefficient is to place the block on the wooden surface and then determine the angle to the horizontal to which the surface has to be tilted for the block to start sliding down the surface (see Section 9.3.1 and Figure 9.5).

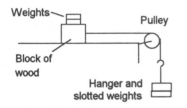

Figure 9.8   *Activity 1*

Problems

1 A constant horizontal force of 50 N acts on a body on a smooth horizontal plane. If the body starts from rest and is observed to move 2.0 m in 1.5 s, what is the mass of the body?

2 What is the tension in a rope passing over a small frictionless pulley if at one end of the rope there is a mass of 200 g and at the other end 100 g? Also, what will be the acceleration of the masses?

3 A man with a mass of 80 kg stands on a platform which itself has a mass of 40 kg. The man pulls the platform, and himself, upwards with an acceleration of 0.6 m/s² by means of pulling on a rope which is attached at one end to the platform and runs over a pulley above his head. With what force does he need to pull on the rope?

4 An object of mass 2.0 kg rests on a smooth horizontal plane and is attached by means of a string passing over a pulley wheel to a

vertically suspended object of mass 2.5 kg. What is (a) the acceleration of the objects and (b) the tension in the string?

5 The locomotive of a train can exert a maximum pull of 160 kN. If the locomotive has a mass of 80 000 kg and the train a mass of 400 000 kg, what will be the maximum acceleration for the train on an incline of 1 in 250 if the resistances to motion amount to 24 kN?

6 A jet of water 50 mm in diameter and with a velocity of 40 m/s strikes a plate fixed at right angles to the jet. Calculate the force acting on the plate if no rebound of water takes place. Density of water is 1000 kg/m³.

7 A gun of mass 150 000 kg fires a shell of mass 1200 kg, giving it a muzzle velocity of 800 m/s. What is the gun's initial recoil velocity?

8 A sphere of mass 150 g is moving with a velocity of 4 m/s along a smooth horizontal plane and collides with a sphere of mass 350 g moving with a velocity of 2 m/s towards it. After the collision the 350 g sphere has a velocity of 1 m/s in the opposite direction. What will be the velocity of the 150 g sphere?

9 A truck of mass 800 kg moves along a straight horizontal track with a velocity of 6 m/s and collides with another truck of mass 2000 kg which is already moving in the same direction with a velocity of 2 m/s. After collision the two trucks remain locked together. What will be their velocity?

10 The hammer of a forging press has a mass of 400 kg and when in use is brought to rest from 10 m/s in 0.02 s when hitting a forging. What is (a) the impulse, (b) the force exerted on the forging?

11 A rough surface is inclined at an angle such that the vertical height increases by 3 m for every 4 m horizontal displacement. A block of mass 50 kg is released from rest at the top of the incline and travels a distance of 10 m down the slope and attains a velocity of 8 m/s. Determine the size of the frictional force.

12 A block of steel of mass 5.0 kg rests on an incline which is at an angle of 30° to the horizontal. If the block just starts to slip under its own weight, what is the coefficient of static friction?

13 What is the minimum horizontal force that must be applied to slide an object of mass 20 kg along a horizontal surface with (a) a constant velocity, (b) a constant acceleration of 0.30 m/s², if the coefficient of kinetic friction is 0.30?

14 A block of mass 3 kg slides down a rough slope which is at an angle to the horizontal such that there is a vertical drop in height of 1 m for every 4 metres along the slope. If the block starts from rest, what will be its speed after it has moved 4 m down the slope if there is a constant frictional force of 4 N?

15 A 2 kg block slides down a rough slope which is at 30° to the horizontal. What is the frictional force, if the block starts from rest and reaches a speed of 2 m/s after travelling 5 m down the slope?

16 A body is projected up an inclined plane which is at 50° to the horizontal. If the body starts with a velocity of 5 m/s at the bottom of the incline, how far up the slope will the body go before coming to rest and then starting to slide back down the plane? The coefficient of friction between the body and the plane is 0.6.

10 D.c. circuits

10.1 Introduction

This chapter is about d.c. circuits that just contain resistors and the basic terms used such as charge, current, voltage, resistance and power.

10.1.1 Charge

If you rub a strip of polyethylene, or a strip of cellulose acetate, or the case of a biro, on your sleeve and hold it above some small scraps of paper, the scraps are attracted to the strip and we say that the strip has acquired electric charge.

If a polyethylene strip is placed in a paper or wire stirrup and suspended on a length of cotton thread (Figure 10.1), each end of the strip rubbed and then the end of a similarly rubbed polyethylene strip brought near to one end of the suspended strip, they are found to repel each other. The charges on the two strips of polyethylene will be the same and thus *like charges repel each other*. If we repeat the experiment with two cellulose acetate strips we obtain the same effect of like charges repelling. However, if we bring a charged polyethylene strip close to a suspended charged acetate strip, or bring a charged acetate strip close to a suspended charged polyethylene strip (Figure 10.2), we find that attraction occurs. The polyethylene and the acetate appear to be charged with different kinds of charge. Thus, *unlike charges attract each other*.

The terms *positive* and *negative* are used to describe the two types of charge produced on the strips; the rubbing causes the polyethylene to become negatively charged and with the acetate strip the rubbing causes it to become positively charged. The rubbing action causing electrons to transfer to or from the rubbing cloth to the strips; with polyethylene the rubbing causes electrons to transfer from the cloth to the polyethylene and so the polyethylene acquires a surplus of electrons and so has a net negative electric charge, whereas with the acetate the rubbing causes electrons to transfer from the acetate to the cloth and so the acetate becomes deficient in electrons and so has a net positive charge.

10.1.2 Electric current in metals

A metal, such as copper, has atoms with outer electrons which are easily detached and so, when such atoms are packed tightly together, the outer electrons experience forces of attraction towards neighbouring atoms and become detached and can move within the metal. These will just drift around in a random manner. However, if a battery is connected across the metal rod, the random movements of the electrons has superimposed on them a general towards the positive terminal. It is this drift of the

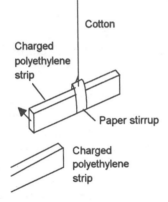

Figure 10.1 *Repulsion between like electric charges*

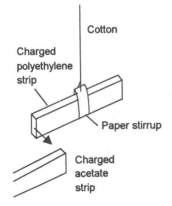

Figure 10.2 *Attraction between unlike electric charges*

electrons that constitutes the electric current in the circuit. *Current* is defined as the rate of movement of charge in a circuit. When there is a current I then the charge Q being moved in a time t is given by $I = Q/t$.

Current has the unit of ampere or amp (A), the unit being defined as that constant current which, when flowing in two straight parallel conductors of infinite length, of negligible cross-section, and placed 1 m apart in a vacuum, produces between the conductors a force of 2×10^{-7} N per metre length (see Section 11.5 for a discussion of the force acting on current carrying conductors). The instrument used to measure current is called an ammeter and is connected in series with the circuit element for which the current is required.

The unit of charge is the coulomb (C) and is defined as the quantity of electricity passing a point in a circuit when a current of 1 A flows for 1 s. In an electric circuit the charge carriers through electrical conductors are electrons. One coulomb of charge is the total charge carried by 6.24×10^{18} electrons.

Example

If there is a current of 4 A through a particular part of a circuit, what will be the charge moved through that part in 1 minute?

Charge moved Q = current $I \times$ time $t = 4 \times 60 = 240$ C.

10.1.3 Conductors and insulators

The term *conductor* is used for a material which readily allows charge to flow through it; the term *insulator* is used for a material which allows very little movement of charge. The metal used for an electrical wire is a conductor, whereas the plastic in which it is sheathed is an insulator.

10.1.4 Electrical energy, e.m.f. and potential difference

When there is a current through a circuit then charge is continuously being moved through it. Energy has to be continually supplied to keep the charge moving. The rate at which this energy is required is called the *power P*. Thus if an energy W is required over a time t then $P = W/t$. The unit of power is the watt (W), one watt being a rate of energy use of 1 joule per second. Thus, for example, a 1 kW electric fire requires 1000 J of electrical energy to be supplied every second.

To establish a flow of charge through a circuit, it is necessary to exert some sort of force on the charge carriers to cause them to move. This is termed the *electromotive force* (e.m.f.). A battery is a source of e.m.f.; chemical energy is transformed in the battery into energy which is used to move the charge carriers round the circuit. If we think of a mechanical analogue, the battery lifts the charge carriers up to the top of a hill, i.e. gives them potential energy, and then the circuit allows them to flow down the hill back to the other terminal of the battery. The e.m.f. is a measure of the potential energy/charge that is given by the battery. The unit of e.m.f. is the volt (V) and an e.m.f. of 1 V is when the battery gives 1 J to each coulomb of charge.

Energy is required to establish a flow of charge through a component. The term *voltage* or *potential difference* is used for the energy required to move a unit charge between two points in a circuit and has the unit of volt (V). Thus if V is the potential difference between two points in a circuit then the energy W required to move a charge Q between the points is given by $V = W/Q$. Thus we can write:

$$P = \frac{W}{t} = \frac{Q}{t} \times \frac{W}{Q} = IV$$

A potential difference of 1 V is said to exist between two points of a conducting wire carrying a constant current of 1 A when the power dissipated between these points is equal to 1 W. The instrument used to measure voltage is termed a voltmeter and is always connected in parallel with the circuit element across which the voltage is required.

Example

What is the electrical power developed between two points in an electrical circuit when there is a current of 2 A between the points and the potential difference between them is 4 V?

$$P = IV = 2 \times 4 = 8 \text{ W.}$$

10.1.5 Resistance

The electrical *resistance* R of a circuit element is the property it has of impeding the flow of electrical current and is defined by the equation:

$$R = \frac{V}{I}$$

where V is the potential difference across an element when I is the current through it. The unit of resistance is the ohm (Ω) when the potential difference is in volts and the current in amps.

It is sometimes more convenient to use the reciprocal value of a resistance, i.e. $1/R$. This reciprocal value is called the *conductance* (G).

$$G = \frac{1}{R} = \frac{I}{V}$$

With the resistance in ohms, the unit for the conductance is the siemen (S). Thus a resistance of 10 Ω is a conductance of $1/10 = 0.10$ S.

Example

An ammeter connected in series with a resistor indicates a current through the resistor of 0.2 A when the voltage across it is 2 V. What is (a) the resistance of the resistor, (b) the power needed to 'push' the current through the resistor, (c) the charge moved through the resistor in 1 minute?

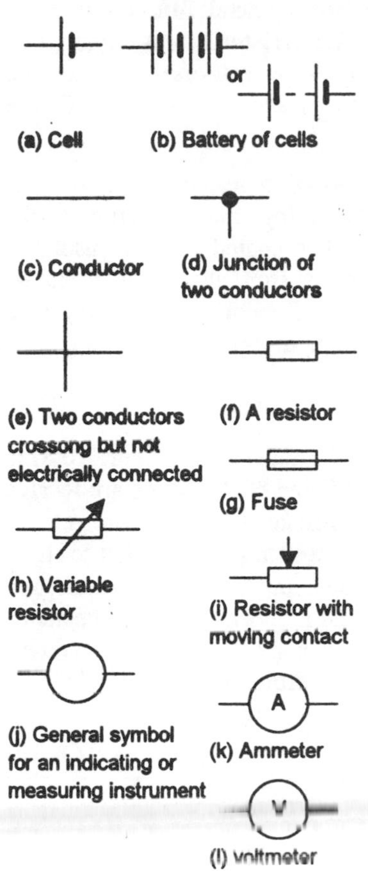

(a) Cell

(b) Battery of cells

(c) Conductor

(d) Junction of two conductors

(e) Two conductors crossong but not electrically connected

(f) A resistor

(g) Fuse

(h) Variable resistor

(i) Resistor with moving contact

(j) General symbol for an indicating or measuring instrument

(k) Ammeter

(l) voltmeter

Figure 10.3 *Basic symbols*

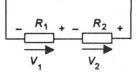

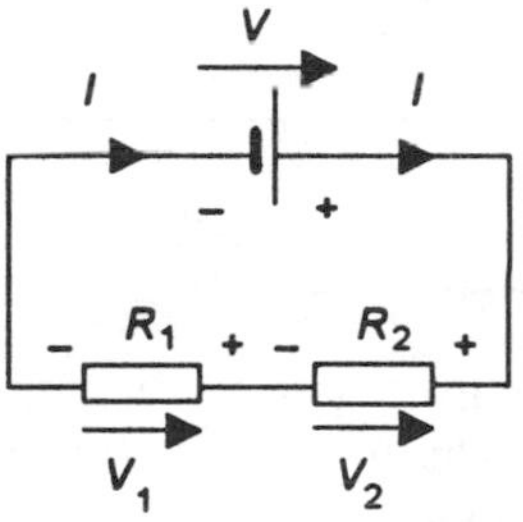

Figure 10.4 *Circuit notation*

(a) Resistance = V/I = 2/0.2 = 10 Ω.

(b) Power = VI = 2 × 0.2 = 0.4 W.

(c) Current is the rate of movement of charge; thus, a current of 0.2 A indicates that 0.2 C are moved in 1 s. Hence, in 1 minute we have 60 × 0.2 = 12 C moved.

Example

What is the conductance of a resistor having a resistance of 50 Ω?

Conductance = 1/resistance = 1/50 = 0.02 S.

10.1.6 Circuit symbols and sign conventions

Figure 10.3 shows the basic symbols used in electrical circuit diagrams. With the cell, the positive plate of the cell is the long line and the shorter fatter line is the negative plate.

The convention adopted with electrical circuits is to indicate the direction of the current in a circuit conductor by an arrow on the conductor (Figure 10.4). Current I is taken as flowing out of the positive plate of a cell (the long line on the cell symbol), round the circuit and back into the negative plate (the short fat line on the symbol), the arrow for current indicating this direction. This is the convention that was originally adopted by those who first investigated electrical currents and who thought that current flow was caused by the motion of positive charge carriers rather than negative charge. Thus they assumed that the current flowed from the positive terminal to the negative terminal and this convention for labelling current direction is still used. Now we know that in conductors the current flow is by electrons, negative charge carriers, and so the direction of the flow of electrons is in the opposite direction to the direction used to label the current in an electrical circuit.

The direction of the potential difference V between two points in a circuit is indicated by an arrow between the points and which is parallel to the conductor (Figure 10.4). The arrow for the potential difference points towards the point which is taken to be more positive. Think of charge flowing down the potential hill of the circuit, with a cell pumping the charge back up to the top of the hill again, the arrow points uphill.

10.2 Resistors

There are a number of types of fixed value resistors.

1 *Carbon composition resistors* are made by mixing finely ground carbon with a resin binder and insulating filler, compressing it into a cylindrical shape, and then firing it (Figure 10.5). Copper leads are provided at each end of the cylindrical shape and the resistor coated with an insulating plastic film. The resistance of the resistor depends on the ratio of carbon to insulating filler.

2 *Film resistors* are made by depositing an even film of a resistive material on a ceramic rod. The materials used may be carbon

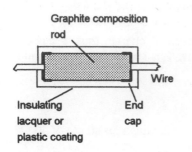

Figure 10.5 *Graphite composition resistor*

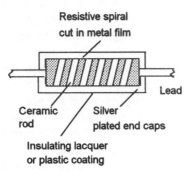

Figure 10.6 *Thin film resistor*

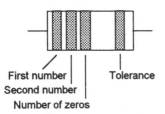

Figure 10.7 *Wire-wound resistor*

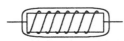

First number
Second number
Number of zeros
Tolerance

Figure 10.8 *Colour coding*

(carbon-film resistors), nickel chromium (metal film resistors), a mixture of metals and glass (metal glaze resistors), or a mixture of a metal and an insulating oxide (metal oxide resistors). The required resistance value is obtained by cutting a spiral in the film (Figure 10.6).

3 *Wire-wound resistors* are made by winding resistor wire, such as nickel chromium or a copper-nickel alloy, onto a ceramic tube (Figure 10.7). The whole resistor is then coated with an insulator. Wire-wound resistors have power ratings from about 1 W to 25 W, the other forms tending to have ratings of about 0.25 W. The *power rating* is the maximum power that can be dissipated by the resistor.

10.2.1 Coding

Small resistors are marked with a series of colour bands (Figure 10.8) to indicate the resistor value and tolerance. Three of the bands are located closer together towards one end of the resistor and with the resistor placed with this end to the left, the bands are then read from left to right. The first band gives the first number of the component value, the second band the second number, the third band the number of zeros to be added after the first two numbers and the fourth band the resistor tolerance. Table 10.1 shows the colours assigned to the different numbers.

Table 3.1 *Colour coding*

Black	Brown	Red	Orange	Yellow
0	1	2	3	4
Green	Blue	Violet	Grey	White
5	6	7	8	9

Tolerance colour

Brown	Red	Gold	Silver	None
1%	2%	5%	10%	20%

In catalogues and circuit drawings it is standard practice to not give decimal points in specifying values but include the prefix for the multiplication factor in place of the decimal point to avoid accidental marks being mistaken for decimal points. This system is also used where values are written on resistors instead of the colour code being used. The symbol R is used for $\times 1$, k for $\times 1000$ and M for $\times 1\,000\,000$. Thus a 4.7 Ω resistor would be written as 4R7, a 4.7 kΩ resistor as 4k7 and a 4.7 MΩ resistor as 4M7. Tolerances can be indicated by adding a letter after the value code: F for ±1%, G for ±2%, J for ±5%, K for ±10%, M for ±20%.

Example

What is the value of a resistor coded, from the left end, yellow, violet, red, silver?

Yellow has the value 4, violet the value 7, red the value 2 and thus the value is 4700 Ω, the tolerance colour silver indicating 10%.

Example

What is the resistance value and tolerance of a resistor specified in a circuit diagram as 6R8M?

6.8 Ω with a tolerance of ±20%.

10.2.2 Preferred values

In order to provide coverage of a wide range of resistor values by means of only a limited number of resistors, a *preferred* range of values of resistances is used by manufacturers of resistors. Resistors are made in various *tolerance bands*, this being the percentage uncertainty with which a manufacturer has stated the nominal value. For example, a resistor stated has having the nominal value of 10 Ω resistor with ±10% (note that often the plus or minus is not stated but merely implied) tolerance band would have a resistance in the range 10 − 10% of 10 = 10 − 1 = 9 Ω to 10 + 10% of 10 = 10 + 1 = 11 Ω.

The preferred values in a particular tolerance band are chosen so that the upper value of one resistor is either equal to or just overlaps with the lower value of the next higher resistor. Thus with the 10% tolerance band, the next resistor above the nominal 10 Ω resistor would be 12 Ω, it having a resistance between about 11 Ω and 13 Ω. Table 10.2 shows examples of nominal preferred resistance values and their tolerances.

Table 10.2 *Preferred resistance values for difference tolerances*

20%	10%	5%	20%	10%	5%
10	10	10	33	33	33
		11			36
	12	12		39	39
		13			43
15	15	15	47	47	47
		16			51
	18	18		56	56
		20			62
22	22	22	68	68	68
		24			75
	27	27		82	82
		30			91

Example

What is the possible resistance range of (a) a 47 Ω resistor with a 5% tolerance, (b) a 330 Ω resistor with a 20% tolerance?

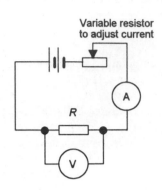

Figure 10.9 *Circuit to obtain current and p.d. values*

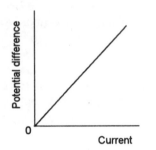

Figure 10.10 *Circuit element obeying Ohm's law*

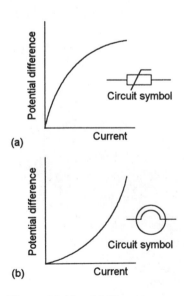

Figure 10.11 *(a) Thermistor, (b) filament lamp*

(a) Since 5% of 47 is 2 (rounded to the nearest whole number), the resistance value lies in the range 47 − 2 = 45 Ω to 47 + 2 = 49 Ω.
(b) Since 20% of 330 is 66, the resistance value lies in the range 330 − 66 = 264 Ω to 330 + 66 = 396 Ω.

10.2.3 Ohm's law

For many resistors, if the temperature does not change, a measurement of the potential difference across a resistor and the current through it (Figure 10.9) shows that the potential difference is proportional to the current through it. A graph of potential difference V against current I is a straight line passing through the origin (Figure 10.10) with a constant of V/I. Hence the gradient is the resistance R. Thus we can write:

$$V = RI$$

as an equation which gives a constant value of R for all values of current. This is called *Ohm's law*. Thus circuit elements which obey Ohm's law have a constant value of resistance which does not change when the current changes. Also, since the power $P = IV$, we can write:

$$P = IV = I^2R = \frac{V^2}{R}$$

Example

A 200 Ω resistor has a power rating of 2 W. What is the maximum current that can be used with the resistor without exceeding the power rating?

Using $P = I^2R$:

$$I = \sqrt{\frac{P}{R}} = \sqrt{\frac{2}{200}} = 0.1 \text{ A}$$

The maximum current is thus 0.1 A or 100 mA.

10.2.4 Non-linear resistance elements

Not all circuit elements give graphs which are straight line for the relationship between the potential difference across them and the current through them. Figure 10.11 shows examples of non-linear relationships that can be obtained with the circuit shown in Figure 10.9. Figure 10.11(a) is the type of graph obtained with a typical thermistor and Figure 10.11(b) is that obtained with a typical filament lamp. With the thermistor, the gradient of the graph decreases as the current increases and so the resistance decreases as the current increases. With the filament lamp, the gradient increases as the current increases and so resistance increases as the current increases.

Thermistors (*therm*ally sensitive re*sistors*) are resistance elements made from mixtures of metal oxides, such as those of chromium, cobalt, iron, manganese and nickel. These materials are semiconductors and have resistances which decrease as the temperature increases. The mixture is formed into elements such as rods, discs or beads and connecting wires attached. The resistance of a thermistor is very sensitive to temperature changes, considerably more than metals, and generally decreases with an increase in temperature. Increasing the current through a thermistor increases the power dissipated in it and hence its temperature, hence the consequential fall in resistance.

With a filament lamp, increasing the current through it raises the temperature of the filament. The material used for a lamp filament is tungsten and this has a resistance which increases as the temperature increases.

10.3 Resistors in series

When circuit elements are in *series* (Figure 10.12) then we have the same rate of flow of charge through each element, i.e. the same current I. Thus the potential difference V_1 across resistance R_1 is IR_1. The potential difference V_2 across R_2 is IR_2. The potential difference V across the series arrangement is thus the sum of the potential differences, i.e.

$$V = V_1 + V_2 = IR_1 + IR_2 = I(R_1 + R_2)$$

We could replace the two resistors by a single, equivalent, resistor R if we have $V = IR$. Hence:

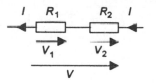

$$R = R_1 + R_2$$

Figure 10.12 *Resistors in series*

For resistors in series the total resistance is the sum of their resistances.

Example

A circuit consists of four resistors in series, they having resistances of 2 Ω, 10 Ω, 5 Ω and 8 Ω. A voltage of 12 V is applied across them. Determine (a) the total resistance, (b) the circuit current, (c) the power dissipated in each resistor, (d) the total power dissipation.

(a) The total resistance is the sum of the four resistances and so we have $R = 2 + 10 + 5 + 8 = 25\ \Omega$
(b) The current I through the circuit is given by $I = V/R = 12/25 = 0.48$ A $= 480$ mA.
(c) The power developed in each resistance is given by $P = I^2R$, the current I being the same through each resistor. Thus, the powers are $0.48^2 \times 2 = 0.46$ W, $0.48^2 \times 10 = 2.30$ W, $0.48^2 \times 5 = 1.15$ W and $0.48^2 \times 8 = 1.84$ W.
(d) The total power developed can be obtained, either by adding together the powers dissipated by each resistor or determining the power dissipated by the equivalent resistor. Thus, $P = 0.46 + 2.30 + 1.15 + 1.84 = 5.75$ W or $P = I^2R = 0.48^2 \times 25 = 5.76$ W. The slight difference in the answers is due to rounding errors.

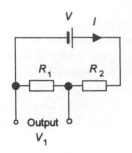

Figure 10.13 *Voltage divider circuit*

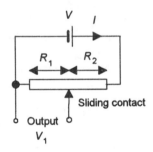

Figure 10.14 *Potentiometer*

10.3.1 Voltage divider circuit

Resistors in series can be used as a simple method of voltage division. Consider two resistors R_1 and R_2 in series, as illustrated in Figure 10.13. A voltage V is applied across the arrangement. The total resistance is $R_1 + R_2$. Thus the current I through the series resistors is:

$$I = \frac{V}{R_1 + R_2}$$

The voltage across resistor R_1 is IR_1. Hence if this voltage is taken as the output voltage, we have:

$$V_1 = \frac{R_1}{R_1 + R_2} V$$

Thus the output voltage is the fraction $R_1/(R_1 + R_2)$ of the input voltage. The result is thus a *voltage-divider circuit*.

It is often necessary to obtain a continuously variable voltage from a fixed supply voltage. For this a resistor with a sliding contact is used, such a circuit element being known as a *potentiometer* (Figure 10.14). The position of the sliding contact determines the ratio of the resistances R_1 and R_2 and hence the output voltage.

Example

A potentiometer has a total track resistance of 50 Ω and is connected across a 20 V supply. What will be the resistance between the slider and the track end when the output voltage taken of between these terminals is 5.0 V?

The potentiometer is effectively two resistors in series connected across the supply voltage (see Figure 10.14), namely the resistance between one end of the potentiometer and the slider and the resistance between the slider and the other end of the potentiometer. The fraction of the output voltage required is 5.0/20 and thus this must be the ratio of the resistance between one end and the slider to the sum of the two resistances, i.e. the total potentiometer resistance. Thus $R/50 = 5.0/20$ and so $R = 12.5$ Ω.

10.4 Resistors in parallel

When circuit elements are in *parallel* (Figure 10.15) then the same potential difference occurs across each element. Since charge does not accumulate at a junction, then the current entering a junction must equal the sum of the current leaving it, i.e. $I = I_1 + I_2$.

For resistance R_1 we have $V = I_1 R_1$ and for resistance R_2 we have $V = I_2 R_2$, thus:

$$I = I_1 + I_2 = \frac{V}{R_1} + \frac{V}{R_2} = V\left(\frac{1}{R_1} + \frac{1}{R_2}\right)$$

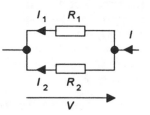

Figure 10.15 *Resistors in parallel*

We could replace the two resistors by a single, equivalent, resistor R if we have $I = V/R$. Hence:

$$\frac{1}{R} = \frac{1}{R_1} + \frac{1}{R_2}$$

We can write this equation in terms of conductances as:

$$G = G_1 + G_2$$

For circuit elements in parallel the total conductance is the sum of their conductances.

With a circuit of resistors in series, the applied voltage is divided across the resistors. With a circuit of resistors in parallel, the current is divided between the parallel branches. A parallel circuit is thus a *current divider circuit*.

Example

A circuit consists of two resistors of 5 Ω and 10 Ω in parallel with a voltage of 2 V applied across them. Determine (a) the total circuit resistance, (b) the current through each resistor.

(a) The total resistance R is given by:

$$\frac{1}{R} = \frac{1}{R_1} + \frac{1}{R_2} = \frac{1}{5} + \frac{1}{10} = \frac{3}{10}$$

Hence $R = 3.3\ \Omega$

(b) The same potential difference will be across each resistor. Thus the currents are $I_1 = V/R_1 = 2/5 = 0.4$ A $= 400$ mA and $I_2 = V/R_2 = 2/10 = 0.2$ A $= 200$ mA.

(c) The total current entering the arrangement can be obtained by either adding the currents through each of the branches or determining the current taken by the equivalent resistance. Thus, $I = 0.4 + 0.2 = 0.6$ A or 600 mA, or $I = V/R = 2/3.3 = 0.61$ A or 610 mA. Rounding errors account for the slight differences in these answers.

10.5 Series–parallel circuits

Many circuits are more complicated than just the simple series or parallel connections of resistors and may have combinations of both series and parallel connected resistors. In general, such circuits are analysed by beginning with the innermost parts of the circuit and block-by-block reducing the elements to single equivalent resistances, before finally simplifying the circuit to just one equivalent resistance.

Example

Determine the total resistance of the circuits shown in Figure 10.16.

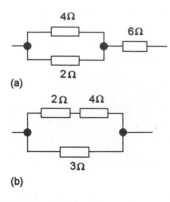

(a)

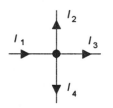

(b)

Figure 10.16   *Example*

(a) The circuit consists of a parallel arrangement of 4 Ω and 2 Ω which is then in series with resistance of 6 Ω. The equivalent resistance for the parallel part of the circuit is given by:

$$\frac{1}{R} = \frac{1}{4} + \frac{1}{2} = \frac{3}{4}$$

and is thus 1.33 Ω. This is in series with 6 Ω and so the total resistance is 7.33 Ω.

(b) The circuit consists of 2 Ω and 4 Ω in series, i.e. an equivalent resistance of 6 Ω. This is then in parallel with 3 Ω. Thus the total resistance is given by:

$$\frac{1}{R} = \frac{1}{3} + \frac{1}{6} = \frac{3}{6}$$

and so $R = 2$ Ω.

10.6 Kirchhoff's laws

There are two fundamental laws used in the analysis of circuits: Kirchhoff's current law and Kirchhoff's voltage law.

10.6.1 Kirchhoff's current law

When there is a current passing through a component, then all the charge that enters it over a period of time must equal all the charge that leaves it in the same time. Thus the rate at which charge enters a component, i.e. the current, must equal the rate at which current leaves the component. This idea can be extended to an electrical circuit; *the total current entering any junction in a circuit must equal the total current leaving it*. This is called *Kirchhoff's current law*. Thus for the junction shown in Figure 10.17, we must have:

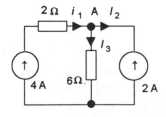

Figure 10.17   *Currents at a junction*

$I_1 = I_2 + I_3 + I_4$

Example

Determine, using Kirchhoff's current law, the current through the 6 Ω resistor in the circuit shown in Figure 10.18.

The circuit has two current sources. For junction A, applying Kirchhoff's current law, $I_1 = I_2 + I_3$. But $I_1 = 4$ A and $I_3 = -2$ A (the current is in the opposite direction to the arrow used for I_2). Thus 4 $= I_2 - 2$ and so $I_2 = 6$ A.

Figure 10.18   *Example*

10.6.2 Kirchhoff's voltage law

Consider the circuit shown in Figure 10.19. We have a voltage source of 10 V and three series resistors. The total resistance is 10 Ω and so the circuit current is 1 A. The potential difference across a resistor is given by the product of the current through it and its resistance and thus the potential difference across the 5 Ω resistor is 5 V, that across the 3 Ω

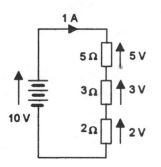

Figure 10.19 *Circuit to illustrate the voltage law*

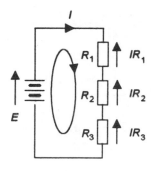

Figure 10.20 *Circuit to illustrate the voltage law*

resistor is 3 V and that across the 2 Ω resistor is 2 A. We see that the applied voltage of 10 V 'drops' 5 V across the first resistor, 3 V across the second resistor and 2 V across the final resistor; when we reach the end we have 'used up' all the applied voltage. The *sum of the voltage drops around the circuit loop is equal to the amount by which the voltage source increased it*. This is the *Kirchhoff's voltage law*.

The direction we move round a loop is quite arbitrary but the usual convention is to go round a loop in the clockwise direction with voltages in the clockwise direction taken as positive and in the opposite direction as negative. Thus moving round the loop in Figure 10.20 in a clockwise direction gives:

$$E - IR_1 - IR_2 - IR_3 = 0$$

Figure 10.21 shows a more complex circuit for which we can consider there are three loops, these being ABCF, CDEF and ABCDEF. We can apply the voltage law to each loop.

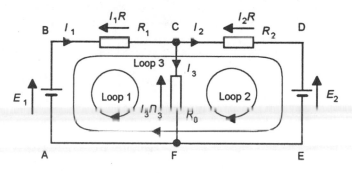

Figure 10.21 *Mesh currents*

For loop 1, the current through R_1 is the current I_1 and so the voltage drop through it is I_1R_1. The current through R_3 is I_3 and so its voltage drop is I_3R_3. Thus, applying Kirchhoff's voltage law to the loop gives:

$$E_1 - I_1R_1 - I_3R_3 = 0$$

For loop 2 we have a current of I_2 through R_2 and so a voltage drop of I_2R_2. Thus applying Kirchhoff's voltage law to this loop gives:

$$-E_2 - I_2R_2 + I_3R_3 = 0$$

For loop 3, applying Kirchhoff's voltage law gives:

$$E_1 - I_1R_1 - I_2R_2 - E_2 = 0$$

We also can apply Kirchhoff's current law to give the relationship between the currents at node C:

$$I_1 = I_2 + I_3$$

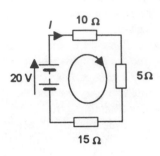

Figure 10.22 *Example*

Example

Use Kirchhoff's voltage law to determine the current in the circuit in Figure 10.22.

The voltage drop across the 10 Ω resistor is $10I$ V, that across the 5 Ω resistor is $5I$ V and across the 15 Ω resistor is $15I$ V. Thus, proceeding clockwise round the loop, Kirchhoff's voltage law gives:

$$20 - 10I - 5I - 15I = 0$$

Hence $30I = 20$ and so $I = 2/3$ A.

Example

Determine, using Kirchhoff's voltage law, the current through the 20 Ω resistor in the circuit shown in Figure 10.23.

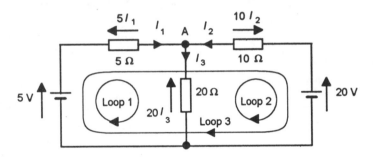

Figure 10.23 *Example*

For loop 1, applying Kirchhoff's voltage law, $5 - 5I_1 - 20I_3 = 0$ and so $I_1 = 1 - 4I_3$. For loop 2, applying Kirchhoff's law, $10I_2 - 20 + 20I_3 = 0$ and so $I_2 = 2 - 2I_3$. Kirchhoff's current law applied to node A gives $I_1 + I_2 = I_3$. Hence, substituting for I_1 and I_2:

$$(1 - 4I_3) + (2 - 2I_3) = I_3$$

and so $I_3 = 0.43$ A. Substituting this back into our earlier equations gives $I_1 = -0.72$ A and $I_2 = 1.14$ A. The minus sign indicates that the currents are in the opposite direction to that indicated in the figure.
For loop 3, applying Kirchhoff's law, $5 - 5I_1 + 10I_2 - 20 = 0$ and so $5 - 5(1 - 4I_3) + 10(2 - 2I_3) - 20 = 0$. Hence $0 = 0$, a useful check that we are applying the law correctly.

10.7 Resistivity

If we add two identical lengths of wire in series we end up with twice the resistance and so the resistance of a length of wire is proportional to its length. If we connect two identical lengths of wire in parallel then the resistance is halved and so, as we are effectively doubling the cross-sectional area of the material in the circuit, the resistance of a length of wire is inversely proportional to the area. The resistance of a wire also depends on the material used for the wire. We can combine these three factors in the equation:

$$R = \frac{\rho L}{A}$$

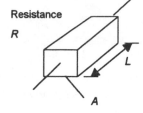

Resistance

R

L

A

Figure 10.24 *Resistivity*

where R is the resistance of a length L of the material of cross-sectional area A (Figure 10.24). The electrical *resistivity* ρ is a measure of the electrical resistance of a material and has the unit of ohm metre. An electrical *insulator*, such as a ceramic, will have a very high resistivity, typically of the order of 10^{10} Ω m or higher. An electrical *conductor*, such as copper, will have a very low resistivity of the order of 10^{-8} Ω m.

The electrical *conductivity* σ is the reciprocal of the resistivity, i.e.

$$\sigma = \frac{1}{\rho} = \frac{L}{RA}$$

The unit of conductivity is thus Ω^{-1} m^{-1} or S m^{-1}. Since conductivity is the reciprocal of the resistivity, an electrical insulator will have a very low conductivity, of the order of 10^{-10} S/m, while an electrical conductor will have a very high conductivity, of the order of 10^{8} S/m.

Pure metals and many metal alloys have resistivities that increase when the temperature increases; some metal alloys do, however, show increases in resistivities when the temperature increases. For insulators, the resistivity increases with an increase in temperature.

Example

What is the resistance of a 3 m length of copper wire at 20°C if it has a cross-sectional area of 1 mm² and the copper has a resistivity of 2×10^{-8} Ω m?

$$R = \frac{\rho L}{A} = \frac{2 \times 10^{-8} \times 3}{10^{-6}} = 0.06 \ \Omega \ m$$

Example

What is the electrical conductance of a 2 m length of nichrome wire at 20°C if it has a cross-sectional area of 1 mm² and an electrical conductivity of 0.9×10^{6} S/m?

$$G = \frac{1}{R} = \frac{\sigma A}{L} = \frac{0.9 \times 10^{6} \times 1 \times 10^{-6}}{2} = 0.45 \ S$$

10.8 Basic measurements

A basic instrument for the measurement of direct current is the *moving coil meter*. The basic meter movement is likely to have a minimum d.c. full scale deflection of between 10 μA and 20 mA. Higher d.c. current ranges, up to typically about 10 A, can be obtained by the use of shunts. These are resistors connected in parallel with the instrument (Figure 10.25). The potential difference across the instrument is I_iR_i. The current

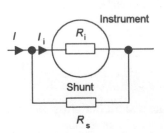

Figure 10.25 *Instrument shunt*

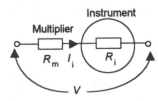

Figure 10.26 *Instrument multiplier*

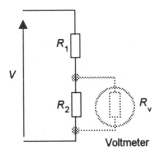

Figure 10.27 *Loading*

through the shunt is $(I - I_i)$ and so the potential difference across the shunt is $(I - I_i)R_s$. Thus, since these potential differences must be the same, we must have $I_iR_i = (I - I_i)R_s$. So the current I which will give a full scale deflection is:

$$I = \frac{I_i(R_i + R_s)}{R_s}$$

Such instruments have linear scales and accuracies up to about ± 0.1% of full scale deflection. They are generally unaffected by stray magnetic fields.

The moving coil meter can be used as a direct voltage voltmeter by measuring the current through a resistance. For the lowest voltage range this resistance is just the resistance of the meter coil. For higher voltage ranges a resistor, termed a multiplier, is connected in series with the meter movement (Figure 10.26). The potential difference V being measured is the sum of the potential differences across the meter resistance R_i and across the multiplier resistance R_m.

$$V = I_i(R_i + R_m)$$

Such instruments typically are used for voltage ranges from about 50 mV to 100 V with an accuracy of up to ± 0.1% of full scale deflection.

An important point, with any voltmeter, is its resistance. This is because it is connected in parallel with a circuit element and acts as a bypass for some of the current. Thus the current through the element is reduced. The voltmeter is said to give a *loading* effect. Consider the circuit shown in Figure 10.27. With no voltmeter connected across R_2 then the true voltage across R_2 is, by voltage division, $VR_2/(R_1 + R_2)$. If now a voltmeter with a resistance R_V is connected across R_2, then we have R_V and R_2 in parallel and so, using the equation for resistances in parallel, a total resistance of $R_VR_2/(R_V + R_2)$. Because this differs from the resistance that was there before the voltmeter was connected, then the voltage V is divided in a different manner and so the voltmeter reading is in error. For a 99% accuracy of reading by the instrument, the resistance of the voltmeter must be at least 99 times larger than that of the element across which it is connected.

For alternating currents and voltages, the moving coil meter can be used with a rectifier circuit. Typically the ranges with alternating currents vary from about 10 mA to 10 A with accuracies up to ± 1% of full scale deflection for frequencies in the region 50 Hz to 10 kHz. With alternating voltages, the ranges are typically from about 3 V to 3 kV with similar accuracies.

Multi-meters give a number of direct and alternating current and voltage ranges, together with resistance ranges (see next section for details of the ohmmeter). A typical meter has full scale deflections for d.c. ranges from 50 μA to 10 A, a.c. from 10 mA to 10 A, direct voltages from 100 mV to 3 kV, alternating voltages 3 V to 3 kV and for resistance 2 kΩ to 20 MΩ. The accuracy for d.c. ranges is about ± 1% of full scale deflection, for a.c. ± 2% of full scale deflection and for resistance ± 3% of the mid-scale reading.

The moving coil meter has the problem of low input impedance and low sensitivity on a.c. ranges. This can be overcome by using an amplifier between the input and the meter to give an *electronic instrument*. The result is an increased sensitivity and a much higher input impedance. Hence there is less chance of loading presenting a problem.

Another form of meter is the *moving iron meter*. Such meters can be used for direct and, because the deflection is proportional to the square of the current, alternating currents. The scale is non-linear, being most cramped at the lower end. Typically such instruments have ranges with full scale deflections of between 0.1 and 30 A, without shunts, and an accuracy of about ± 0.5% of full scale deflection.

Example

A moving coil meter has a resistance of 10 Ω and gives a full scale deflection with 15 mA. What shunt resistance will be required if the meter is to give a full scale reading with 5 A?

The potential difference across the meter equals that across the shunt, i.e. $I_i R_i = (I - I_i)R_s$. Thus:

$$10 \times 10^{-3} \times 10 = (5 - 10 \times 10^{-3})R_s$$

Hence the resistance of the shunt is 0.0200 Ω.

Example

A circuit consists of two 1 MΩ resistors in series with a voltage of 50 V applied across them. What will be the reading indicated by a voltmeter of resistance 50 kΩ when connected across one of the resistors?

Before the voltmeter is connected the voltage across one of the resistors is, as a result of voltage division, 25 V. When the voltmeter is connected across a resistor the combined resistance R is given by:

$$\frac{1}{R} = \frac{1}{10^6} + \frac{1}{50 \times 10^3}$$

Thus $R = 47.6$ kΩ. Hence the voltage is now, by voltage division:

$$V = 50 \frac{47.6 \times 10^3}{10^6 + 47.6 \times 10^3} = 2.3 \text{ V}$$

There is thus error in using a voltmeter of such low resistance.

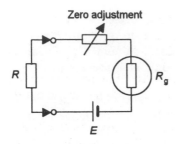

Figure 10.28   *Ohmmeter*

10.8.1 Measurement of resistance

Figure 10.28 shows one form of a basic *ohmmeter* circuit. A battery of e.m.f. E is connected in series with the meter resistance R_g, a zero adjustment resistor and the resistance R being measured. With R equal to zero, i.e. the terminals of the ohmmeter short circuited, the zero adjustment resistor R_z is adjusted so that the meter gives a full scale deflection, this corresponding to the 0 mark on the resistance scale. When R is across the instrument terminals then the total resistance in the circuit is increased and so the current I reduces from the zero value and can be used as a measure of the resistance R:

$$I = \frac{E}{R_g + R_z + R}$$

This type of circuit is used in multimeters for the measurement of resistance.

Example

An ohmmeter circuit, of the form indicated in Figure 10.28, is used to measure a resistance R. The moving coil meter used has a resistance of 50 Ω and a full scale deflection of 10 mA. The ohmmeter battery has an e.m.f. of 1.0 V. What should be the value of the zero adjustment resistance for there to be zero current when $R = 0$ and what then will be the value of R which will give the mid-scale reading on the meter?

When $R = 0$ then the current is the full scale value of 1.0 mA and so $1.0 = 10 \times 10^{-3}(50 + R_z)$. Thus $R_z = 50$ Ω. The mid-scale current reading for the meter is 5 mA. Hence, $1.0 = 5 \times 10^{-3}(50 + 50 + R)$ and so $R = 100$ Ω.

Activities

1 Investigate how the current through electrical components is related to the potential difference across them. Consider (a) A wire wound or carbon composition resistor with a value between 1 and 10 Ω. Either use a variable voltage supply (0 to 6 V d.c.) or a 6 V fixed d.c. voltage supply and a variable resistor (as in Figure 10.9). You will need a 0 to 1 A d.c. ammeter and a 0 to 5 V voltmeter. (b) A filament lamp such as a 6 V, 0.06 A lamp. Either use a variable voltage supply (0 to 6 V d.c.) or a 6 V fixed d.c. voltage supply and a variable resistor (as in Figure 10.9). You will need a 0 to 1 A d.c. ammeter and a 0 to 5 V voltmeter. (c) A thermistor. Consider what voltage supply and meters you might require for the thermistor given to you (you need to know from its specification its nominal resistance at 25°C and its power rating); you might have to carry out some preliminary experimental work to determine what you require. Because of the wide variation in resistance that can occur, you might like to use a multi-range meter.

Problems

1 The following results were obtained from measurements of the voltage across a lamp at different currents. Is Ohm's law obeyed?

Voltage in V	2.0	4.0	6.0	7.8	9.6
Current in A	0.05	0.10	0.15	0.20	0.25

2 An electrical lamp filament dissipates 20 W when the current through it is 0.10 A. What is (a) the potential difference across the lamp and (b) its resistance at that current?

3 Within what range will the resistances lie for resistors specified as being (a) 20 Ω, 5 % tolerance, (b) 68 Ω, 10 % tolerance, (c) 47 Ω, 20 % tolerance?

4 What are the nominal resistance and range of resistance values for resistors with the following colour codes: (a) blue, grey, black, gold, (b) yellow, violet, orange, silver, (c) blue, grey, green, gold.

5 Write (a) the standard abbreviations used for resistors with resistances of 4.7 kΩ and 5.6 Ω and (b) the values represented by the abbreviations (a) 6M8, (b) 4R7.

6 A 100 Ω resistor has a power rating of 250 mW. What should be the maximum current used with the resistor?

7 A 270 Ω resistor has a power rating of 0.5 W. What should be the maximum potential difference applied across the resistor?

8 A 10 Ω resistor is required for use in a circuit where the potential difference across it will be 4 V. What should its power rating be?

9 Resistors of resistances 20 Ω, 30 Ω and 50 Ω are connected in series across a d.c. supply of 20 V. What is the current in the circuit and the potential difference across each resistor?

10 The current to a light-emitting diode (LED) of resistance 250 Ω has to be limited to 5 mA. What resistor should be connected in series with the LED when it is connected to a 5 V d.c. supply?

11 A potential divider circuit is to be used to reduce a 24 V d.c. supply to 5 V. What resistors should be used if the circuit current should not exceed 10 mA?

12 Determine the power dissipated in each resistor and the total power dissipated when a voltage of 10 V is applied across (a) a series-connected (b) a parallel-connected pair of resistors with resistances of 20 Ω and 50 Ω.

13 Calculate the equivalent resistance of three resistors of 5 Ω, 10 Ω and 15 Ω which are connected in (a) series, (b) parallel.

14 Three resistors of 15 kΩ, 20 kΩ and 24 kΩ are connected in parallel. What is (a) the total resistance and (b) the power dissipated if a voltage of 24 V is applied across the circuit?

15 A circuit consists of two resistors in parallel. The total resistance of the parallel arrangement is 4 Ω. If one of the resistors has a resistance of 12 Ω, what will be the resistance of the other one?

16 A circuit consists of two resistors in series. If they have resistances of 4 Ω and 12 Ω, what will be (a) the circuit current and (b) the potential difference across each resistor when a potential difference of 4 V is applied across the circuit?

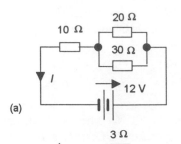

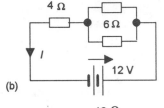

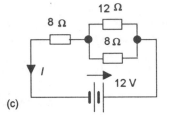

Figure 10.29   *Problem 21*

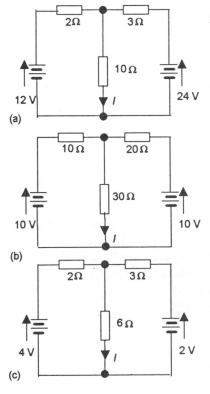

Figure 10.30 *Problem 24*

17 A circuit consists of two resistors in parallel. If they have resistances of 4 Ω and 12 Ω, what will be (a) the current through each resistor and (b) the total power dissipated when a potential difference of 4 V is applied across the circuit?

18 A circuit consists of two parallel-connected resistors of 2.2 kΩ and 3.9 kΩ in series with a 1.5 kΩ resistor. With a supply voltage of 20 V, what is the current drawn from the supply?

19 A circuit consists of three parallel-connected resistors of 330 Ω, 560 Ω and 750 Ω in series with a 800 Ω resistor. If the supply voltage connected to the circuit is 12 V, what are the voltages across the parallel resistors and the series resistor?

20 Three resistors of 6 Ω, 12 Ω and 24 Ω are connected in parallel across a voltage supply. What fraction of the supply current will flow through each resistor?

21 Determine the total resistances of the circuits shown in Figure 10.29 and the circuit currents I.

22 A circuit consists of a resistor of 8 Ω in series with an arrangement of another 8 Ω resistor in parallel with a 12 Ω resistor. With a 24 V input to the circuit,what will be the current taken from the supply?

23 A circuit consists of a resistor of 2 Ω in series with an arrangement of three parallel resistors, these having resistances of 4 Ω, 10 Ω and 20 Ω. If the voltage input to the circuit is 10 V, what will be the current through the 2 Ω resistor?

24 Determine, using Kirchhoff's laws, the current I in each of the circuits shown in Figure 10.30.

25 A moving coil meter has a resistance of 5 Ω and gives a full scale deflection with 1 mA. Determine (a) the shunt resistance required if it is to give a full scale reading with 0.1 A, and (b) the multiplier resistance if it is to give a full scale deflection with 10 V.

26 A moving coil meter has a resistance of 10 Ω and gives a full scale deflection with 10 mA. Determine (a) the shunt resistance required if it is to give a full scale reading with 5 A, and (b) the shunt multiplier resistance if it is to give a full scale reading with 100 V.

27 A circuit consists of two resistors, with resistances of 20 kΩ and 10 kΩ, in series. A voltage of 100 V is applied to the circuit. What reading will be indicated by a voltmeter of resistance of 50 kΩ connected across the 10 kΩ resistor?

28 An ohmmeter, of the form shown in Figure 10.28, uses a basic meter movement of resistance 50 Ω and having a full scale deflection of 1 mA. The meter is to be used with a shunt. With a 3 V battery, what will be the shunt resistance and the zero adjustment resistance if the mid-scale reading of the instrument is to occur when the measured resistance R is 2 kΩ?

29 If aluminium has a resistivity of 2.5×10^{-8} Ω m, what will be the resistance of an aluminium wire with a length of 1 m and a cross-sectional area of 2 mm² and what is the conductivity of the aluminium?

30 What length of copper wire, of diameter 1.5 mm, is needed to give a resistance of 0.3 Ω if the resistivity is 1.7×10^{-8} Ω m?

11 Magnetism

11.1 Introduction

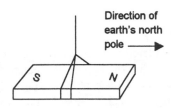

Figure 11.1 *Suspended magnet*

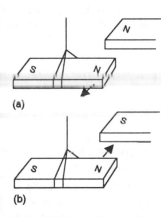

Figure 11.2 *(a) Repulsion, (b) attraction*

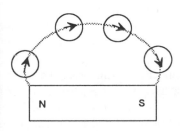

Figure 11.3 *Use of compass needle to plot directions of magnetic field*

This chapter reviews the concept of a magnetic field and then takes a look at the principles of electromagnetic induction, the force on current carrying conductors in magnetic fields and the principles of motors, generators, transformers and moving coil meters.

11.1.1 Magnetic fields

If a permanent bar magnet is suspended, as in Figure 11.1, so that it is free to swing in a horizontal plane, it will rotate until it lines up with one end pointing towards the earth's north pole and the other towards the south. The end of the magnetic which points towards the north is termed the *north pole* and the end that points towards the south is the *south pole*. If the north pole of another magnet is brought near to the north pole of a suspended magnet (Figure 11.2), repulsion occurs; if the south pole of another magnet is brought near to the north pole of the suspended magnet, attraction occurs. *Like poles repel each other, unlike poles attract each other*.

A permanent magnet is said to produce a *magnetic field* in the space around it. Thus it is the magnetic field in the space around the magnetic which interacts with the suspended magnet and causes it to be attracted or repulsed. The magnetic field pattern in the space surrounding a permanent magnet can be shown by sprinkling iron filings around it or by means of a compass needle being used to plot the directions of the field (Figure 11.3). The direction of the magnetic field at a point is defined as the direction of the force acting on an imaginary north pole placed at that point and is thus the direction in which the north pole of a compass needle points. Figure 11.4 shows examples of the field patterns produced by permanent bar magnets.

11.1.2 Characteristics of magnetic field lines

Magnetic lines of force can be thought of as being lines along which something flows. The thing that flows is called *flux* (Φ). Such lines can be considered to have the following properties:

1 The direction of a line of magnetic flux at any point is given by the north pole of a compass needle placed at that point.

2 The lines of magnetic flux never intersect. If lines did intersect then we would have, at the point of intersection, two possible field directions and a compass needle would not know which way to point.

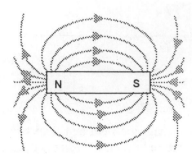

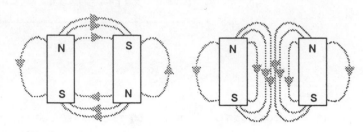

Figure 11.4 *Magnetic fields*

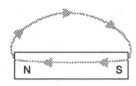

Figure 11.5 *Closed path*

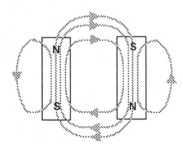

Figure 11.6 *Magnetic flux lines*

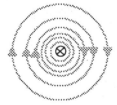

Figure 11.7 *Current-carrying conductor*

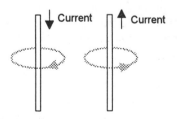

Figure 11.8 *Field directions*

3 Each line of magnetic flux can be considered to form a closed path. This means that a line of flux emerging from the N pole of a magnet and, passing through the surrounding space back to the S pole is assumed to continue through the magnet back to the N pole to complete the closed loop (Figure 11.5).

4 Lines of magnetic flux are like stretched elastic cords, always trying to shorten themselves. This is illustrated by the two bar magnets in Figure 11.6. The lines in 'attempting to shorten themselves' cause the magnets to be attracted to each other.

11.1.3 Magnetic field produced by a current-carrying conductor

When a current flows through a conductor, a magnetic field is produced in the space around the conductor. For a single wire the field pattern is concentric circles centred on the wire; Figure 11.7 shows the field pattern for the current passing at right angles into the plane of this page. For the current passing at right angles out of the plane of this page, the directions of the field are in the opposite direction (Figure 11.8). A useful way of remembering the direction of the field round a wire is the *corkscrew rule*. If a right-handed corkscrew is driven along the wire in the direction of the current, then the corkscrew rotates in the direction of the magnetic field.

With a single turn of wire, the magnetic field pattern is just the result of the addition of the fields due to each side of the wire loop which constitutes (Figure 11.9) the turn. With the solenoid, i.e. a coil with many turns and a length significantly larger than its diameter, the magnetic fields due to each turn add up to give a magnetic field with lines of force entering one end of the coil and leaving from the other (Figure 11.10(a)). The field pattern is just like that for a permanent bar magnet and as a result we can think of it having magnetic poles. Looking at the end elevation of a solenoid, then that end is a north pole if the current flow is anticlockwise and a south pole if clockwise (Figure 11.10(b)). The magnetic field from a solenoid may be increased in strength by inserting a rod of iron inside the solenoid.

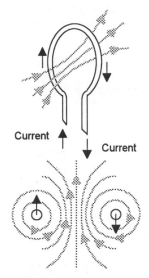

Current ↑ ↓ Current

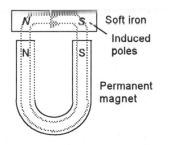

Figure 11.9   *Single loop*

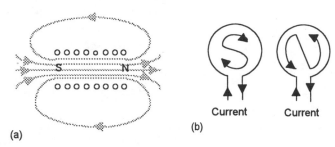

(a)

(b)

Current   Current

Figure 11.10   *Solenoid*

11.1.4 Magnetic induction

If a piece of unmagnetised soft iron is placed near a permanent magnet it becomes attracted to it (Figure 11.11). The lines of magnetic flux due to the permanent magnet pass through the soft iron and we can explain the force of attraction as being due to the flux lines trying to shorten themselves. An alternative way of explaining the attraction is in terms of magnetic poles being induced in the soft iron and then attraction between the unlike poles of the permanent magnet and the induced poles in the soft iron.

11.2 Electromagnetic induction

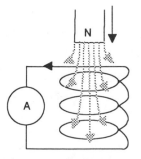

Soft iron
Induced poles

Permanent magnet

Figure 11.11 *Magnetic induction*

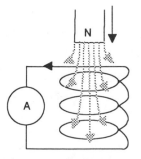

Figure 11.12   *Electromagnetic induction*

Consider what happens if we move a magnet towards a coil along its axis, as illustrated in Figure 11.12. The meter indicates a current and thus we conclude that an e.m.f. is being induced in the coil. Think of the pole of the magnet being rather like a garden water sprinkler spraying out water. As the magnet moves towards the coil so more of the 'sprayed out water' passes through the turns of the coil. An e.m.f. is only induced when the amount of flux linked by the coil, i.e. passing through the turns of the coil, is changing. Immediately the magnet stops moving, the induced e.m.f. ceases.

If the magnet is kept stationary and the coil moved towards the magnet, an e.m.f. is induced. When the coil is stationary there is no e.m.f. Thus, for the magnet and the coil, when one moves relative to the other then there is a change in the flux linked and so an e.m.f.

> *An e.m.f. is induced in a coil of wire when the magnetic flux linked by it changes.*

The faster we move a magnet towards a stationary coil, or a coil towards a stationary magnet, then bigger the size of the induced e.m.f.

> *The size of the induced e.m.f. is proportional to the rate of change of flux linked by the coil. This is known as Faraday's law.*

If we move the magnet towards the coil there is an induced e.m.f.; if we move the magnet away from the coil there is an e.m.f but it is in the

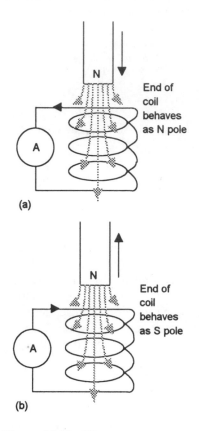

(a)

(b)

Figure 11.13 *Electromagnetic induction*

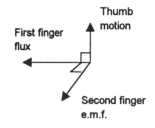

Figure 11.14 *Right-hand rule*

opposite direction. The direction of the induced e.m.f. in the coil is always in such a direction as to produce a current which sets up magnetic fields which tend to neutralise the change in magnetic flux linked by the coil which caused it.

> *The direction of the induced e.m.f. is such as to oppose the change producing it. This is known as Lenz's law.*

This means, for example, that when the north pole of the magnet approaches the coil (Figure 11.13(a)) that the current produced in the coil is in such a direction as to make the end of the coil nearest the magnet a north pole, so opposing the approach of the magnet. If the north pole of the magnet is being moved away from the coil (Figure 11.13(b)) then the current induced in the coil is in such a direction as to make the end of the coil nearest the magnet north pole a south pole, so opposing the removal of the magnet.

A useful way of determining the direction of the induced e.m.f. in a conductor is *Fleming's right-hand rule* (Figure 11.14). If the thumb, the forefinger and the second finger of the right hand are set at right angles to each other, the thumb points in the direction of the movement of the conductor relative to the flux, the first finger in the direction of the magnetic flux and the second finger in the direction of the induced e.m.f.

We can represent Faraday's law and Lenz's law as the relationship:

induced e.m.f. $e \propto -$ (rate of change of flux Φ with time t)

The minus sign indicates that the induced e.m.f. is in such a direction as to oppose the change producing it. We can put the constant of proportionality as 1 and write rate of change of flux as $d\Phi/dt$:

$$e = -\frac{d\Phi}{dt}$$

The unit of flux is the weber (Wb). If the flux linked changes by 1 Wb/s then the induced e.m.f. is 1 V.

For coil with N turns, each turn will produce an induced e.m.f. and so the total m.m.f. will be the sum of those due to each turn and thus:

$$e = -N\frac{d\Phi}{dt}$$

11.2.1 Magnitude of the induced e.m.f.

The amount of magnetic flux passing at right angles through a unit area is called the *magnetic flux density B*, i.e.

$$B = \frac{\Phi}{A}$$

The unit of flux density is the tesla (T), when the flux is in webers (Wb) and the area A in square metres. If the flux does not pass at right angles through the area but makes some angle θ to the axis at right angles to the

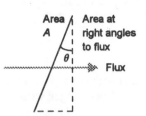

Figure 11.15 *Flux passing through an area*

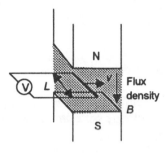

Figure 11.16 *Changing the flux linked*

area (Figure 11.15), then since the area at right angles to the flux is now $A \cos \theta$, the flux density is:

$$B = \frac{\Phi}{A \cos \theta}$$

An e.m.f. is induced in a conductor when it moves in a magnetic field. Consider a conductor of length L moving through a magnetic field with a velocity v, as illustrated in Figure 11.16. In a time t the conductor will move a distance vt and the area through which the flux passes will have been changed by Lvt. Thus the flux linked will change by $BLvt$ in a time t. The rate of change of flux linked with time is thus BLv and there will be an induced e.m.f. e of:

$$e = BLv$$

Example

What is the average e.m.f. induced in a coil of 100 turns if the flux linked by it changes from zero to 0.05 Wb in a time of 10 s?

The change of flux linked by each turn is 0.05 Wb in 10 s, and thus by the 100 turns is 100×0.05 Wb in 10 s. This is an average rate of change of flux linked of 0.5 Wb/s. Hence the induced e.m.f. is 0.5 V.

Example

What is the flux passing through an area of 1 cm² if the flux density is 0.2 T?

Using $B = \Phi/A$, we have $\Phi = BA = 0.2 \times 1 \times 10^{-4} = 2 \times 10^{-5}$ Wb.

Example

A conductor of length 0.01 m moves at right angles to a magnetic field of flux density 0.1 T with a velocity of 4 m/s. What will be the e.m.f. induced between the ends of the conductor?

E.m.f $e = BLv = 0.1 \times 0.01 \times 4 = 4$ mV.

11.3 Generators

One way of changing the flux linked by a coil when in a magnetic field is to rotate the coil so that the angle between the field direction and the plane of the coil changes (Figure 11.17(a)). When the plane of the coil is at right angles to a field of flux density B then the flux linked per turn of wire is a maximum and given by BA, where A is the area of the coil. When the plane of the coil is parallel to the field then there is no flux linked by the coil. When the axis of the coil is at an angle θ to the field direction then the flux linked per turn is $\Phi = BA \cos \theta$. The flux linked thus changes as the coil rotates. If the coil rotates with an angular

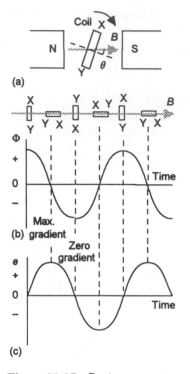

(a)

(b)

(c)

Figure 11.17 *Basic generator*

velocity ω then its angle θ at a time t is ωt and so the flux linked per turn varies with time according to:

$$\Phi = BA \cos \omega t$$

Figure 11.17(b) shows how the flux linked varies with time. The induced e.m.f. produced per turn is the rate at which the flux changes with time. It is thus the gradient of the graph of flux against time. Thus when the gradient is zero there is no induced e.m.f., when the gradient is a maximum the induced e.m.f. is a maximum (because of Lenz's law the induced e.m.f. is $-d\Phi/dt$). The result of considering the gradients is that the induced e.m.f. is given by a sine graph (Figure 11.17(c)), i.e.

$$e = E_{max} \sin \omega t$$

Thus rotating the coil in a magnetic field has led to a basic alternating voltage *generator*.

We could have obtained the above equation by differentiating $\Phi = BA \cos \omega t$ to give for the rate of change of flux with time $d\Phi/dt = -BA \sin \omega t$ and thus, since $e = -d\Phi/dt$, obtain

$$e = BA\omega \sin \omega t = E_{max} \sin \omega t$$

where the maximum e.m.f. E_{max} is $BA\omega$.

11.4 Induced e.m.f with two coils

To induce an e.m.f. in a coil A we need to have the magnetic flux linking its turns changing. One way we can do this is by moving a permanent magnet along the axis of the coil A (Figure 11.18(a)) However, rather than using a permanent magnet to produce the magnetic field, we could pass a current through a coil to give a magnetic field. Thus we could induce an e.m.f. in coil A by moving a current-carrying coil B rather than a permanent magnet (Figure 11.18(b)). An alternative to moving coil B in order to give a changing magnetic field is to change the current in coil B, e.g. by supplying it with alternating current (Figure 11.18(c)).

11.4.1 Transformers

A *transformer* basically consists of two coils, called the primary coil and the secondary coil, wound on the same core of magnetic material (Figure 11.19). An alternating voltage is applied to the primary coil. This produces an alternating current in that coil and so an alternating magnetic flux in the core. The secondary coil is wound on the same core and the flux produced by the primary coil links its turns so the alternating flux induces an alternating e.m.f. in the secondary coil. An alternating e.m.f. in the primary coil has been transformed into an alternating e.m.f. in the secondary coil.

Because the core is made of a magnetic material, to a reasonable approximation, the flux linking each turn of the secondary coil is the same as the flux linking each turn of the primary coil. Because the flux is

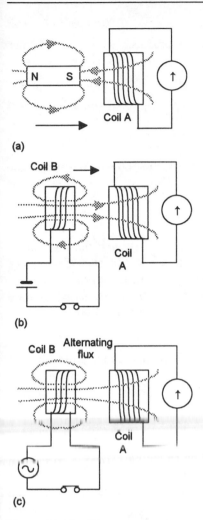

(a)

(b)

(c)

Figure 11.18 *Electromagnetic induction*

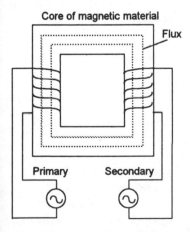

Figure 11.19 *Basic transformer*

changing there must be an induced e.m.f. produced in each turn of each coil. The e.m.f. induced per turn of either the secondary or primary coils must be the same since the same rate of change of flux occurs. The total induced e.m.f. for a coil is thus proportional to the number of turns of that coil. Thus we can write:

$$\frac{\text{induced e.m.f. in primary}}{\text{induced e.m.f. in secondary}} = \frac{N_1}{N_2}$$

where N_1 is the number of turns for the primary coil and N_2 the number of turns for the secondary coil. When the secondary coil is on open-circuit, i.e. there is no load connected to the coil, then the voltage between the terminals of the coil is the same as the induced e.m.f. If there is no load then there is no current and so no energy taken from the secondary coil. This means, if no energy is wasted, that no energy is taken from the primary coil. This can only be the case if the induced e.m.f. is equal to and opposing the input voltage. Thus if we take V_1 as the input voltage and V_2 the output voltage, we can write:

$$\frac{V_1}{V_2} = \frac{N_1}{N_2}$$

If V_2 is less than V_1 the transformer is said to have a step-down voltage ratio, if V_2 is more than V_1 a step-up voltage ratio.

Now consider what happens when there is a resistive load and a secondary current flows. Because there is a current flowing through a resistance, then power is dissipated. This current in the secondary coil produces its own alternating flux in the core, this then resulting in an alternating e.m.f. being induced in the primary coil. Consequently this produces a current in the primary coil. If the power losses in a transformer are negligible, then when there is a current in the primary coil the power supplied to the primary coil must equal the power taken from the secondary coil. Thus:

$$I_1V_1 = I_2V_2$$

where I_1 is the current in the primary coil and I_2 that in the secondary coil. We can rearrange this as $V_1/V_2 = I_2/I_1$ and so:

$$\frac{V_1}{V_2} = \frac{I_2}{I_1} = \frac{N_1}{N_2}$$

Hence:

$$I_1N_1 = I_2N_2$$

The product of the current through a coil and its number of turns is called its *ampere-turns*. Thus, the number of ampere-turns for the secondary winding equals the number of ampere-turns for the primary winding.

Example

What will be the secondary voltage produced with a transformer having 400 primary turns and 50 secondary turns, when there is an alternating voltage input of 240 V?

Using $V_1/V_2 = N_1/N_2$, then $V_2 = V_1N_2/N_1 = 240 \times 50/400 = 40$ V.

Example

A transformer has a step-down voltage ratio of 6 with a primary coil alternating voltage input of 240 V. What will be the primary and secondary currents when a lamp dissipating 40 W is connected across the second coil?

The power output from the secondary is 40 W and thus, assuming that there are no transformer power losses, the input to the primary coil must be equal to 40 W. Since $P = I_1V_1$, then the primary current I_1 is 40/240 = 0.167 A. The secondary current can be obtained by using $I_1N_1 = I_2N_2$. Thus $I_2 = (N_1/N_2)I_1 = 6 \times 0.167 = 1.00$ A.

11.5 Force on a current-carrying conductor

A wire carrying a current produces a magnetic field in the region around it and is thus able to exert forces on magnets in that region. Conversely, a wire carrying a current when placed in a magnetic field experiences a force. Figure 11.20 shows how such a force might be demonstrated. Thick bare copper wire is used to make two horizontal rails and a movable crossbar that can slide on the rails. With the magnet giving lines of magnetic flux at right angles to the crossbar, when an electric current is switched on the crossbar moves along the rails. Thus a force is acting on the current-carrying conductor.

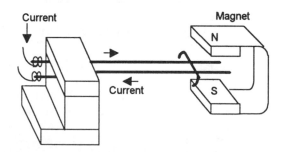

Figure 11.20 *Force on a current-carrying conductor in a magnetic field*

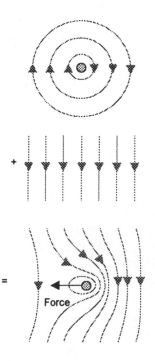

Figure 11.21 *Catapult field*

The current in the movable crossbar gives circular lines of magnetic flux around the wire. The magnet gives almost straight lines of magnetic flux across the gap between the poles. When we have both magnetic fields together, the fields add together to give the result shown in Fig. 11.21; the result is what can be termed a catapult field. The distorted field lines act like a stretched elastic string bent out of the straight and which endeavours to straighten out and so exerts a force on the conductor. The force does *not* act along the wire carrying the current, nor does it act along the magnetic field; it acts in a direction which is perpendicular to both the current and the magnetic field.

> *If a wire is placed in a magnetic field so that it is at right angles to the field, then when a current passes through the wire a force acts on the wire. This force is in a direction at right angles to both the current and the magnetic field.*

A rule that helps in remembering the directions is *Fleming's left-hand rule* (Figure 11.22). If the thumb, forefinger and second finger of the left hand are set at right angles to each other, then the forefinger indicates the flux direction, the thumb the motion (or force) and the second finger the current direction. Only if there is a component of the magnetic field at right angles to the wire is there any force.

Suppose that as a result of the force acting on the conductor in Figure 11.23 it moves with a velocity v. An e.m.f. e will be induced in the conductor, with e being equal to BLv. If the source e.m.f. is E then the net e.m.f. in the circuit supplying the current I is $E - e$. If the circuit has a resistance R then

$$E - BLv = IR$$

Multiplying throughout by I, then

$$EI \quad BILv - I^2R$$

EI is the power supplied by the source and I^2R is the power dissipated in the circuit by the current. Thus $BILv$ must represent the power developed by the force used to move the conductor. But the power developed by a force F moving its point of application with a velocity v is Fv (see Section 2.7). Hence $Fv = BILv$ and so:

$$F = BIL$$

Example

A wire has a length of 100 mm in a magnetic field of flux density 1.0 T. What is the force on the wire when it carries a current of 2 A and is (a) at right angles to the field, (b) at 30° to the field and (c) in the same direction as the field.

(a) $F = BIL = 1.0 \times 2 \times 0.10 = 0.20$ N
(b) The flux density at right angles to the wire is $B \sin \theta$, hence $F = 1.0 \sin 30° \times 2 \times 0.10 = 0.10$ N.
(c) The flux density at right angles to the wire is 0, hence the force is 0.

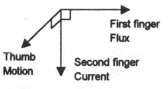

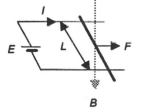
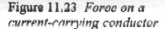

Figure 11.22 *Fleming's left-hand rule*

Figure 11.23 *Force on a current-carrying conductor*

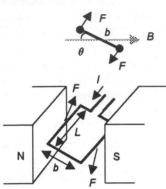

Figure 11.24 *Coil in a magnetic field*

11.5.1 Force on a current carrying coil

Consider the forces acting on a single turn, current carrying rectangular coil in a magnetic field (Figure 11.24). The sides of the coil have a length L and breadth b. The plane of the coil is at an angle θ to the direction of the field which has a uniform flux density B. Each of the sides of length L of the coil has a flux density component $B \sin \theta$ at right angles to it and so will experience a force F, with:

$$F = (B \sin \theta)IL$$

Using Fleming's left-hand rule, we can determine the directions of these forces to be as shown in the figure. These forces are in such directions as to rotate the coil about the horizontal axis (note that the reason we are not interested in the forces on the sides of length b is that these forces do not cause rotation and are in opposite directions, cancelling each other out) . The turning moment or torque T about this axis is:

$$T = F\tfrac{1}{2}b + F\tfrac{1}{2}b = Fb = BILb \sin \theta$$

But Lb is the area A of the coil. Hence:

$$T = BIA \sin \theta$$

If there are N turns on the coil, then each turn will experience the above torque and so the total torque will be

$$T = NBIA \sin \theta$$

The maximum torque will be when $\sin \theta = 1$, i.e. $\theta = 90°$ and the coil is at right angles to the field.

Example

What is the torque experienced by a rectangular coil with 50 turns of length 100 mm and breadth 50 mm if it carries a current of 200 mA in a magnetic field having a flux density of 0.6 T at 45° to the plane of the coil?

Using the equation $T = NBIA \sin \theta$ gives:

$$T = 50 \times 0.6 \times 0.200 \times 100 \times 10^{-3} \times 50 \times 10^{-3} \sin 45°$$

$$= 0.021 \text{ N m.}$$

11.5.2 D.c. motor

The above is the basic principle involved in *d.c. motors*. The motor is essentially just a coil in a magnetic field. Figure 11.25 shows the basic principle. When a current passes one way through the coil it is acted on by a torque which causes it to rotate. However, when the coil has rotated

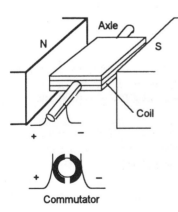

Figure 11.25 *Basic elements of a d.c. motor*

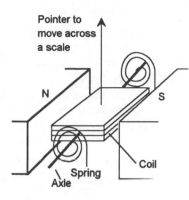

Figure 11.26 *Basic elements of a moving-coil ammeter*

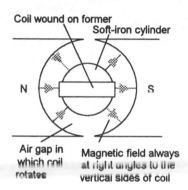

Figure 11.27 *Radial field*

to the vertical position the torque drops to zero. If the coil overshoots this point then the torque acts in the reverse direction to return the coil back to the vertical position and so the rotation ceases. To keep it rotating the current has to be reversed. The current is fed to the coil via a commutator attached to the axle and as this rotates it reverses the current as the coil passes through the vertical position.

11.5.3 Moving coil galvanometer

The *moving coil galvanometer* follows the same principle. Figure 11.26 shows the basic principle involved. With a current through the coil, a torque acts on it and causes it to rotate. The coil moves against springs. These supply a restoring torque which is proportional to the angle ϕ through which they have been twisted. We can write this restoring torque as $c\phi$, where c is a constant for the springs. Thus the coil rotates to an angle at which the restoring force exerted by the springs just balances the torque produced by the current in the coil; the deflection of the coil is thus a measure of the current.

With a more realistic form of moving coil meter, the magnet pole pieces are so designed that the flux density is always at right angles to the sides of the coil (Figure 11.27) and so the torque acting on the coil with a current I is always $NBIA$. The coil rotates until $c\phi = NBIA$. Thus the angular deflection ϕ of the coil is:

$$\phi = \left(\frac{NBA}{c}\right)I$$

The term in the brackets is a constant for the galvanometer and so the angular deflection is proportional to the current.

Activities

1 Determine the magnetic field patterns for two bar magnets in a line end-to-end with (a) like poles facing each other, (b) opposite poles facing each other.
2 Examine a moving coil meter, draw a diagram of its construction and explain the reasons behind the form and materials used, and the principles of its operation.

Problems

1 What is the flux density when a flux of 2×10^{-4} Wb passes through an area of 2 cm^2 at (a) right angles, (b) 45° to it?
2 What is the flux density inside a solenoid with a cross-sectional area of 1 cm^2 if the flux passing through it is 20 µWb?
3 The flux density in the air gap between the north and south poles of a horseshoe magnet is 2 T. If each of the pole surfaces has an area of 2 cm^2, what is the flux passing between the poles?
4 What is the average e.m.f. induced in a coil of 120 turns if the flux linked changes from 0.05 Wb to 0.10 Wb in a time of 4 s?
5 A coil with 100 turns is linked by a flux of 0.20 Wb. If the direction of the flux is reversed in a time of 5 ms, what is the average e.m.f. induced in the coil?

6 A conductor of length 500 mm moves with a velocity of 20 m/s through a uniform magnetic field of flux density 1.0 T, the field being at right angles to the conductor. What is the e.m.f. induced in the conductor if the direction of motion is (a) at right angles to the direction of the field, (b) at an angle of 30° to the field?

7 A conductor of length 15 mm moves with a velocity of 20 m/s through a uniform magnetic field at an angle of 60° to the magnetic field direction. If the magnetic field has a flux density of 0.002 T, what is the e.m.f. induced in the conductor?

8 A conductor of length 500 mm moves at right angles to its length with a velocity of 40 m/s in a uniform magnetic field of flux density 1 Wb/m². What is the e.m.f. induced in the conductor when the direction of motion is (a) perpendicular to the field, (b) at 30° to the direction of the field?

9 What will be the secondary voltage produced by a transformer which has 400 primary turns and 100 secondary turns, when there is an alternating voltage input to the primary of 240 V?

10 A transformer has a step-down voltage ratio of 5 to 1. What voltage and current must be supplied to the primary coil if the transformer is to supply 50 V at 20 A at the secondary?

11 What is the force acting per metre length of a conductor carrying a current of 2.0 A if it is (a) at right angles, (b) at 60°, (c) parallel to a magnetic field having a flux density of 0.50 T?

12 A conductor has a length of 200 mm and is at right angles to a magnetic field of flux density 0.7 T. What is the force acting on it when it carries a current of 1.5 A?

13 A square coil with sides of length 100 mm has 50 turns of wire. The plane of the coil is at an angle of 45° to a magnetic field of flux density 0.4 T. What is the torque experienced by the coil when a current of 2 A flows through it?

14 A moving coil galvanometer has a rectangular coil with breadth 10 mm and length 15 mm situated in a radial magnetic field of flux density 0.5 T. The suspension of the coil has a torsional constant of 10^{-8} N m/rad. How many turns will the coil need to have if it is to give a deflection of 90° when the current is 10 μA?

12 Circuit analysis

12.1 Introduction

This chapter is concerned with the analysis of circuits containing just resistors and d.c. sources and follows on from the discussion of such circuits in Chapter 10.

The fundamental laws we use in circuit analysis are Kirchhoff's laws. The laws can be stated as (see Section 10.6 for a discussion of them):

1 *Kirchhoff's current law states that at any junction in an electrical circuit, the current entering it equals the current leaving it.*

2 *Kirchhoff's voltage law states that around any closed path in a circuit, the sum of the voltage drops across all the components is equal to the sum of the applied voltage rises.*

The techniques used in this chapter for d.c. circuit analysis with circuits containing just resistors are:

1 Circuit reduction techniques for resistors in series and parallel, involving the current divider rule and the voltage divider rule.

2 Kirchhoff's laws with node and mesh analysis.

3 Superposition theorem.

4 Norton's theorem.

5 Thévenin's theorem.

12.2 Series and parallel resistors

Circuit elements are said to be connected in *series* when each element carries the same current as the others; they are in *parallel* with one another when the same voltage appears across each of the elements. A series–parallel circuit is one that contains combinations of series- and parallel-connected components. A technique for solving such circuits is to systematically determine the equivalent resistance of series or parallel connected resistors and so reduce the analysis problem to a very simple circuit. This was illustrated in Chapter 10 and this reviews the principles and then extends them to more difficult problems.

For series resistors the equivalent resistance is the sum of the resistances of the separate resistors (Section 10.3):

$$\text{equivalent resistance } R_e = R_1 + R_2 + \ldots$$

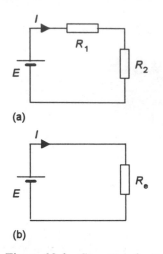

(a)

(b)

Figure 12.1 *Circuit reduction*

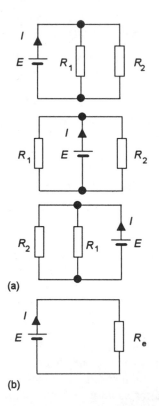

(a)

(b)

Figure 12.2 *(a) Versions of the same parallel circuit, (b) the equivalent circuit*

where R_1, R_2, etc. are the resistances of the separate resistors. As an illustration of the reduction technique, consider the series circuit shown in Figure 12.1(a). The equivalent resistance R_e of the two resistors R_1 and R_2 is $R_1 + R_2$ and thus we can reduce the circuit to that in Figure 12.1(b). The circuit current I can then be obtained for this reduced circuit, being E/R_e and hence is $E/(R_1 + R_2)$.

For resistors in parallel the equivalent resistance R_e is given by (see Section 10.4):

$$\frac{1}{R_e} = \frac{1}{R_1} + \frac{1}{R_2} + \ldots$$

and hence, for two resistors:

$$R_e = \frac{R_1 R_2}{R_1 + R_2}$$

As an illustration of the circuit reduction technique involving parallel resistors, consider the circuit shown in Figure 12.2(a). This circuit is shown in a number of alternative but equivalent forms, all involving a battery of e.m.f. E and two parallel resistors. The equivalent resistance R_e for the two parallel resistors is given by the above equation as $R_e = R_1 R_2/(R_1 + R_2)$. Thus the current I drawn from the voltage source E is $E/R_e = E(R_1 + R_2)/R_1 R_2$.

For a circuit consisting of both series and parallel components, we can use the above techniques to simplify each part of the circuit in turn and so obtain a simple equivalent circuit. As an illustration, consider the circuit shown in Figure 12.3(a). As a first step we can reduce the two parallel resistors to their equivalent, thus obtaining circuit (b) with $R_p = R_2 R_3/(R_2 + R_3)$. We then have R_1 in series with R_p and so can obtain the equivalent resistance $R_e = R_1 + R_2 R_3/(R_2 + R_3)$ and circuit (c). Thus the current drawn from the voltage source E is $I = E/[R_1 + R_2 R_3/(R_2 + R_3)]$.

Figure 12.4(a) gives another illustration. As a first step we can reduce the two series resistors to their equivalent, thus obtaining circuit (b) with $R_s = R_1 + R_2$. We then have R_s in parallel with R_3 and so can obtain the equivalent resistance $R_e = R_s R_3/(R_s + R_3) = (R_1 + R_2)R_3/(R_1 + R_2 + R_3)$. Thus the current $I = E(R_1 + R_2 + R_3)/[(R_1 + R_2)R_3]$.

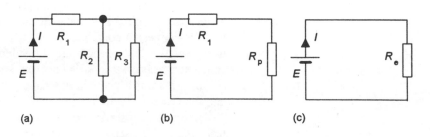

(a)   **(b)**   **(c)**

Figure 12.3 *(a) Series–parallel circuit, (b) first reduction, (c) second reduction*

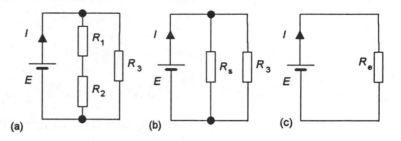

Figure 12.4 *(a) Series–parallel circuit, (b) first reduction, (c) second reduction*

Example

Determine the current I taken from the voltage source in the circuit given by Figure 12.5.

For the two resistors in parallel we have an equivalent resistance R_p given by:

$$\frac{1}{R_p} = \frac{1}{60} + \frac{1}{30} = \frac{1+2}{60} = \frac{1}{20}$$

Hence the circuit reduces to that in Figure 12.6(a). For the two resistors in series, the equivalent resistance is $20 + 20 = 40\ \Omega$. We now have the circuit shown in (b) and so I is $E/R_e = 24/40 = 0.6$ A.

Example

Determine the current I taken from the voltage source in the circuit given by Figure 12.7.

The equivalent resistance for the two series resistors is $8 + 4 = 12\ \Omega$. This gives the simpler circuit shown in Figure 12.8(a). The equivalent resistance for the two parallel resistors is $12 \times 6/(12 + 6) = 4\ \Omega$. This gives Figure 12.8(b). The equivalent resistance for the two series resistors is $8 + 4 = 12\ \Omega$. Hence we end up with the equivalent circuit shown in Figure 12.8(c). The current I is thus $24/12 = 2$ A.

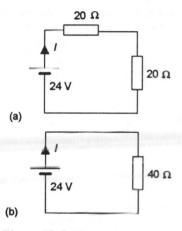

Figure 12.5 *Example*

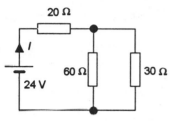

(a)

(b)

Figure 12.6 *Example*

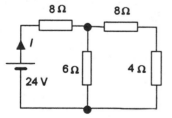

Figure 12.7 *Example*

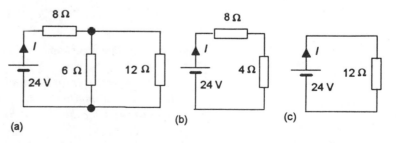

(a)

(b)

(c)

Figure 12.8 *Example*

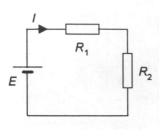

Figure 12.9 *Voltage drops in a series circuit*

12.2.1 Voltage and current division

Often in circuit analysis it is necessary to find the voltage drops across one or more series resistors. Consider the circuit shown in Figure 12.9. The current I is given by:

$$I = \frac{E}{R_1 + R_2}$$

Thus the voltage drop V_1 across resistance R_1 of IR_1 is:

$$V_1 = E\frac{R_1}{R_1 + R_2}$$

and the voltage drop V_1 across resistance R_2 of IR_2 is:

$$V_2 = E\frac{R_2}{R_1 + R_2}$$

Note that the above equations indicate that the voltage drop across a resistor is proportional to the ratio of that resistance to the total resistance of the circuit. We have *voltage division* with the supply voltage being divided into bits across each resistor according to its resistance as a fraction of the total resistance. The voltage division rule can be stated as:

The voltage across any resistance in a series circuit is the source voltage multiplied by the ratio of that resistance to the total resistance of the circuit.

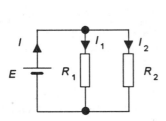

Figure 12.10 *Current division with parallel circuit*

For a parallel circuit we have a *current division* rule. Consider the circuit shown in Figure 12.10. The equivalent resistance for the two parallel resistors is $R_e = R_1R_2/(R_1 + R_2)$ and so $E = IR_e = IR_1R_2/(R_1 + R_2)$. The potential drop across each resistor is the same, namely E. Thus the current I_1 through resistor R_1 is:

$$I_1 = \frac{E}{R_1} = \frac{IR_e}{R_1} = I\frac{R_1R_2}{R_1(R_1 + R_2)} = I\frac{R_2}{R_1 + R_2}$$

The current I_2 through resistor R_2 is:

$$I_2 = \frac{E}{R_2} = \frac{IR_e}{R_2} = I\frac{R_1R_2}{R_2(R_1 + R_2)} = I\frac{R_1}{R_1 + R_2}$$

Thus we can state a current division rule as:

With two resistors in parallel, the current in each resistor is the total current multiplied by the fraction of the resistance of the opposite resistor divided by the sum of the two resistances.

When we have a series–parallel circuit then we can use the voltage division rule to determine the voltage drop across each group of series

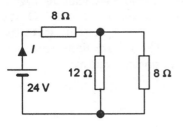

Figure 12.11   *Example*

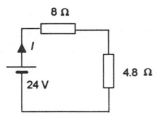

Figure 12.12   *Example*

elements and the current division rule to determine the current through each branch of parallel elements.

Example

Determine the voltage drops across, and the current through, each of the resistors in the circuit given in Figure 12.11.

The equivalent resistance R_p of the pair of parallel resistors is:

$$R_p = \frac{12 \times 8}{12 + 8} = 4.8 \ \Omega$$

We can thus draw the equivalent circuit of Figure 12.12. The voltage drop across the 8 Ω resistor is given by the voltage divider rule as:

$$V_8 = 24\frac{8}{8 + 4.8} = 15 \ V$$

The voltage drop across the parallel resistors is given by the voltage divider rule as:

$$V_p = 24\frac{4.8}{8 + 4.8} = 9 \ V$$

We can check these values by the use of Kirchhoff's voltage law which states that: around any closed path in a circuit that the sum of the voltage drops and voltages rises from sources is zero. Thus we have 15 + 9 = 24 V.

The total circuit resistance is the sum of the series resistances in the equivalent circuit in Figure 12.12, i.e. 8 + 4.8 = 12.8 Ω. Thus the circuit current I is 24/12.8 = 1.875 A. This is the current through the 8 Ω resistor. It is the current entering the parallel arrangement and being divided between the two resistors. Using the current divider rule, the current through the 12 Ω resistor is:

$$I_{12} = 1.875\frac{8}{12 + 8} = 0.75 \ A$$

and the current through the 8 Ω resistor is:

$$I_8 = 1.875\frac{12}{12 + 8} = 1.125 \ A$$

We can check these values by using Kirchhoff's current law: the current entering a junction equals the current leaving it. The current entering the parallel arrangement is 1.875 A and the current leaving it is 0.75 + 1.125 = 1.875 A.

Example

Determine the voltage drops across, and the current through, each of the resistors in the circuit given in Figure 12.13.

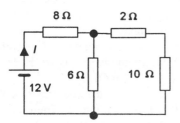

Figure 12.13   *Example*

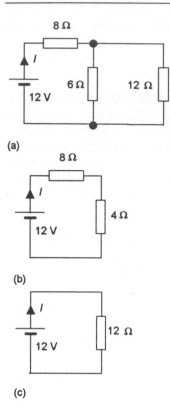

(a)

(b)

(c)

Figure 12.14   *Example*

For the series arrangement of the 2 Ω and 10 Ω, the equivalent resistance is 12 Ω. This gives the simplified circuit of Figure 12.14(a). For the parallel arrangement of the 12 Ω and 6 Ω, the equivalent resistance is 12 × 6/(12 + 6) = 4 Ω. This gives the simplified circuit of Figure 12.14(b). The voltage drop across the 8 Ω resistor is thus given by the voltage division rule as:

$$V_8 = 12\frac{8}{8+4} = 8 \text{ V}$$

The voltage drop across the parallel arrangement is:

$$V_P = 12\frac{4}{8+4} = 4 \text{ V}$$

We can check this with Kirchhoff's voltage law: 8 + 4 = 12 V. For the series element of the parallel arrangement we have 4 V across 2 Ω in series with 10 Ω. Hence, using the voltage divider rule:

$$V_2 = 4\frac{2}{2+10} = 0.67 \text{ A}$$

$$V_{10} = 4\frac{10}{2+10} = 3.33 \text{ A}$$

We can check this with Kirchhoff's voltage law: 0.67 + 3.33 = 4 V.

For the currents through the resistors we need the total circuit current I. The simplified circuit of Figure 12.14(b) has 8 Ω and 4 Ω in series and so we can obtain the simplified circuit of Figure 12.14(c). The circuit current is thus I = 12/(8 + 4) = 1.0 A. This is the current through the 8 Ω resistor. It is also the current entering the parallel arrangement. Using the current divider rule for Figure 12.14(a):

$$I_6 = 1.0\frac{12}{6+12} = 0.67 \text{ A}$$

$$I_{12} = 1.0\frac{6}{6+12} = 0.33 \text{ A}$$

12.3 Kirchhoff's laws   One way we can use Kirchhoff's laws to analyse circuits is to write equations for the currents at every node in the circuit and write equations for every loop in the circuit. We then have to solve the resulting set of simultaneous equations. A *node* is a point in a circuit where two or more devices are connected together, i.e. it is a junction at which we have current entering and current leaving. A *loop* is a sequence of circuit elements that form a closed path. For the circuit shown in Figure 12.15, there are four different nodes a, b, c and d and three loops L1, L2 and L3. Loop 1 is through a, b and d, loop 2 is through b, c and d and loop 3 is round the outer elements of the circuit, i.e. a, b, c and d. There are, however, two methods that can be used to reduce the number of

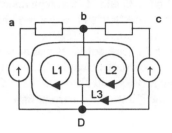

Figure 12.15 *Nodes and Loops*

simultaneous equations that have to be solved, these being node analysis and mesh analysis.

12.3.1 Node analysis

Node analysis uses Kirchhoff's current law to evaluate the voltage at each principal node in a circuit. A *principal node* is a point where three or more elements are connected together. Thus in Figure 12.15, just b and c are principal nodes. One of the principal nodes is chosen to be a reference node so that the potential differences at the other nodes are with reference to it; in Figure 12.15 we might choose d to be the reference node. Kirchhoff's current law is then applied to each non-reference node. The procedure is thus:

1 Draw a labelled circuit diagram and mark the principal nodes.

2 Select one of the principal nodes as a reference node.

3 Apply Kirchhoff's current law to each of the non-reference nodes, using Ohm's law to express the currents through resistors in terms of node voltages.

4 Solve the resulting simultaneous equations. If there are n principal nodes there will be $(n - 1)$ equations.

5 Use the derived values of the node voltages to determine the currents in each branch of the circuit.

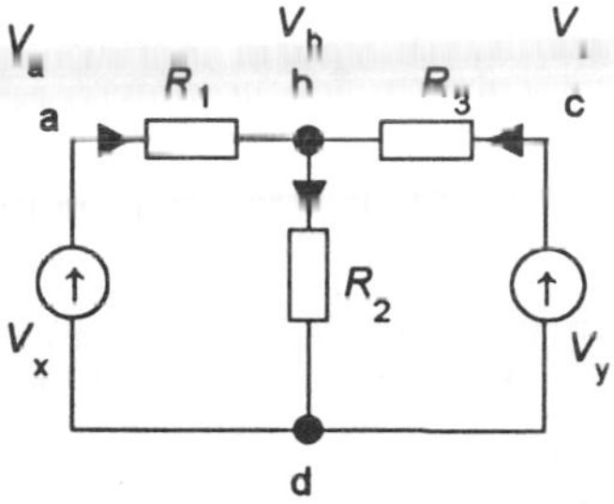

Figure 12.16 *Node analysis*

As an illustration of the application of the above method of circuit analysis, consider the circuit shown in Figure 12.16. There are four nodes a, b, c and d, of which b and d are principal nodes. If we take node d as the reference node, then the voltages V_a, V_b and V_c are the node voltages relative to node d. This means that the potential difference across resistor R_1 is $(V_a - V_b)$, across resistor R_2 is V_b and across R_3 is $(V_c - V_b)$. Thus the current through R_1 is $(V_a - V_b)/R_1$, through resistor R_2 is V_b/R_2 and through R_3 is $(V_c - V_b)/R_3$. Thus, applying Kirchhoff's current law to node b gives:

$$\frac{V_a - V_b}{R_1} + \frac{V_c - V_b}{R_3} = \frac{V_b}{R_2}$$

But $V_a = V_x$ and $V_c = V_y$ and so:

$$\frac{V_x - V_h}{R_1} + \frac{V_y - V_b}{R_3} = \frac{V_b}{R_2}$$

Hence the voltage at node b can be determined and hence the currents in each branch of the circuit.

Example

Determine the currents in each branch of the circuit shown in Figure 12.17.

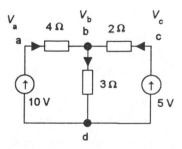

Figure 12.17 *Example*

The nodes are a, b, c and d with nodes b and d being principal nodes. Node d is taken as the reference node. If V_a, V_b and V_c are the node voltages relative to node d then the potential difference across the 4 Ω resistor is $(V_a - V_b)$, across the 3 Ω resistor is V_b and across the 2 Ω resistor is $(V_c - V_b)$. Thus the current through the 4 Ω is $(V_a - V_b)/4$, through the 3 Ω resistor is $V_b/3$ and through the 2 Ω resistor is $(V_c - V_b)/2$. Thus, applying Kirchhoff's current law to node b gives:

$$\frac{V_a - V_b}{4} + \frac{V_c - V_b}{2} = \frac{V_b}{3}$$

But $V_a = 10$ V and $V_c = 5$ V and so:

$$\frac{10 - V_b}{4} + \frac{5 - V_b}{2} = \frac{V_b}{3}$$

$$\frac{2(10 - V_b) + 4(5 - V_b)}{8} = \frac{V_b}{3}$$

$$60 - 6V_b + 60 - 12V_b = 8V_b$$

Thus $V_b = 4.62$ V. The potential difference across the 4 Ω resistor is thus $10 - 4.62 = 5.38$ V and so the current through it is $5.38/4 = 1.35$ A. The potential difference across the 3 Ω resistor is 4.62 V and so the current is $4.62/3 = 1.54$ A. The potential difference across the 2 Ω resistor is $5 - 4.62 = 0.38$ V and so the current is $0.38/2 = 0.19$ A.

12.3.2 Mesh analysis

A *mesh* is a loop which does not contain any other loops within it. Thus for Figure 12.18 loops 1 and 2 are meshes but loop 3 is not. Mesh analysis involves defining a current as circulating round each mesh. The same direction must be chosen for each mesh current and the usual convention is to make all the mesh currents circulate in a clockwise direction. Thus for Figure 12.18 we would define a current I_1 as circulating round mesh 1 and a current I_2 round mesh 2. Having specified mesh currents, Kirchhoff's voltage law is then applied to each mesh. The procedure is thus:

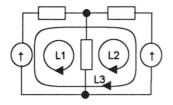

Figure 12.18 *Meshes and loops*

1   Label each of the meshes with clockwise mesh currents.

2   Apply Kirchhoff's voltage law to each of the meshes, the potential differences across each resistor being given by Ohm's law in terms of the current through it and in the opposite direction to the current. The current through a resistor which borders just one mesh is the mesh current; the current through a resistor bordering two meshes is the algebraic sum of the mesh currents through the two meshes.

3   Solve the resulting simultaneous equations to obtain the mesh currents. If there are *n* meshes there will be *n* equations.

4   Use the results for the mesh currents to determine the currents in each branch of the circuit.

Note that mesh analysis can only be applied to planar circuits, these being circuits that can be drawn on a plane so that no branches cross over each other.

As an illustration of the above method of circuit analysis, consider the circuit shown in Figure 12.19. There are three loops ABCF, CDEF and ABCDEF but only the first two are meshes. We define currents I_1 and I_2 as circulating in a clockwise direction in these meshes.

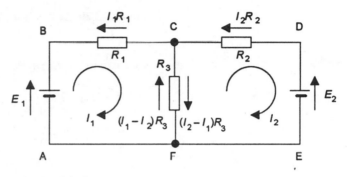

Figure 12.19 *Mesh currents*

Consider mesh 1. The current through R_1 is the mesh current I_1. The current through R_2, which is common to the two meshes, is the algebraic sum of the two mesh currents, i.e. $I_1 - I_2$. Thus, applying Kirchhoff's voltage law to the mesh gives

$$E_1 - I_1R_1 - (I_1 - I_2)R_2 = 0$$

For mesh 2 we have a current of I_2 through R_3 and a current of $(I_2 - I_1)$ through R_2. Thus applying Kirchhoff's voltage law to this mesh gives:

$$-E_2 - I_2R_2 - (I_2 - I_1)R_2 = 0$$

We thus have the two simultaneous equations for the two meshes.

Example

Determine, using mesh analysis, the current through the 20 Ω resistor in the circuit shown in Figure 12.20.

There are two meshes and we define mesh currents of I_1 and I_2 as circulating round them. For mesh 1, applying Kirchhoff's voltage law,

$$5 - 5I_1 - 20(I_1 - I_2) = 0$$

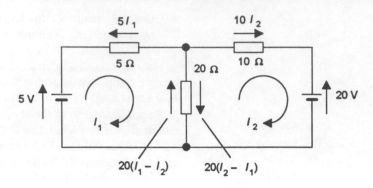

Figure 12.20 *Example*

This can be rewritten as:

$$5 = 25I_1 - 20I_2$$

For mesh 2, applying Kirchhoff's voltage law:

$$-10I_2 - 20 - 20(I_2 - I_1) = 0$$

This can be rewritten as:

$$20 = 20I_1 - 30I_2$$

We now have a pair of simultaneous equations. Multiplying the equation for mesh 1 by 4 and subtracting from it five times the equation for mesh 2 gives:

$$20 = 100I_1 - 80I_2$$
$$\text{minus } \underline{100 = 100I_1 - 150I_2}$$
$$-80 = 0 + 70I_2$$

Thus $I_2 = -1.14$ A and, back substituting this value in one of the mesh equations, $I_1 = -0.71$ A. The minus signs indicate that the currents are in the opposite directions to those indicated in the figure. The current through the 20 Ω resistor is thus, in the direction of I_1, $-0.71 + 1.14 = 0.43$ A.

12.4 Superposition theorem

Before considering this theorem we need to explain the term 'linear circuit'. If some relationship is linear then we have an equation relating the variables y and x of the form $y = mx$, where m is a constant. A resistor that obeys Ohm's law has a linear relationship between voltage V and current I of $V = RI$, assuming the resistance R is a constant. For such a linear relationship, if we have, for a resistance of 4 Ω, a current of 1 A then the voltage is 4 V. If the current is 2 A then the voltage is 8 V. If the current is 3 A then the voltage is 12 V. But suppose we have two

With a single
source:
1 A gives 4 V,
2 A gives 8 V,
3 A gives 12 V

With two
sources:
1 giving 1 A,
2 giving 2 A,
the voltage
is the sum of
that given by
each source
and so 12 V

Figure 12.21 *Linearity*

sources supplying current through the resistor (Figure 12.21), one a current of 1 A and the other a current of 2 A. The total current is 3 A and so the voltage is 12 V. But this is the same as considering each source as acting alone and adding the voltages given by each. We can superpose the results due to each source. This is a fundamental property of a linear component, or indeed a linear circuit.

If we have a linear circuit containing more than one source the *superposition theorem* states that:

> *The voltage and current response of a linear circuit to a number of independent sources is the sum of the responses obtained by applying each independent source once with the other independent sources set equal to zero.*

The contributions of sources are made equal to zero by replacing ideal voltage sources by short circuits, i.e. making $E = 0$, and ideal current sources by open circuits, i.e. making $I = 0$.

The procedure for analysing a circuit by using superposition is thus:

1 Select any one source in the circuit and make all the others zero.

2 Calculate the voltage or current in a component arising when just this one source is present.

3 Repeat steps 1 and 2 for each of the sources when all the others are made zero.

4 Add together the computed values obtained from analysing the circuit with each source acting alone. The sum is the voltage or current that would be obtained when all the sources were simultaneously present.

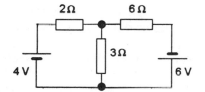

Figure 12.22 *Example*

The following examples illustrate the use of the above procedure. Note that you cannot calculate the power at each step and then superimpose the powers to give the power that would occur when all the sources were simultaneously present. To determine the power you would need to determine the currents at each step and hence the current when all sources were simultaneously present, then calculate the power.

Example

Determine the currents in the branches of the circuit shown in Figure 12.22.

First, we select the 4 V source and make the 6 V source zero, this then giving the circuit shown in Figure 12.23. We now have a 2 Ω resistor in series with a parallel arrangement of 3 Ω and 6 Ω. The parallel arrangement has an equivalent resistance of $3 \times 6/(3 + 6) = 2$ Ω. The total circuit resistance is then $2 + 2 = 4$ Ω and so the circuit current $4/4 = 1.0$ A. Thus the current through the 2 Ω resistor is 1.0 A and, by current division, that through the 3 Ω resistor is 1.0

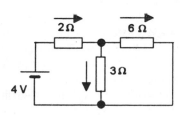

Figure 12.23 *Example*

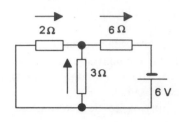

Figure 12.24   *Example*

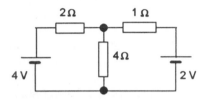

Figure 12.25   *Example*

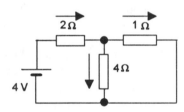

Figure 12.26   *Example*

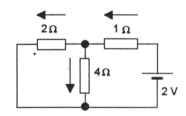

Figure 12.27   *Example*

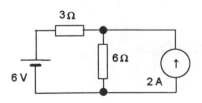

Figure 12.28   *Example*

× 6/9 = 0.67 A and through the 6 Ω resistor is 1.0 × 3/9 = 0.33 A. The directions of the currents are shown on the Figure.

Now we put the 4 V source equal to zero and consider the circuit with just the 6 V source (Figure 12.24). We then have a 6 Ω resistor in series with a parallel arrangement of 3 Ω and 2 Ω. The parallel arrangement has an equivalent resistance of 3 × 2/5 = 1.2 Ω. The total circuit resistance is thus 6 + 1.2 = 7.2 Ω and so the circuit current is 6/7.2 = 0.83 A. This is the current through the 6 Ω resistor. By current division, the current through the 3 Ω resistor is 0.83 × 2/5 = 0.33 A and through the 2 Ω is 0.83 × 3/5 = 0.50 A. The directions of the currents are shown in the Figure.

Taking into account the directions of the currents, due to both sources the current through the 2 Ω resistor = 1.0 + 0.5 = 1.5 A, the current through the 3 Ω resistor = 0.67 − 0.33 = 0.34 A and the current through the 6 Ω resistor = 0.83 + 0.33 = 1.16 A.

Example 6

Determine the currents in the branches of the circuit shown in Figure 12.25 and the power through the 4 Ω resistor.

With the 2 V source put equal to zero we have the circuit in Figure 12.26. We thus have 2 Ω in series with a parallel arrangement of 1 Ω and 4 Ω. The total circuit resistance is thus 2 + (1 × 4/5) = 2.8 Ω and the circuit current is 4/2.8 = 1.43 A. The current through the 2 Ω resistor is thus 1.43 A and, by current division, the current through the 1 Ω resistor is 1.43 × 4/5 = 1.14 A and through the 4 Ω resistor is 1.43 × 1/5 = 0.29 A. The current directions are as shown in the Figure.

With the 4 V source put equal to zero we have Figure 12.27. We thus have 1 Ω in series with a parallel arrangement of 2 Ω and 4 Ω. The total circuit resistance is 1 + (2 × 4/6) = 2.33 Ω and the circuit current is 2/2.33 = 0.86 A. The current through the 1 Ω resistor is thus 0.86 A and, by current division, the current through the 4 Ω resistor is 0.86 × 2/6 = 0.29 A and through the 2 Ω resistor is 0.86 × 4/6 = 0.57 A. Because these are both positive values, the current directions are as shown in the Figure.

Taking into account the directions of the currents, due to both sources gives the current through the 2 Ω resistor as 1.43 − 0.57 = 0.86 A, the current through the 4 Ω resistor as 0.29 + 0.29 = 0.58 A and the current through the 1 Ω resistor as 1.14 − 0.86 = 0.28 A.

The power through the 4 Ω resistor is $I^2R = 0.58^2 × 4 = 1.35$ W. Note that it is *not* $0.29^2 × 4 + 0.29^2 × 4 = 0.67$ W.

Example

Determine the current through the 6 Ω resistor for the circuit shown in Figure 12.28.

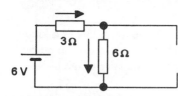

Figure 12.29 *Example*

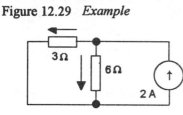

Figure 12.30 *Example*

This circuit has a voltage source and a current source. Firstly we consider just the voltage source and put the current source to zero. This means replacing the current source by an open-circuit to give the circuit shown in Figure 12.29. The total circuit resistance is then $3 + 6 = 9 \ \Omega$ and hence the current is $6/9 = 0.67$ A. The current direction is as shown in the Figure.

Now we put the voltage source to zero, obtaining the circuit shown in Figure 12.30. Applying the current divider rule, the current through the 6 Ω resistor is $2 \times 3/(3 + 6) = 0.67$ A. The current directions are as shown in the Figure.

Taking account of the current directions, superposition gives the current through the 6 Ω resistor as the sum of the current values obtained when each source was considered alone and so we have the current $= 0.67 + 0.67 = 1.34$ A.

12.5 E.M.F. and internal resistance

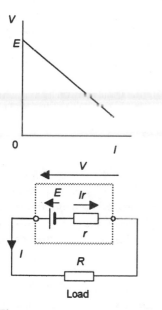

Figure 12.31 *Voltage–current relationship for a voltage source*

If the potential difference V between the terminals of a voltage source, such as a battery, is measured then the value obtained is found to depend on the current I taken from the battery. The greater the current taken the lower the potential difference. Figure 12.31 shows the type of result obtained. The value of the potential difference for a battery when no current is taken is called the *electromotive force* (e.m.f. E) and for a power supply is the *no load output voltage*.

The voltage source can be considered to consist of an ideal voltage source in series with an internal resistance, the ideal voltage source being one with no internal resistance and giving at all currents the same voltage, namely the e.m.f. E. When a current I occurs, then the potential difference V between the terminals of the source is E minus the potential drop across the internal resistance r, i.e.

$$V = E - Ir$$

When voltage sources such as cells are connected in series, then the total e.m.f. is the sum of the e.m.f.s of the separate sources and the total internal resistance is the sum of the internal resistances of each source, they are after all just resistances in series. Series connection is thus a way of getting a larger e.m.f. but at the expense of a higher internal resistance. When voltage sources with the same e.m.f. are connected in parallel, then the total e.m.f. is the e.m.f. of one source and the total internal resistance r_{tot} is that of the parallel connected internal resistances. Thus if each source has the same internal resistance r, for three cells we have:

$$\frac{1}{r_{tot}} = \frac{1}{r} + \frac{1}{r} + \frac{1}{r} = \frac{3}{r}$$

The internal resistance is thus one-third that of a single cell. For n cells we have $r_{tot} = r/n$. Parallel connection is thus a way of getting a lower internal resistance for the same e.m.f.

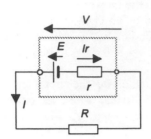

Figure 12.32 *Example*

Example

When a 2 Ω resistor is connected across the terminals of a battery a current of 1.5 A occurs. When a 4 Ω resistor is used the current is 1.0 A. Determine the e.m.f. and internal resistance of the battery.

The circuit is as shown in Figure 12.32. The total circuit resistance is $R + r$ and so $E = I(R + r)$. Initially we have $E = 1.5(2 + r) = 3.0 + 1.5r$ and then $E = 1.0(4 + r) = 4.0 + 1.0r$. Hence:

$$3.0 + 1.5r = 4.0 + 1.0r$$

and so $r = 2.0$ Ω. Substituting this into either of the initial equations gives $E = 6.0$ V.

12.6 Thévenin's theorem

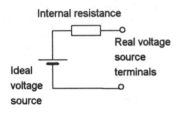

Figure 12.33 *Equivalent circuit for a voltage source*

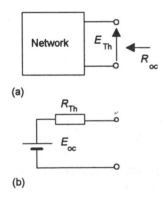

Figure 12.34 *(a) The network, (b) its equivalent*

An ideal voltage source is one that gives a constant e.m.f. regardless of what current is drawn from it, i.e. whatever load resistance is connected between its terminals. A real voltage source gives an e.m.f. which depends on the current drawn and can be considered to be an ideal voltage source in series with its internal resistance (Figure 12.33). We can use such an equivalent when we connect some network of components between the two terminals of a voltage source.

Often we connect one network of components to the two terminals of another network, e.g. a speaker between the two terminals of an amplifier. To solve such problems we use, in the same way as above with an equivalent circuit for a voltage source, an equivalent circuit for the amplifier. The equivalent circuit for any two-terminal network containing a voltage or current source is given by *Thévenin's theorem*:

Any two-terminal network (Figure 12.34(a)) containing voltage or current sources can be replaced by an equivalent circuit consisting of a voltage equal to the open-circuit voltage of the original circuit in series with the resistance measured between the terminals when no load is connected between them and all independent sources in the network are set equal to zero (Figure 12.34(b).

If we have a linear circuit, to use Thévenin's theorem, we have to divide it into two circuits, A and B, connected at a pair of terminals. We can then use Thévenin's theorem to replace, say, circuit A by its equivalent circuit. The open-circuit Thévenin voltage for circuit A is that given when circuit B is disconnected and the Thévenin resistance for A is the resistance looking into the terminals of A with all its independent sources set equal to zero. Figure 12.35 illustrates this with the next but one Example showing how it is applied.

Example

Determine the equivalent Thévenin circuit for the network of Figure 12.36.

1. Identify the two parts A and B of the circuit and separate by terminals

2. Separate A from B

3. Replace A by its Thévenin equivalent

4. Reconnect circuit B and carry out the analysis

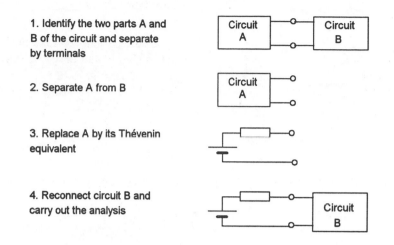

Figure 12.35 *Step-by-step approach for circuit analysis*

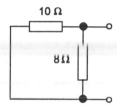

Figure 12.36 *Example*

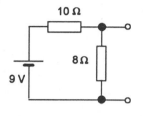

Figure 12.37 *Example*

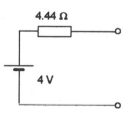

Figure 12.38 *Example*

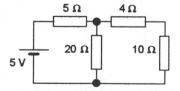

Figure 12.39 *Example*

The voltage between the open-terminals is that across the 8 Ω and is thus the open-circuit voltage. Using the voltage division rule, the voltage across the 8 Ω is:

$$E_{oc} = 9\left(\frac{8}{8+10}\right) = 4 \text{ V}$$

The Thévenin resistance is calculated by looking back into the circuit from the terminals and putting the voltage source to zero. Note that if the sources are not ideal they should be replaced by the ideal source with its internal resistance and so the voltage source is then replaced by its internal resistance. Putting the voltage source to zero in Figure 12.36 gives the circuit shown in Figure 12.37. This has an equivalent resistance of:

$$R_{Th} = \frac{10 \times 8}{10 + 8} = 4.44 \, \Omega$$

Thus the equivalent circuit is that given in Figure 12.38.

Example

Using Thévenin's theorem, determine the current through the 10 Ω resistor in the circuit given in Figure 12.39.

Since we are interested in the current through the 10 Ω resistor we identify it as network B and the rest of the circuit as network A, connecting them by terminals (Figure 12.40). We then separate A from B (Figure 12.41) and determine the Thévenin equivalent for it. The open-circuit voltage is that across the 20 Ω resistor. Using the voltage division rule:

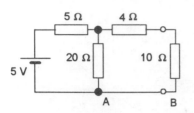

Figure 12.40 *Example*

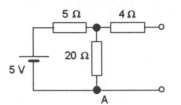

Figure 12.41 *Example*

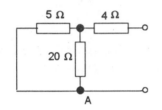

Figure 12.42 *Example*

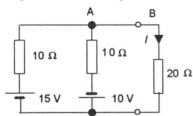

Figure 12.45 *Example*

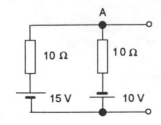

Figure 12.46 *Example*

Figure 12.47 *Example*

$$E_{Th} = 5\frac{20}{20+5} = 4 \text{ V}$$

The resistance looking into the terminals when voltage source is equal to zero is that of 4 Ω in series with a parallel arrangement of 5 Ω and 20 Ω (Figure 12.42).

$$R_{Th} = 4 + \frac{20 \times 5}{20+5} = 8 \text{ Ω}$$

Thus the equivalent Thévenin circuit is as shown in Figure 12.43 and when network B is connected to it we have the circuit shown in Figure 12.44. Hence the current through the 10 Ω resistor is:

$$I_{10} = \frac{4}{8+10} = 0.22 \text{ A}$$

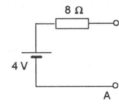

Figure 12.43 *Example*

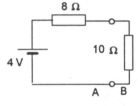

Figure 12.44 *Example*

Example

Using Thévenin's theorem, determine the current I in the circuit shown in Figure 12.45.

We can redraw the circuit to isolate the 20 Ω load resistance, as in Figure 12.46, and identify it as network B with the rest of the circuit becoming network A. We can then apply Thévenin's theorem to A (Figure 12.47). The open-circuit voltage is 15 V minus the potential drop across its series 10 Ω resistor, or 10 V minus the potential drop across its series 10 Ω resistor. The two e.m.f.s add and so the current in the network is (15 + 10)/(10 + 10) = 1.25 A. Thus the open-circuit potential is:

$$E_{oc} = 15 - (10 \times 1.25) = 2.5 \text{ V}$$

The resistance between the terminals when the voltage sources are put equal to zero is:

$$R_{Th} = \frac{10 \times 10}{10+10} = 5 \text{ Ω}$$

We can thus replace network A by the Thévenin circuit and obtain the circuit shown in Figure 12.48. The current I is thus 2.5/(5 + 20) = 0.1 A.

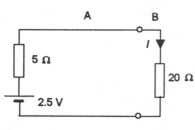

Figure 12.48 *Example*

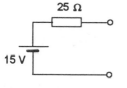

Figure 12.50 *Example*

12.7 Norton's theorem

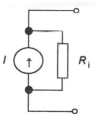

Figure 12.51 *Current source*

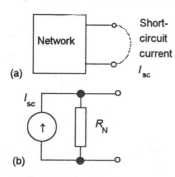

Figure 12.52 *(a) Network, (b) Norton equivalent*

Example

Determine the Thévenin network for the network in Figure 12.49.

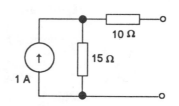

Figure 12.49 *Example*

The open-circuit voltage is that across the 15 Ω resistor, there being no current through the 10 Ω resistor and so no potential drop across it. Thus:

$$E_{th} = 1 \times 15 = 15 \text{ V}$$

When the current source is put to a zero value, i.e. open-circuit, we have 15 Ω resistor in parallel with the 10 Ω resistor and so:

$$R_{th} = 15 + 10 = 25 \ \Omega$$

The Thévenin equivalent network is thus as in Figure 12.50.

A current source can be regarded as an ideal current source in parallel with an internal resistance (Figure 12.51). An ideal current source is one that supplies a constant current regardless of the load resistance connected to its terminals; a real current source supplies a current which depends on the load resistance. In a similar way to Thévenin's theorem, we can have an equivalent circuit for any two-terminal network containing voltage or current sources in terms of an equivalent network of a current source shunted by a resistance. This is known as *Norton's theorem*:

Any two-terminal network containing voltage or current sources can be replaced by an equivalent network consisting of a current source, equal to the current between the terminals when they are short-circuited, in parallel with the resistance measured between the terminals when there is no load between them and all independent sources in the network are set equal to zero (Figure 12.52).

If we have a linear circuit we have to divide it into two circuits, A and B, connected at a pair of terminals (Figure 12.53). We can then use Norton's theorem to replace, say, circuit A by its equivalent circuit. The short-circuit Norton current for circuit A is that given when circuit B is disconnected and the Norton resistance for A is the resistance looking into the terminals of A with all its independent sources set equal to zero.

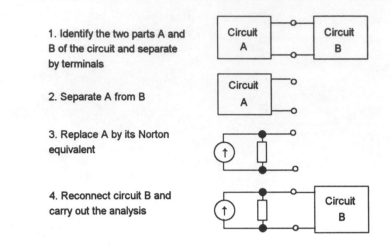

1. Identify the two parts A and B of the circuit and separate by terminals

2. Separate A from B

3. Replace A by its Norton equivalent

4. Reconnect circuit B and carry out the analysis

Figure 12.53 *Step-by-step approach for circuit analysis*

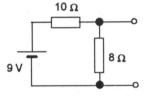

Figure 12.54   *Example*

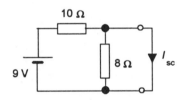

Figure 12.55   *Example*

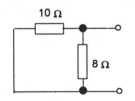

Figure 12.56 *Example*

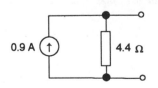

Figure 12.57   *Example*

Example

Determine the Norton equivalent network for the circuit in Figure 12.54.

When the output terminals are short-circuited (Figure 12.55), the circuit resistance is 10 Ω and so the short-circuit current is:

$$I_{sc} = \frac{9}{10} = 0.9 \text{ A}$$

The Norton resistance is obtained by making independent sources equal to zero and then determining the resistance between the terminals (Figure 12.56). This resistance is thus:

$$R_N = \frac{10 \times 8}{10 + 8} = 4.4 \text{ Ω}$$

Thus the equivalent Norton circuit is as shown in Figure 12.57.

Example

Using Norton's theorem, determine the current I through the 20 Ω resistor in Figure 12.58. Note this is the same circuit as solved earlier by Thévenin's theorem.

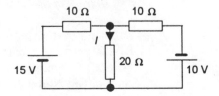

Figure 12.58   *Example*

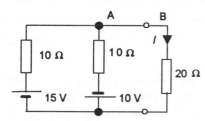

Figure 12.59 *Example*

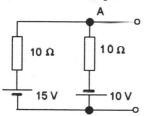

Figure 12.60 *Example*

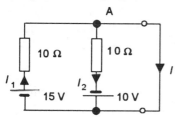

Figure 12.61 *Example*

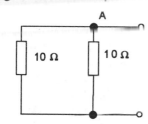

Figure 12.62 *Example*

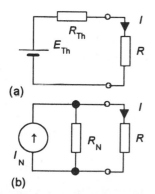

Figure 12.65 *(a) Thévenin,*
(b) Norton networks

We can redraw the circuit in the form shown in Figure 12.59 as two connected networks A and B. Network B is selected to be the 20 Ω resistor through which we require the current. We then determine the Norton equivalent circuit for network A (Figure 12.60). Short-circuiting the terminals gives the circuit shown in Figure 12.61. The short circuit current will be the sum, taking into account directions, of the currents from the two branches of the circuits containing the voltage sources:

$$I_{sc} = I_1 - I_2$$

The current $I_1 = 15/10 = 1.5$ A, since the other branch of the network is short-circuited, and $I_2 = 10/10 = 1.0$ A. Thus $I_{sc} = 0.5$ A. The Norton resistance is given by that across the terminals when all the sources are set to zero (Figure 12.62). It is thus:

$$R_N = \frac{10 \times 10}{10 + 10} = 5 \, \Omega$$

Thus the Norton equivalent circuit is that shown in Figure 12.63. Hence when we put this with network B (Figure 12.64) we can obtain the current I by the use of the current divider rule as:

$$I = 0.5 \times \frac{5}{5 + 20} = 0.1 \, \text{A}$$

The result is the same as that obtained in the earlier Example by the use of the Thévenin theorem.

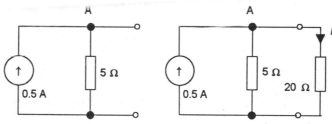

Figure 12.63 *Example*   Figure 12.64 *Example*

12.7.1 Relationship between Thévenin and Norton networks

For a two-terminal network we can determine Thévenin and Norton equivalent circuits. The two must themselves be equivalent. The Thévenin and Norton circuits in Figure 12.65 must give the same voltage across and current through the load resistance R. Thus for the same load current:

$$\overset{\text{Thévenin}}{} \quad \overset{\text{Norton}}{}$$
$$I = \frac{E_{Th}}{R + R_{Th}} = I_N \times \frac{R_N}{R + R_N}$$

Hence we must have:

$$E_{th} = I_N R_N \text{ and } R_{th} = R_N$$

Thus if we wanted to obtain a Thévenin or a Norton equivalent network we determine the resistance between the terminals when all independent sources are put equal to zero, this giving R_{th}, which equals R_N. We might then determine the Thévenin voltage and from it, using the above equation, determine the Norton current, or vice versa.

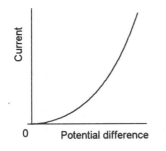

Example

Determine the Norton equivalent network for the network in Figure 12.66 by first determining the Thévenin equivalent and then converting it to the Norton equivalent. In an earlier Example the Norton equivalent was directly obtained.

Figure 12.66 *Example*

The Thévenin and Norton resistance is obtained by putting the voltage source to zero and hence is $10 \times 8/(10 + 8) = 4.4$ Ω. The Thévenin voltage is the open-circuit voltage and is thus $9 \times 8/(10 + 8) = 4$ V. Thus the Norton current is $E_{th}/R_N = 4/4.4 = 0.9$ A.

12.8 Non-linear circuits

Resistors which obey Ohm's law have the current through them directly proportional to the potential difference across them, a graph of current against potential difference being a straight line. Devices with such characteristics are said to be *linear*. There are, however, circuit components which have *non-linear* characteristics, e.g. diodes and transistors. For such components, manufacturers normally provide graphs of the current–potential difference characteristic. Typically such a characteristic is of the form shown in Figure 12.67.

We can use a graphical technique to analyse circuits involving a non-linear component. Consider the circuit in Figure 12.68 where there is a linear resistor in series with a non-linear component. Applying Kirchhoff's voltage law then:

Figure 12.67   *Non-linear characteristics*

$$V_s - V_R - V_D = 0$$

where V_D is the potential difference across the non-linear device and V_R the potential difference across the resistor R. But $V_R = IR$ and so:

$$V_s - IR - V_D = 0$$

We can write this equation in the form:

$$I = -\left(\frac{1}{R}\right)V_D + \frac{V_s}{R}$$

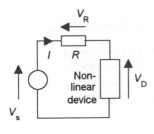

Figure 12.68   *Non-linear device in a circuit*

A graph of the current I against V_D is a straight line of slope $-(1/R)$ and intercept with the I axis of V_s/R. When $I = 0$ then $V_D = V_s$ and thus the

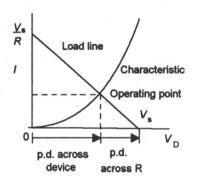

Figure 12.69 *Use of the load line*

intercept with the V_D axis is V_s. This graph is termed the *load line* as the resistor R can be considered as the load on the device.

If we superimpose this load line graph on the characteristic graph for the non-linear device (Figure 12.69), then the intersection between the two graphs gives the *operating point* of the circuit and enables us to read off the circuit current value and the voltage across the device.

Example

Figure 12.70(a) shows the characteristic of a diode. Determine the current in the circuit when the diode is connected in series with a 100 Ω resistance and the circuit supplied with a voltage of 2 V.

The load line is a line drawn between the values $I = V_s/R = 2/100 = 0.020$ A $= 20$ mA and $V_D = V_s = 2$ V. Figure 12.70(b) shows this load line superimposed on the diode characteristic. The operating point is thus at a current of 12.5 mA.

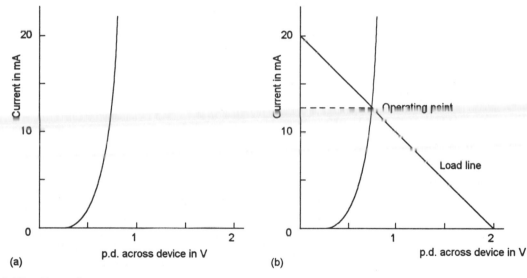

Figure 12.70 *Example*

Problems

1 Determine by simplifying the circuits the total resistance and the total circuit drawn from the voltage source in each of the series–parallel arrangements of resistors shown in Figure 12.71.

2 Determine the output voltage V_o for the circuit given in Figure 12.72.

3 Determine, by simplifying the circuits, the current through and the voltage drop across each of the resistors in the series–parallel circuits shown in Figure 12.73.

4 Determine the potential difference across the 10 Ω resistor in the circuit shown in Figure 12.74.

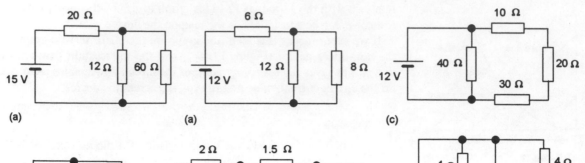

(a)   (a)   (c)

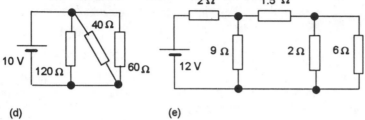

(d)   (e)   (f)

Figure 12.71 *Problem 1*

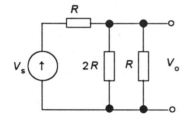

Figure 12.72   *Problem 2*

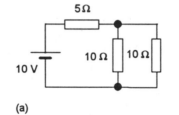

(a)

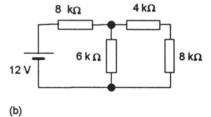

(b)

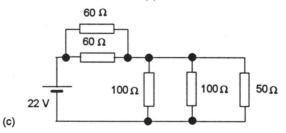

(c)

Figure 12.73 *Problem 3*

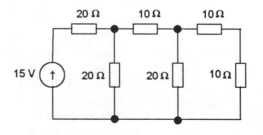

Figure 12.74 *Problem 4*

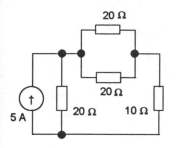

Figure 12.75 *Problem 5*

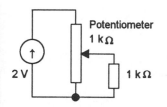

Figure 12.76 *Problem 6*

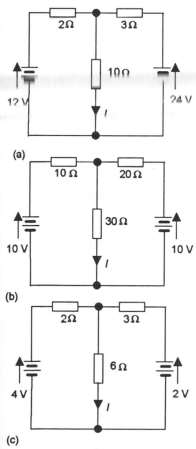

(a)

(b)

(c)

Figure 12.77 *Problem 7*

5 Determine the current through the 10 Ω resistor in the circuit shown in Figure 12.75.

6 Determine the current through the 1 kΩ resistor in Figure 12.76 when the slider of the potentiometer is located at its midpoint.

7 Determine, using (a) node analysis and (b) mesh analysis, the current I in each of the circuits shown in Figure 12.77.

8 Use mesh analysis to determine the currents through the resistors in the circuit shown in Figure 12.78.

9 A battery consists of four cells in series, each having an e.m.f. of 1.5 V and an internal resistance of 0.1 Ω. What is the e.m.f. and internal resistance of the battery?

10 A battery has an e.m.f. of 6 V and an internal resistance of 1.0 Ω. What will be the potential difference between its terminals when the current drawn is 2.0 A?

11 A battery supplies a current of 2.0 A when a 1.5 Ω resistor is connected across its terminals and 0.8 A when the resistor is increased to 4.0 Ω. What is the e.m.f. and internal resistance of the battery?

12 Four batteries, each of e.m.f. 2.0 V and internal resistance 0.2 Ω are connected in parallel. What is the total e.m.f. and internal resistance of the arrangement?

13 Using the superposition theorem determine the current flowing through each of the resistors in the circuits given in Figure 12.79.

14 Using the superposition theorem, determine the current I in the circuits given in Figure 12.80.

15 Determine the Thévenin equivalent networks for the networks shown in Figure 12.81.

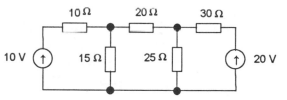

Figure 12.78 *Problem 8*

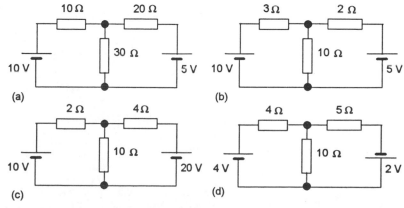

(a)

(b)

(c)

(d)

Figure 12.79 *Problem 13*

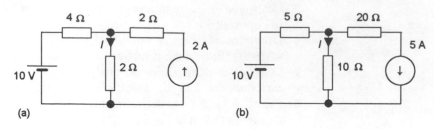

(a)   (b)

Figure 12.80   *Problem 14*

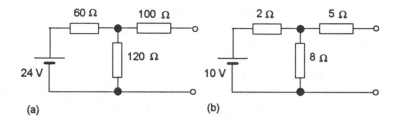

(a)   (b)

Figure 12.81   *Problem 15*

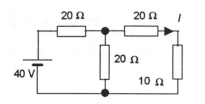

Figure 12.82   *Problem 16*

16   Using Thévenin's theorem determine the current *I* for the circuit shown in Figure 12.82.

17   Determine the Norton equivalent networks for the networks shown in Figure 12.81.

18   Using Norton's theorem determine the current *I* for the circuits shown in Figure 12.82.

19   Figure 12.83 shows the characteristic of a diode. Determine the current in the circuit when the diode is connected in series with (a) a 100 Ω resistance, (b) a 120 Ω resistance, and the circuit supplied with a voltage of 2 V.

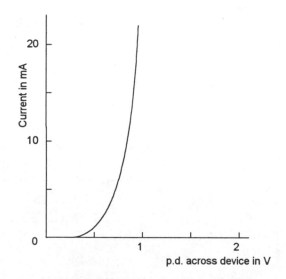

Figure 12.83   *Problem 19*

13 Capacitance

13.1 Introduction

This chapter is about capacitors, these being components which are widely used in electronic circuits. The concept of an electric field is reviewed and the basic principles and use of capacitors considered.

13.1.1 Electric fields

If we pick up an object and then let go, it falls to the ground. We can explain this by saying that there are attractive gravitational forces between two masses, the earth and the object. There is another way of explaining this. We can say that the earth produces in its surroundings a *gravitational field* and that when the object is in that field it experiences a force which causes it to fall. A mass is thus said to produce a gravitational field. Other masses placed in that field experience forces.

A charged body is said to produce an *electric field* in the space around it. Any other charged body placed in the electric field experiences a force. The direction of the electric field at a point is defined as being the direction the force would be if a positive charge was placed at the point. The field can be visualised by drawing lines representing the directions of the field, these lines being called *lines of force*. Figure 13.1 shows the field patterns for isolated positive and negative charges.

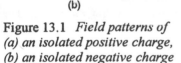

(a)

(b)

Figure 13.1 *Field patterns of (a) an isolated positive charge, (b) an isolated negative charge*

13.2 Capacitor

For parallel, oppositely charged, conducting plates the lines of force in the central area between the plates are straight lines at right angles to the plates (Figure 13.2). Also, if we measure the force on a charged body we find that between the plates the force is constant, indicating a constant electric field. The electric field is, with the exception of near the edges of the plates, constant between the plates and at right angles to them.

If the plates are connected to a direct voltage supply such as a battery (Figure 13.3), the battery pushes electrons through the circuit onto one of the plates and removes them from the other plate. The result is that the plate that has gained electrons becomes negatively charged and the plate that has lost electrons becomes positively charged. In this situation there is a potential difference between the plates.

The electric field strength between the plates depends on the potential difference V between the plates and their separation d. For a given plate separation, the larger the potential difference the larger the electric field. For a given potential difference, the larger the plate separation the smaller the electric field. However, a constant value of (potential difference)/(the distance between the plates) gives a constant electric field. We thus define the *electric field strength E* as:

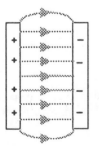

Figure 13.2 *Field pattern between parallel oppositely charged plates*

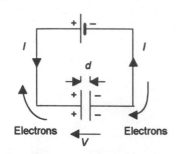

Figure 13.3   *Charging a capacitor*

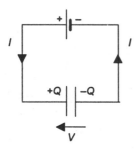

Figure 13.4   *Charging a capacitor*

$$E = \frac{V}{d}$$

The unit of electric field strength is volt/metre. Note that the electric field strength can also be defined as the force per unit charge $E = F/Q$ and the above equation derived. However, since we are more concerned with potential differences it seems more appropriate to define E as the potential gradient.

Example

What is the electric field strength between a pair of parallel conducting plates, 2 mm apart, when the potential difference between them is 200 V?

$E = V/d = 200/(2 \times 10^{-3}) = 1.0 \times 10^5$ V/m.

13.2.1 Capacitance

When a pair of parallel conducting plates are connected to a d.c. supply and a potential difference produced between them, one of the plates becomes positively charged and the other negatively charged (Figure 13.4). This arrangement of two parallel conducting plates is called a *capacitor* and the material between the plates is the *dielectric*. The amount of charge Q on a plate depends on the potential difference V between the plates. Q is found to be directly proportional to V. The constant of proportionality is called the *capacitance C*. Thus:

$$Q = CV$$

The unit of capacitance is the farad (F), when V is in volts and Q in coulombs. Note that a capacitance of 1 F is a very large capacitance and more usually capacitances will be microfarads (μF), i.e. 10^{-6} F, or nanofarads (nF), i.e. 10^{-9} F, or picofarads (pF), i.e. 10^{-12} F.

Example

What are the charges on the plates of an 8 μF capacitor when the potential difference between its plates is 120 V?

$Q = CV = 8 \times 10^{-6} \times 120 = 960 \times 10^{-6} = 960\ \mu$C and so the charges on the plates are +960 μC and −960 μC.

13.3 Capacitors in series and parallel

Consider three capacitors in *series*, as in Figure 13.5. The potential difference V across the arrangement will be the sum of the potential differences across each capacitor. Thus:

$$V = V_1 + V_2 + V_3$$

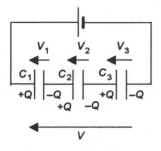

Figure 13.5 *Capacitors in series*

In order to have the same current through all parts of the series circuit we must have the same charge flowing onto and off the plates of each capacitor. Thus each capacitor will have the same charges of $+Q$ and $-Q$. Dividing the above equation throughout by Q gives:

$$\frac{V}{Q} = \frac{V_1}{Q} + \frac{V_2}{Q} + \frac{V_3}{Q}$$

But $C_1 = Q/V_1$, $C_2 = Q/V_2$ and $C_3 = Q/V_3$. Hence, if we replaced the three series-connected capacitors by a single equivalent capacitor with a capacitance given by $C = Q/V$, we must have:

$$\frac{1}{C} = \frac{1}{C_1} + \frac{1}{C_2} + \frac{1}{C_3}$$

The reciprocal of the equivalent capacitance of series-connected capacitors is the sum of the reciprocals of their individual capacitances.

Consider the voltage drops across series capacitors. For two capacitors C_1 and C_2 in series we have, for the first capacitor, $Q = C_1V_1$ and, for the second capacitor, $Q = C_2V_2$. Therefore, $C_1V_1 = C_2V_2$ and we can write:

$$\frac{V_2}{V_1} = \frac{C_1}{C_2}$$

and we have voltage division.

Consider three capacitors in *parallel*, as in Figure 13.6. Because they are in parallel, the potential difference V across each capacitor will be the same. The charges on each capacitor will depend on their capacitances. If the total charge shared between the capacitors is Q, then:

$$Q = Q_1 + Q_2 + Q_3$$

Dividing throughout by V gives:

$$\frac{Q}{V} = \frac{Q_1}{V} + \frac{Q_2}{V} + \frac{Q_3}{V}$$

But $C_1 = Q_1/V$, $C_2 = Q_2/V$ and $C_3 = Q_3/V$. Hence, if we replaced the three parallel-connected capacitors by a single equivalent capacitor with a capacitance given by $C = Q/V$, we must have:

$$C = C_1 + C_2 + C_3$$

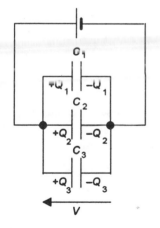

Figure 13.6 *Capacitors in parallel*

The equivalent capacitance of parallel-connected capacitors is the sum of their individual capacitances.

Example

What is the capacitance of a 2 μF capacitor and 4 μF capacitor when connected in (a) series, (b) parallel?

(a) $1/C = 1/2 + 1/4 = 3/4$, hence $C = 1.33$ μF.
(b) $C = 2 + 4 = 6$ μF.

Example

If two capacitors having capacitances of 6 μF and 10 μF are connected in series across a 200 V supply, determine (a) the p.d. across each capacitor and (b) the charge on each capacitor.

(a) Let V_1 and V_2 be the p.d.s across the 6 μF and 10 μF capacitors respectively. As the charges on each capacitor will be the same, then $V_1 = Q/(6 \times 10^{-6})$ and $V_2 = Q/(10 \times 10^{-6})$. Thus $V_1/V_2 = 10/6$. Since $V_1 + V_2 = 200$ then $(10/6)V_2 + V_2 = 200$ and so $V_2 = 75$ V and $V_1 = 125$ V.
(b) $Q = C_1 V_1 = 6 \times 10^{-6} \times 125 = 750$ μC.

13.4 Capacitance of a parallel plate capacitor

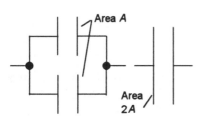

Figure 13.7 *Capacitors in parallel and equivalent capacitor*

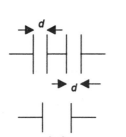

Figure 13.8 *Capacitors in series and equivalent capacitor*

The capacitance of a pair of parallel conducting plates depends on the plate area A, the separation d of the plates and the medium, i.e. the dielectric, between the plates. To see how these factors determine the capacitance, consider the effect of doubling the area of the plates as being effectively the combining of two unit area capacitors in parallel (Figure 13.7). For two identical capacitors in parallel, the total capacitance is the sum of the individual capacitances and so the capacitance is doubled. Thus doubling the area doubles the capacitance. The capacitance is proportional to the area A:

$$C \propto A$$

Now consider the effect of doubling the separation of the plates. This can be thought of as combining two identical capacitors in series (Figure 13.8). The reciprocal of the total capacitance is then $2/C$ and so the capacitance is halved by doubling the separation. The capacitance is thus inversely proportional to the separation:

$$C \propto \frac{1}{d}$$

Hence we can write:

$$C = \varepsilon \frac{A}{d}$$

where ε is the constant or proportionality and is the factor, called the *absolute permittivity*, which depends on the material between the plates (Figure 13.9), this being termed the *dielectric*. A more usual way of writing the equation is, however, in terms of how much greater the permittivity of a material is than that of a vacuum.

$$C = \varepsilon_r \varepsilon_0 \frac{A}{d}$$

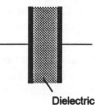

Figure 13.9 *The dielectric*

where $\varepsilon = \varepsilon_r \varepsilon_0$ with ε_r being called the *relative permittivity* of a material and ε_0 the *permittivity of free space*. The permittivity of free space has the value 8.85×10^{-12} F/m. The relative permittivity is often referred to as the *dielectric constant*. For a vacuum the relative permittivity is 1, for dry air close to 1, for plastics about 2 to 3, for oxide films such as aluminium or tantalum oxide about 7 to 30, and for ceramics it can be many thousands.

Example

What are the capacitances of a parallel plate capacitor which has plates of area 0.01 m² and 1 mm with a dielectric of (a) air with relative permittivity 1, (b) mica with relative permittivity 7?

(a) $C = \varepsilon_r \varepsilon_0 \dfrac{A}{d} = \dfrac{1 \times 8.85 \times 10^{-12} \times 0.01}{1 \times 10^{-3}} = 8.85 \times 10^{-11} = 88.5$ pF.

(b) The capacitance will be 7 times that given in (a) and so 619.5 pF.

13.4.1 Capacitance of a multi-plate capacitor

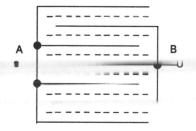

Figure 13.10 *Multi-plate capacitor*

Suppose a capacitor to be made up of n parallel plates, alternate plates being connected together as in Figure 13.10. Let A = the area of one side of each plate, d = the thickness of the dielectric and ε_r its relative permittivity. The figure shows a capacitor with seven plates, four being connected to A and three to B. There are six layers of dielectric sandwiched by the plates and so, consequently, the useful surface area of each set of plates is $6A$. For n plates, the useful area of each set is thus $(n-1)A$. Therefore:

$$C = \frac{\varepsilon_r \varepsilon_0 (n-1)A}{d}$$

Example

A capacitor is made with seven identical plates connected as in Figure 13.10 and separated by sheets of mica having a thickness of 0.3 mm and a relative permittivity of 6. If the area of one side of each plate is 0.05 m², calculate the capacitance.

$$C = \frac{6 \times 8.85 \times 10^{-12} \times (7-1) \times 0.05}{0.0003} = 5.31 \times 10^{-8} \text{ F}$$

13.4.2 Dielectric and electric fields

The term *dipole* is used for atoms or groups of atoms that effectively have a positive charge and a negative charge separated by a distance (Figure 13.11). These may be permanent dipoles because of an uneven distribution of charge in a molecule or temporally created as a result of the electric field distorting the arrangement of the electrons in an atom. When an electric field is applied to a material containing dipoles, i.e. a

Figure 13.11 *Dipole*

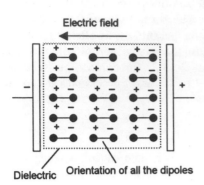

Figure 13.12   *Dipoles in field*

dielectric, the dipoles become reasonably lined up with the field. Figure 13.12 illustrates this, the material being between the plates of a parallel plate capacitor and the electric field produced by a potential difference applied between the plates. The result of using a dielectric between the plates of the capacitor, whether the dipoles are permanent or temporary, is thus to give some alignment of the dipoles with the electric field and the charge on the plates is partially cancelled by the charge on the dipoles adjacent to the plates. Thus there is a net smaller charge Q on the plates for the same potential difference V and so a higher capacitance.

13.4.3 Dielectric strength

For the dielectric in a capacitor, if the electric field becomes large enough it can cause electrons to break free from the dielectric atoms and so render the dielectric conducting. When this happens the capacitor can no longer maintain charge on the plates. The maximum field strength that a dielectric can withstand is called the *dielectric strength*. Dry air has a dielectric strength of about 3×10^6 V/m, ceramics about 10×10^6 V/m, plastics about 20×10^6 V/m, mica about 40×10^6 V/m.

Example

What is the maximum potential difference that can be applied to a parallel plate capacitor having a mica dielectric of thickness 0.1 mm if the dielectric strength of the mica is 40×10^6 V/m?

$$V_{max} = E_{max} d = 40 \times 10^6 \times 0.001 = 4 \text{ kV}$$

13.5 Forms of capacitors

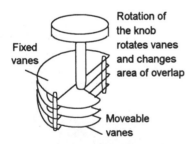

Figure 13.13   *Air capacitor*

The following are types of capacitors commonly available:

1 *Variable air capacitors*
These are parallel plate capacitors with air as the dielectric and are generally multi-plate (Figure 13.13). They consist of a set of plates which can be rotated and so moved into or out of a set of fixed plates, so varying the area of overlap of the plates and hence the area of the capacitor. Such capacitors have maximum capacitances of the order of 1000 pF.

2 *Paper capacitors*
These consist of a layer of waxed paper, relative permittivity about 4, sandwiched between two layers of metal foil, the whole being wound into a roll (Figure 13.14) and either sealed in a metal can or encapsulated in resin. An alternative form is metallised paper in which the foil is replaced by thin films of metal deposited on the paper. Such capacitors have capacitances between about 10 nF and 10 μF. Working voltages are up to about 600 V.

3 *Plastic capacitors*
These have a similar construction to paper capacitors, consisting of layers of plastic film, e.g. polystyrene with a relative permittivity of

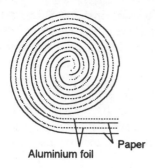

Paper

Aluminium foil

Figure 13.14 *Paper capacitor*

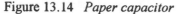

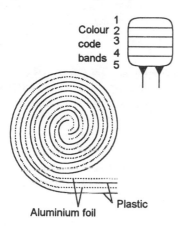

Colour
code
bands

1
2
3
4
5

Aluminium foil

Plastic

Figure 13.15 *Plastic film capacitor*

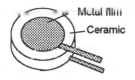

Metal film

Ceramic

Figure 13.16 *Ceramic disc capacitor*

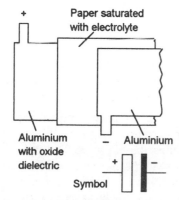

Paper saturated
with electrolyte

Aluminium
with oxide
dielectric

− Aluminium

+ −

Symbol

Figure 13.17 *Aluminum electrolytic capacitor*

2.5, between layers of metal foil (Figure 13.15) or metallised plastic film with metal films being deposited on both sides of a sheet of plastic, e.g. polyester with a relative permittivity of 3.2. Polystyrene film capacitors have capacitances from about 50 pF to 0.5 µF and working voltages up to about 500 V, metallised polyester about 50 pF to 0.5 µF up to about 400 V. The capacitor value may be marked on it or a colour code used (Table 13.1). Such a capacitor is compact and widely used in electronic circuits.

Table 13.1 *Colour code, values in pF*

Colour	Band 1 1st no.	Band 2 2nd no.	Band 3 no of 0s	Band 4 tolerance	Band 5 Max. voltage
Black		0	None	20%	
Brown	1	1	1		100 V
Red	2	2	2		250 V
Orange	3	3	3		
Yellow	4	4	4		400 V
Green	5	5	5	5%	
Blue	6	6	6		
Violet	7	7	7		
Grey	8	8	8		
White	9	9	9	10%	

4   *Mica capacitors*
These generally consist of thin sheets of mica, relative permittivity about 5 to 7, about 2.5 µm thick and coated on both sides with silver. They have capacitances ranging from about 5 pF to 10 nF with a working voltage up to 600 V. These are used in situations where high stability is required.

5   *Ceramic capacitors*
These can be in tube, disc or rectangular plate forms, being essentially a plate of ceramic silvered on both sides (Figure 13.16). Depending on the ceramic used, they have capacitances in the range 5 pF to 1 µF, or more, and working voltages up to about 1 kV.

6   *Electrolytic capacitors*
One form of electrolytic capacitor consists of two sheets of aluminium foil separated by a thick absorbent material, e.g. paper, impregnated with an electrolyte such as ammonium borate, with the whole arrangement being rolled up and put in an aluminium can (Figure 13.17). Electrolytic action occurs when a potential difference is connected between the plates and results in a thin layer of aluminium oxide being formed on the positive plate. This layer forms the dielectric. Another form uses tantalum instead of aluminium, with tantalum oxide forming the dielectric. The electrolytic capacitor must always be used with a d.c. supply and always connected with the correct polarity. This is because if a reverse voltage is used, the dielectric layer will be removed and a

large current can occur with damage to the capacitor. Electrolytic capacitors, because of the thinness of the dielectric, have very high capacitances. Aluminium electrolytic capacitors have capacitances from about 1 µF to 100 000 µF, tantalum ones 1 µF to 2000 µF. Working voltages can be as low as 6 V.

13.6 Capacitors in circuits

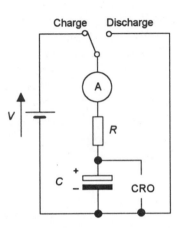

Figure. 13.18 *Capacitor charge and discharge circuit*

Consider the action of a capacitor, e.g. a 20 µF electrolytic capacitor (note that the capacitor has a polarity indicating that it can only be connected one way in the circuit), in a circuit in series with a resistor, e.g. 1 MΩ (Figure 13.18). A cathode ray oscilloscope is connected across the capacitor in order to indicate the potential difference across that component and a centre-zero microameter used to monitor the current.

When the two-way switch is closed to the charge position, the deflection on the meter rises immediately to its maximum value and then falls off to zero (Figure 13.19) as the capacitor is charged by electrons flowing onto one plate of the capacitor to give it an excess of electrons and so a negative charge and, at the same time, electrons flowing away from the other plate of the capacitor to leave it with a deficiency of electrons and hence a positive charge. The current decreases with time as more and more charge accumulates on the capacitor plates, until eventually the current drops to zero and the capacitor is fully charged. At the same time as the current is decreasing, the CRO shows that the p.d. across the capacitor is increasing, reaching a maximum value when the capacitor is fully charged.

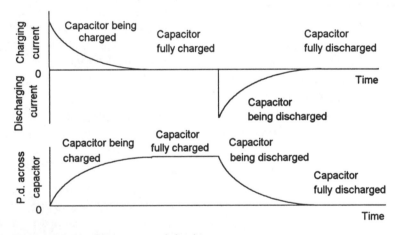

Figure 13.19 *Charging and discharging a capacitor*

During the charging, when the capacitor is fully charged and the current has dropped to zero, there is no current through the resistor and hence no p.d. across the resistor and thus the entire applied p.d. *V* is across the capacitor. When the two-way switch is moved to the discharge position, i.e. a path between the plate with the excess of negative charge and the plate with the deficiency, i.e. the positively charged plate, the current rises immediately to the same maximum as during the charging but in the reverse direction; it then decreases with time (Figure 13.19).

The CRO shows that the p.d. across the capacitor drops with time from its initially fully charged value to become zero when the capacitor is fully discharged.

If the experiment is repeated with more resistance in the charge and discharge circuits, it takes longer to charge or discharge a capacitor.

13.6.1 Capacitor in d.c. and a.c. circuits

When a capacitor is in a d.c. circuit, once it has become fully charged then the current in the circuit is zero and so it 'blocks' the passage of a d.c. current. However, with an a.c. circuit, we can think of the continual changes in the direction of the applied current as charging the capacitor, then discharging it, then charging it, then discharging it, and so on. During charging and discharging, there is a current in a capacitor circuit. Thus, with a.c. we have an alternating current in the capacitor circuit. A capacitor does 'not block' an a.c. current. Thus, if we have a signal which might be a mixture of a direct current and an alternating current and we put a capacitor in the circuit, the d.c. component will be blocked off and only the a.c. component transmitted.

13.7 Energy stored in a charged capacitor

Consider a constant current I being applied to charge an initially uncharged capacitor C for a time t. Since current is the rate of movement of charge then the charge moved onto one of the plates and off the other plates will be It. If this movement of charge results in the p.d. across the capacitor rising from 0 to V then the charge $= It = CV$ and thus $I = VC/t$. The average p.d. across the capacitor during the charging is ½V and so the average power to capacitor during charging $= I \times \tfrac{1}{2}V$. The energy supplied to capacitor during charging = average power × time and, hence stored by it, is thus:

$$\text{energy stored} = \tfrac{1}{2}IVt = \tfrac{1}{2}(VC/t) \times Vt = \tfrac{1}{2}CV^2$$

Since $Q = CV$, we can also write the above equation in the two forms:

$$\text{energy stored} = \tfrac{1}{2}CV \times V = \tfrac{1}{2}QV$$

$$\text{energy stored} = \tfrac{1}{2}C\left(\frac{Q}{C}\right)^2 = \frac{Q^2}{2C}$$

Example

What is the energy stored in a 10 μF capacitor when the voltage between its terminals is 20 V?

$$\text{Energy} = \tfrac{1}{2}CV^2 = \tfrac{1}{2} \times 10 \times 10^{-6} \times 20^2 = 0.002 \text{ J}$$

Example

A 50 μF capacitor is charged from a 20 V supply. After being charged, it is immediately connected in parallel with an uncharged

30 μF capacitor. Determine (a) the p.d. across the combination, (b) the total energy stored in the capacitors before and after the capacitors are connected in parallel.

(a) Since $Q = CV$, then the charge $Q = 50 \times 10^{-6} \times 20 = 0.001$ C. When the capacitors are connected in parallel, the total capacitance is $50 + 30 = 80$ μF. The charge of 0.001 coulomb is then divided between the two capacitors. Hence, using $Q = CV$ we have $V = Q/C = 0.001/80 \times 10^{-6} = 12.5$ V.
(b) When the 50 μF capacitor is charged to a p.d. of 20 V the stored energy $= \frac{1}{2} \times 50 \times 10^{-6} \times 20^2 = 0.01$ J. With the capacitors in parallel the total stored energy $= \frac{1}{2} \times 80 \times 10^{-6} \times 12.5^2 = 0.00625$ J.

Connecting the uncharged capacitor in parallel with the charged capacitor results in a reduction in the energy stored in the capacitors from 0.01 J to 0.00625 J. This loss of energy appears mainly as heat in the wires as a result of the circulating current responsible for equalising the p.d.s.

Activities

1 Use the circuit shown in Figure 13.18 to investigate the action of a capacitor in a d.c. circuit. You might initially try a 20 μF electrolytic capacitor in series with 1 MΩ resistor and then see the effect of doubling the resistance to 2 MΩ. Note that with the oscilloscope, the a.c./d.c. switch should be on the d.c. setting.

Problems

1 What is the electric field strength between a pair of parallel plates 3 mm apart if the potential difference between the plates is 120 V?

2 A capacitor of capacitance 10 μF has a potential difference of 6 V between its plates. What are the charges on the capacitor plates?

3 What size capacitor is required if it is to have 24 μC of charge on its plates when a 6 V supply is connected across it?

4 What is the potential difference between the plates of a 4 μF capacitor when there is charge of 16 μC on its plates?

5 A 2 μF capacitor and a 4 μF capacitor are connected in series across a 12 V d.c. supply. What will be the resulting (a) charge on the plates of each capacitor and (b) the potential difference across each?

6 What will be the total capacitances when three capacitors, 2 μF, 4 μF and 8 μF, are connected in (a) series, (b) parallel?

7 Three capacitors, 1 μF, 2 μF and 3 μF, are connected in parallel across a 10 V d.c. supply. What will be (a) the total capacitance, (b) the resulting charge on each capacitor, (c) the resulting potential difference across each capacitor?

8 Three capacitors, 20 μF, 30 μF and 60 μF, are connected in series across a 36 V d.c. supply. What will be (a) the total equivalent capacitance, (b) the charge on the plates of each capacitor and (c) the potential differences across each capacitor?

9 Three capacitors, 50 pF, 100 pF and 220 pF, are connected in parallel across a 10 V d.c. supply. What will be (a) the total

capacitance, (b) the resulting charge on each capacitor, (c) the resulting potential difference across each capacitor?

10 A parallel plate capacitor has plates of area 200 cm², separated by a dielectric of thickness 1.0 mm and having a relative permittivity of 5. What is the capacitance?

11 A 2 μF capacitor is to be made from rolled up sheets of aluminium foil separated by paper of thickness 0.1 mm and relative permittivity 6. What plate area is required?

12 A 50 μF electrolytic capacitor has aluminium plates of area 600 cm² with an oxide dielectric of relative permittivity 20. What is the thickness of the dielectric?

13 A 1 μF capacitor is made up of two strips of metal foil, width 100 mm, separated by a paper dielectric of thickness 0.02 mm and relative permittivity 3.0. What will be the length of each strip of metal foil?

14 A parallel plate capacitor has two rectangular plates, each 10 mm by 20 mm, separated by a 0.2 mm thick sheet of mica. What is the capacitance? Take the relative permittivity of the mica to be 5.

15 A parallel plate capacitor has 7 metal plates connected as shown in Figure 13.10 and separated by sheets of mica of relative permittivity 5, thickness 0.2 mm and having an area on each side of 0.040 m². What is the capacitance?

16 A parallel plate capacitor has 9 metal plates connected as shown in Figure 13.10 and separated by sheets of mica of relative permittivity 5 and having an area on each side of 0.012 m². Determine the thickness of mica required to give a capacitance of 200 pF.

17 What is the maximum voltage that can be applied to a parallel plate capacitor with a dielectric of thickness 0.1 mm and dielectric strength 16×10^6 V/m?

18 An electrolytic capacitor has an oxide dielectric of thickness 1 μm and dielectric strength 6×10^6 V/m. What is the maximum voltage that can be used with this capacitor?

19 What energy is stored in a 100 μF capacitor when it has a potential difference of 12 V between its plates?

20 What size capacitor is required if it is to store energy of 0.001 J when it is fully charged by a 10 V d.c. supply?

21 Three capacitors of 2 μF, 3 μF and 6 μF are connected in series across a 10 V d.c. supply. What will be (a) total capacitance, (b) the charge on each capacitor, (c) the potential difference across each capacitor, (d) the energy stored by each capacitor?

22 A 10 μF capacitor is charged until the potential difference across its plates is 5 V. It is then connected across an uncharged 4 μF capacitor. Calculate the stored energy before and after they are connected together.

23 A parallel plate capacitor is formed by plates each of area 100 cm² spaced 1 mm apart in air. What is the capacitance? If a voltage of 1 kV is applied to the plates, what is the energy stored in the capacitor?

14 Semiconductors

14.1 Introduction

Chapter 10 opened with a discussion of charge and an explanation of current flow in metals. This chapter extends the discussion of the mechanism of current flow to consider its flow in semiconductors and discuss the basic principles involved in the action of the semiconductors devices of diodes and bipolar transistors.

In terms of their electrical conductivity, materials can be grouped into three categories, namely conductors, semiconductors and insulators. Conductors have electrical resistivities of the order of 10^{-6} Ω m, semiconductors about 1 Ω m and insulators 10^{10} Ω m. The higher the resistivity of a material the lower is its conductivity. Conductors are metals with insulators being polymers or ceramics. Semiconductors include silicon, germanium and compounds such as gallium arsenide.

14.2 Current flow

In discussing electrical conduction in materials, a useful picture is of an atom as consisting of a positively charged nucleus surrounded by its electrons. The electrons are bound to the nucleus by electric forces of attraction. The force of attraction is weaker the further an electron is from the nucleus. The electrons in the furthest orbit from the nucleus are called the *valence electrons* since they are the ones involved in the bonding of atoms together to form compounds.

Metals can be considered to have a structure of atoms with valence electrons which are so loosely attached that they easily drift off and can move freely between the atoms. Typically a metal will have about 10^{28} free electrons per cubic metre. Thus, when a potential difference is applied across a metal, there are large numbers of free electrons able to respond and give rise to a current. We can think of the electrons pursuing a zigzag path through the metal as they bounce back and forth between atoms (Figure 14.1). An increase in the temperature of a metal results in a decrease in the conductivity. This is because the temperature rise does not result in the release of any more electrons but causes the atoms to vibrate and scatter electrons more, so hindering their progress through the metal.

Insulators, however, have a structure in which all the electrons are tightly bound to atoms. Thus there is no current when a potential difference is applied because there are no free electrons able to move through the material. To give a current, sufficient energy needs to be supplied to break the strong bonds which exist between electrons and insulator atoms. The bonds are too strong to be easily broken and hence

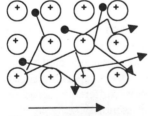

Direction of electron flow

Figure 14.1 *Electric current in a metal*

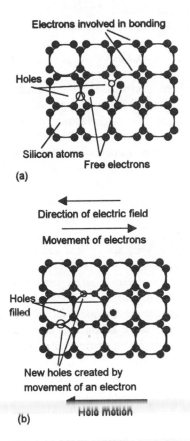

Figure 14.2 *(a) Holes and free electrons in silicon, (b) current when an electric field is applied*

normally there is no current. A very large temperature increase would be necessary to shake such electrons from the atoms.

Semiconductors, e.g. silicon, can be regarded, at a temperature of absolute zero, as insulators in having no free electrons, all the valence electrons being involved in bonding and so not free to move. However, the energy needed to remove a valence electron from an atom is not very high and at room temperature there has been sufficient energy supplied by the thermal agitation resulting from the temperature for some electrons to have broken free. Thus the application of a potential difference will result in a current. Increasing the temperature results in more electrons being shaken free and hence an increase in conductivity. At about room temperature, a typical semiconductor will have about 10^{16} free electrons per cubic metre and 10^{16} atoms per cubic metre with missing electrons.

When a semiconductor atom, such as silicon, loses an electron, we can consider there to be a hole in its valence electrons (Figure 14.2(a)) in that the atom is now one electron short. When electrons are made to move as a result of the application of a potential difference, i.e. an electric field, they can be thought of as hopping from a valence site into a hole in a neighbouring atom, then to another hole, etc. Not only do electrons move through the material but so do the holes, the holes moving in the opposite direction to the electrons. We can think of the above behaviour in the way shown in Figure 14.2(b). One way of picturing this behaviour is in terms of a queue of people at, say, a bus-stop. When the first person gets on the bus a hole appears in the queue between the first and second person. Then the second person moves into the hole, which now moves to between the second and third person. Thus as people move up the queue, the hole moves down the queue.

Because, when there is an electric field, holes move in the opposite direction to electrons we can think of the holes as behaving as though they have a positive charge.

14.2.1 Doping

With a pure semiconductor, each hole is produced by the release of an electron and so we have an equal number of electrons and holes. Thus an electric current with such a material can be ascribed in equal parts to the movement of electrons and holes. Such a semiconductor is said to be *intrinsic*.

The balance between the number of electrons and holes can be changed by replacing some of the semiconductor atoms by atoms of other elements. This process is known as *doping* and typically about one atom in every ten million might be replaced.

The silicon atom has four electrons which participate in bonding and so if atoms of an element having five electrons which participate in bonding, e.g. phosphorus, are introduced, since only four of the electrons can participate in bonds with silicon atoms we have a spare electron with each dopant atom (Figure 14.3). These extra electrons easily break free and become available for conduction. Thus we now have more free

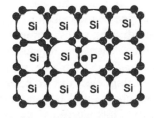

Figure 14.3 *Doping with phosphorus*

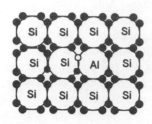

Figure 14.4 *Doping with aluminium*

electrons than holes available for conduction. Such a doped material is said to be *n-type*, the n inducting that the conduction is predominantly by negative charge carriers, i.e. electrons. Arsenic, antimony and phosphorus are examples of elements added to silicon to give n-type semiconductors.

If an element having atoms with just three electrons which participate in bonding is added to silicon, then all its three electrons will participate in bonds with silicon atoms. However, there is a deficiency of one electron and so one bond with a silicon atom is incomplete (Figure 14.4). A hole has been introduced. We thus have more holes than free electrons. Electrical conduction with such a doped semiconductor is predominantly by holes, i.e. positive charge carriers, and so such materials are termed *p-type*. Boron, aluminium, indium and gallium are examples of elements that are added to silicon to give p-type semiconductors.

Such doped semiconductors, n-type and p-type, are termed *extrinsic*. This signifies that they have a *majority charge carrier* and a *minority charge carrier*. With n-type material, the majority charge carrier is electrons and for p-type it is holes. Typically, doping replaces about one in every ten million atoms, i.e. 1 in 10^7. Since there are about 10^{28} atoms per cubic metre, about 10^{21} dopant atoms per cubic metre will be used. Each dopant atom will donate one electron or provided one hole. Thus there will be about 10^{21} electrons donated or holes provided per cubic metre. Since intrinsic silicon has about 10^{16} conduction electrons and holes per cubic metre, the doping introduces considerably more charge carriers and swamps the intrinsic charge carriers. The majority charge carriers are thus considerably in excess of the minority charge carriers.

Thus when a potential difference is connected across a piece of semiconductor, if it is n-type then the resulting current is largely the result of the movement of electrons in the conduction band. If it is p-type, the current is largely the result of movement of holes in the valence band.

14.2.2 The pn junction

Consider what happens if we put an n-type semiconductor in contact with a p-type semiconductor. Before contact we have two materials that are electrically neutral, i.e. in each the amount of positive charge equals the amount of negative charge. However, in the n-type semiconductor we have mostly electrons available for conduction and in the p-type material we have mostly holes available for conduction. When the two materials are in contact then electrons from the n-type semiconductor can diffuse across the junction and into holes in the p-type semiconductor (Figure 14.5). We can also consider the holes to be diffusing across the junction in the opposite direction. Because electrons leave the n-type semiconductor it is losing negative charge and so ends up with a net positive charge. Because the p-type material is gaining electrons it becomes negatively charged. Atoms which have lost charge and become either positive or negative charged are termed *ions*. Thus electrons and holes diffuse across the junction until the build-up of charge on each material is such as to prevent further charge movement. We end up with a layer

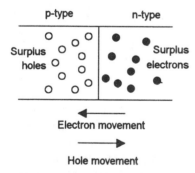

Figure 14.5 *p and n put in contact*

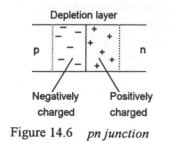

Figure 14.6 *pn junction*

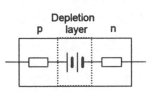

Figure 14.7 *Equivalent circuit for pn junction*

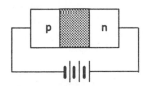

(a) Reverse bias

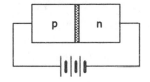

(b) Forward bias

Figure 14.8 *Effect of bias*

on either side of the junction which is short of the necessary charges to give neutrality and we call such a layer the *depletion layer*. The result is shown in Figure 14.6. This separation of charge in the vicinity of the junction gives a barrier potential across the depletion layer and a result which is similar to a battery (Figure 14.7). At room temperature, with doped silicon the barrier potential is approximately 0.7 V and with doped germanium about 0.3 V.

When an external potential difference is connected across a pn junction, if it is connected so that the p-type side is made more negative than the n-type side, i.e. so termed *reverse bias*, then the depletion layer becomes larger as more positive ions are created in the n region and more negative ions in the p region (Figure 14.8(a)). The bias voltage adds to the barrier potential of the unbiased junction. As a consequence there is only a very small current across the junction since there are only few charge carriers with enough energy to overcome the potential barrier at the junction. When the junction is connected to an external potential difference so that the p-type side is made positive with respect to the n-type side, i.e. the so termed *forward bias*, the depletion layer is made narrower (Figure 14.8(b)). This is because the external potential difference provides the n-region electrons with enough energy to move across the junction where they combine with p-region holes. As electrons leave the n-region then more flow in from the negative terminal of the battery and so there is a circuit current. Thus, we can easily get electrons to flow through the circuit when they flow in the direction from p to n, in the reverse direction they are opposed by the barrier potential at the junction and so current flow in just one direction is allowed.

When the pn junction is reverse biased, the depletion layer acts as an insulator between layers of oppositely charged ions and thus acts like a parallel plate capacitor. Since the depletion layer widens when the reverse biased voltage is increased, the capacitance decreases; the capacitance is thus a function of the size of the applied bias voltage.

14.3 Junction diodes

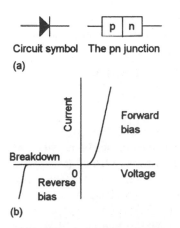

Circuit symbol   The pn junction

(a)

Figure 14.9 *Junction diode*

The *junction diode* is a device involving a pn junction and has a high resistance to the flow of current in one direction and a low resistance in the other. Figure 14.9(a) shows the diode symbol.

The current–voltage characteristic of a junction diode is the graph of the current through the device plotted against the voltage applied across it. Figure 14.9(b) shows the general form of the characteristic of a junction diode. With forward bias there is no significant current until a particular voltage is reached. With reverse bias there is no significant current until breakdown occurs. Figure 14.10 shows typical forward and reverse bias characteristics for silicon and germanium diodes. For the silicon diode the forward current does not noticeably increase until the forward bias voltage is greater than about 0.6 V. For the germanium diode the forward current does not noticeably increase until the forward bias voltage is greater than about 0.2 V.

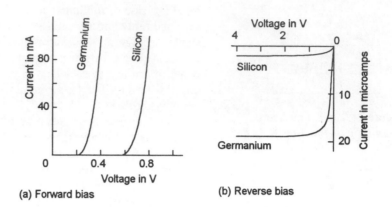

(a) Forward bias

(b) Reverse bias

Figure 14.10   *Characteristics*

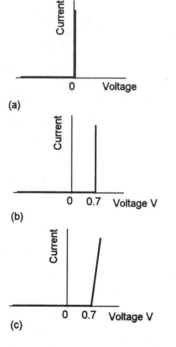

Figure 14.11   *Diode models*

14.3.1 Diodes in circuits

A simple model we can use for a junction diode in a circuit is that of a switch which is on when the diode is forward biased and off when reverse biased; the voltage drop across the diode when forward biased is neglected. The characteristic assumed for such a diode is thus as shown in Figure 14.11(a).

A more refined model assumes that there is a constant voltage drop when forward biased (Figure 14.11(b)). For a silicon diode we might assume a value of 0.7 V.

Another model assumes there is a voltage drop which is not constant but increases with the current (Figure 14.11(c)). We can think of the diode as being like a battery with a series resistance, the value of the resistance depending on the current in the circuit. The *d.c. resistance* of a diode at a particular voltage is the value of the voltage divided by the value of the current; however, in selecting the value of the resistance to be used, we need to consider the voltage at which the diode will be operating and thus use a different resistance value called the *a.c. resistance* r_{ac} (or *dynamic resistance*). This is the reciprocal of the slope of the characteristic at a particular voltage and is thus:

$$r_{ac} = \frac{\delta V}{\delta I}$$

where δV is the change in voltage and δI is the resulting change in current.

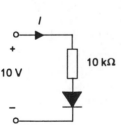

Figure 14.12   *Example*

Example

Determine the current I in the circuit of Figure 14.12 when the voltage is 10 V assuming the diode can be modelled by (a) a switch, (b) a constant voltage drop of 0.7 V.

(a) The current $I = 10/10\ 000 = 1$ mA.
(b) The current $I = (10 - 0.7)/10\ 000 = 0.93$ mA.

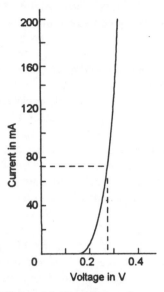

Figure 14.13   *Example*

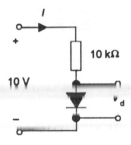

Figure 14.14   *Example*

Example

Estimate (a) the d.c. resistance and (b) the a.c. resistance of the diode giving the characteristic shown in Figure 14.13 at a voltage of 0.28 V.

(a) The current at 0.28 V is 66 mA and so the d.c. resistance is 0.28/0.066 = 4.2 Ω.

(b) The graph is almost straight line at 0.28 V so that we can obtain a reasonable estimate of the a.c. resistance by considering the change in current produced by a small incremental change in voltage (it must be small since the graph is not a perfectly straight line). We thus find values of the currents at voltages which are small equal increments either side of the 0.28 V. At 0.27 V the current is 60 mA and at 0.29 V it is 86 mA. Thus a change of voltage of 0.02 V produces a change in current of 26 mA and so the a.c. resistance is 0.02/0.026 = 0.77 Ω.

Example

For the circuit shown in Figure 14.14, the diode has an a.c. resistance of 50 Ω at the nominal current of 1 mA. What will be the change in the voltage drop across the diode when the supply voltage changes by ±1 V.

The circuit is a potential divider and so a change of 1 V will give a voltage change across the diode v_d of:

$$v_d = \frac{r_d}{R + r_d} v = \frac{50}{10\,000 + 50} \times 1 = 5.4 \text{ mA}$$

Thus the voltage drop changes by ±5.4 mA.

14.3.2 Diode data sheets

The term *signal diode* is used for all diodes which have been designed for use in circuits where large currents and/or large voltages do not occur. The term *power diode* is used for diodes that are generally used as rectifiers, i.e. the conversion of alternating current to direct current.

Manufacturer's data sheets give detailed information about the devices. Typically they include maximum ratings, electrical characteristics, mechanical data and graphs of various parameters. For example, the data sheet for a signal diode is likely to include:

Maximum continuous reverse voltage, e.g. 50 V
Maximum repetitive peak reverse voltage, e.g. 50 V
Maximum repetitive peak forward current, e.g. 150 mA
Maximum forward current (d.c.), e.g. 75 mA
Storage temperature, e.g. −65 to +200°C
Maximum junction temperature, e.g. 200°C
Characteristics at 25° as tabulated values and graphs
Diode capacitance, e.g. at zero reverse voltage and 1 MHz, < 3 pF.

14.4 Transistors

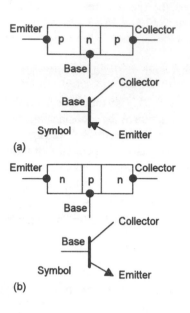

Figure 14.15 *(a) pnp, (b) npn*

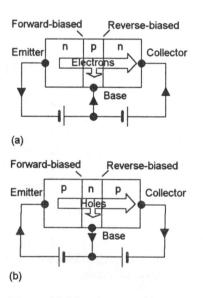

Figure 14.16 *(a) npn, (b) pnp*

The *bipolar junction transistor* has three doped semiconductor regions to give two pn junctions (Figure 14.15) and is available in two forms, one (npn) having two n-type regions separated by a p-region and the other (pnp) with two p-regions separated by an n-region. The three regions are called the *base, emitter* and *collector*. The base is the middle of the three regions and is much narrower than the other two regions.

The npn transistor contains two pn junctions and is normally operated with the emitter–base junction forward biased and the collector–base junction reverse biased (Figure 14.16(a)). The forward bias on the emitter–base junction will cause current to flow across the junction. The current will consist of electrons injected from the emitter into the base and holes injected from the base into the emitter. However, the emitter is more heavily doped than the base and so there is a much higher density of electrons in the emitter than holes in the base. As a consequence the current is predominately due to electrons. Since the base is very thin, most of the electrons moving into it from the emitter will get clean through it without combining with holes and so to the collector and give a collector current. A base current occurs as a result of holes injected from the base region into the emitter region and also holes that have to be supplied by the external circuit in order to replace the holes lost from the base by combining with electrons. However, because the base is lightly doped and thin, the base current I_B is but a small fraction of the emitter current, typically about $0.01I_E$.

The pnp transistor contains two pn junctions and is normally operated with the emitter–base junction forward biased and the collector–base junction reverse biased (Figure 14.16(b)). Unlike the npn transistor, current is mainly due to holes. Because the emitter is more heavily doped than the base, there is a much higher density of holes in the emitter than electrons in the base. As a consequence, the current from the emitter to the base is mainly holes with little movement of electrons. Since the base is very thin, most of the holes moving into it from the emitter will get clean through it without combining with electrons and so to the collector and give a collector current. A base current occurs as a result of electrons injected from the base region into the emitter region and also electrons that have to be supplied by the external circuit in order to replace the electrons lost from the base by combining with holes. However, because the base is lightly doped and thin, the base current I_B is but a small fraction of the emitter current, typically about $0.01I_E$.

With both the npn and pnp transistors, if we apply Kirchhoff's current law then the emitter current I_E must equal the base current I_B plus the collector current I_C:

$$I_E = I_B + I_C$$

and so typically we have $I_E = 0.01I_E + I_C$ and so $I_C = 0.99I_E$.

The ratio of the output current of a transistor to its input current, in the absence of an a.c. signal, is termed the *d.c. current gain h_{FB}*:

$$h_{FB} = -\frac{I_C}{I_E}$$

The minus sign is because the collector and emitter currents flow in opposite directions. Thus the above typical data gives h_{FB} as −0.99. Note that, in general, we use h as the gain symbol and when the suffix is in capital letters we are referring to d.c. operation and when in lower case we are dealing with a.c. operation.

With both the npn and pnp transistors the emitter–base junction is forward biased and thus presents a low resistance to signals, while the collector–base junction is reverse biased and so is high resistance. Thus the output current is through a high resistance and there is a large voltage and power gain.

14.4.1 Transistor connections

A transistor may be connected in a circuit in three different ways with, in each case, one terminal common to both the input and output. In all connections, the base–emitter junction is always forward biased and the collector–base junction is always reverse biased.

1   *Common-base connection*

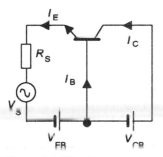

Figure 14.17   *Common-base*

Figure 14.17 shows the connections for a common-base amplifier. An alternating source giving a signal V_S with internal resistance R_S is connected in series with V_{EB} to provide the input to the transistor. V_{EB} applies the voltage necessary to ensure the emitter–base junction is always forward biased. The *d.c. current gain* h_{FB} is:

$$h_{FB} = \frac{I_C}{I_E}$$

With an a.c. input, during the negative half-cycles of the signal the forward bias is increased and the emitter current increases. During the positive half-cycles of the signal, the forward bias is decreased and the emitter current decreases. The collector current is proportional to the emitter current and so varies in the same way as the applied signal. The collector–base circuit is shown with just a bias battery V_{CB}. With this having negligible internal resistance, the collector circuit is effectively short-circuited. The *short-circuit current gain* h_{fb} is defined as the ratio of a change in collector current δI_C to the change in emitter current δI_E producing it when the collector–base voltage is maintained constant:

$$h_{fb} = \frac{\delta I_C}{\delta I_E}$$

Since the collector current is always less than the emitter current (see previous section), then h_{fb} is always less than 1; a typical value is 0.995.

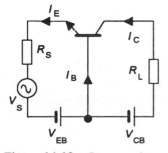

Figure 14.18   *Common-base*

Because the reverse bias collector–base junction has such a high resistance, the inclusion of a resistor R_L in the collector–base circuit (Figure 14.18) has little effect on the current gain. Thus we can take an output voltage from across the load resistor and obtain a voltage gain. The common-base connected transistor is rarely used because,

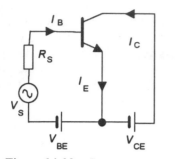

Figure 14.19 *Common-emitter*

although the voltage gain can be large, the current gain is less than one.

2 *Common-emitter connection*

The term common-emitter signifies that the emitter is common to both the input and output circuits (Figure 14.19). This is the most widely used form since it gives both current and voltage gain. The emitter–base junction is forward-biased by V_{BE} and the collector–base junction is reverse-biased by $(V_{CE} - V_{BE})$. The input current is the base current and the output is the collector current. Thus we can define the *d.c. current gain* h_{FE} as:

$$h_{FE} = \frac{I_C}{I_B}$$

With an a.c. input, during the negative half-cycles of the signal the forward bias is decreased and the emitter current decreases. During the positive half-cycles of the signal, the forward bias is increased and the emitter current increases. The collector current is proportional to the emitter current and so varies in the same way as the applied signal. Applying Kirchhoff's current law, we have $I_C = I_E - I_B$ and the change in the collector current δI_C equals the change in emitter current δI_E minus the change in base current δI_B.

$$\delta I_C = \delta I_E - \delta I_B$$

With $h_{fb} = \delta I_C / \delta I_E$ then we can write $h_{fc} \delta I_E = \delta I_E - \delta I_B$ and so:

$$\frac{\delta I_E}{\delta I_B} = \frac{1}{h_{fc} - 1}$$

The collector–emitter circuit is shown with just a bias battery V_{CB}. With this having negligible internal resistance, the collector circuit is effectively short-circuited. The *short-circuit current gain* h_{fe} is defined as the ratio of a change in collector current δI_C to the change in base current δI_B producing it when the collector–emitter voltage is maintained constant and so:

$$h_{fe} = \frac{\delta I_C}{\delta I_B} = \frac{h_{fb} \delta I_E}{\delta I_B} = \frac{h_{fb}}{1 - h_{fb}}$$

A typical value for the short-circuit current gain h_{fb} of a common-base connected transistor is 0.995 and so h_{fe} is typically 199. Thus the common-emitter connection gives considerable current gain. With a load resistor R_L connected in the output circuit (Figure 14.20), the current gain will be less than the short-circuit current gain, decreasing as R_L increases.

If the input resistance of the transistor is R_{IN} then, with the a.c. signal in the input circuit of V_S, the a.c. current input to the transistor is:

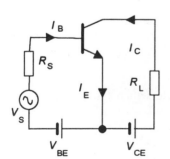

Figure 14.20 *Common-emitter circuit with load*

$$I_B = \frac{V_S}{R_S + R_{IN}}$$

The a.c. voltage input to the transistor is $V_{IN} = I_B R_{IN}$. The output current is $I_C = h_{fe} I_B = h_{fe} V_{IN}/R_{IN}$. The output current flows through the load resistor and develops the output voltage V_{OUT} across it. Thus $V_{OUT} = I_C R_L = R_L h_{fe} V_{IN}/R_{IN}$. The voltage gain A_v is thus:

$$A_v = \frac{V_{OUT}}{V_{IN}} = \frac{h_{fe} R_L}{R_{IN}}$$

Since h_{fe} is fairly large and the load resistance is usually greater than the input resistance, there is a voltage gain.

The input power to the transistor is $P_{IN} = I_B{}^2 R_{IN}$ and the output power is $P_{OUT} = I_C{}^2 R_L = (h_{fe} I_B)^2 R_L$. The power gain A_p is thus:

$$A_p = \frac{P_{OUT}}{P_{IN}} = \frac{(h_{fe} I_B)^2 R_L}{I_B^2 R_{IN}} = \frac{h_{fe}^2 R_L}{R_{IN}} = A_v h_{fe}$$

Since h_{fe} is the current gain, the power gain is the product of the voltage gain and the current gain.

Example

A transistor has an emitter current of 2.5 mA and a collector current of 2.4 mA. What is (a) the base current, (b) the current gain when connected as common-base, (c) the current gain when connected as common-emitter?

(a) Applying Kirchhoff's current law gives a base current of 2.5 − 2.4 = 0.1 mA.
(b) The input is the emitter current and the output the collector current and so the current gain is 2.4/2.5 = 0.96.
(c) The input is the base current and the output is the collector current and so the current gain is 2.4/0.1 = 24.

Example

A transistor is connected as common-emitter with a load resistance of 5 kΩ in the collector circuit. If h_{fe} is 100 and the transistor has an input resistance of 1 kΩ, what are the voltage and power gains?

$$A_v = \frac{V_{OUT}}{V_{IN}} = \frac{h_{fe} R_L}{R_{IN}} = \frac{100 \times 5000}{1000} = 500$$

$$A_p = \frac{P_{OUT}}{P_{IN}} = A_v h_{fe} = 500 \times 100 = 50\,000$$

3 *Common-collector connection*
Figure 14.21 shows a transistor connected with the collector common to both input and output circuits, the base current being the

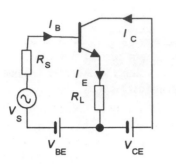

Figure 14.21 *Common-collector*

input current and the emitter current the output current. The *short-circuit current gain* h_{fc} is defined as, when V_{CE} is constant:

$$h_{fc} = \frac{\delta I_E}{\delta I_B}$$

Since $I_B = I_E - I_C$ then:

$$h_{fc} = \frac{\delta I_E}{\delta I_E - \delta I_C} = \frac{\delta I_E}{\delta I_E - h_{fb}\delta I_E} = \frac{1}{1 - h_{fb}}$$

Since we have $h_{fe} = h_{fb}/(1 - h_{fb})$ then, when rearranged, we have $h_{fb} = h_{fe}/(1 + h_{fe})$ which, when substituted in the above equation gives:

$$h_{fc} = \frac{1}{1 - h_{fe}/(1 + h_{fe})} = h_{fe} + 1$$

Thus the short-circuit current gain of a transistor connected with common-emitter is approximately equal to the short-circuit current gain of the same transistor when connected with common-emitter. When a load is connected into the circuit, the current gain is reduced by an amount determined by the size of the load.

With common-collector, the input resistance is considerably greater than the load resistance and thus a voltage gain of less than 1 is obtained. Thus the main use of the common-collector form of connection is to act as a buffer between a high impedance load and a low impedance load.

Table 14.1 summarises the main characteristics of the three modes of connection for a transistor.

14.4.2 Transistor characteristics

A number of current–voltage graphs are used to describe the input–output behaviour of a transistor when in a circuit, these graphs being termed the *static characteristics*. There are four sets of characteristics for each form of connection of a transistor: the input characteristic, the transfer characteristic, the output characteristic and the mutual characteristic.

Table 14.1 *Characteristics of modes of connection*

Form of connection	Short-circuit current gain	Voltage gain	Input resistance	Output resistance
Common-base	h_{fb}	Good	Low: 30 to 100 Ω	High: 100 to 1000 kΩ
Common-emitter	$h_{fe} = \dfrac{h_{fb}}{1 - h_{fb}}$	Better than common-base	Medium: 800 to 5000 Ω	Medium: 10 to 50 kΩ
Common-collector	$h_{fc} = \dfrac{1}{1 - h_{fb}} = h_{fe} + 1$	1 or less	High: 5 to 500 kΩ	Low: 50 to 1000 Ω

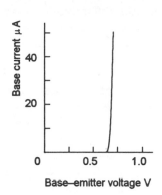

Figure 14.22   *Input characteristic*

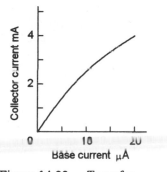

Figure 14.23   *Transfer characteristic*

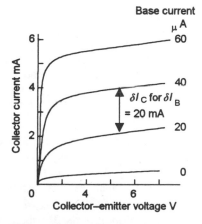

Figure 14.24   *Output characteristic*

For a bipolar junction transistor in common-emitter connection, the input parameters are the base–emitter voltage V_{BE} and the base current I_B. The output parameters are the collector–emitter voltage V_{CE} and the collector current I_C. The characteristics are thus:

1   *Input characteristic*
This shows how the base current varies with changes in the base–emitter voltage when the collector–emitter voltage is kept constant. Figure 14.22 shows a typical characteristic. The base–emitter voltage is almost constant at 0.7 V for a silicon transistor. When the transistor is used as an amplifier we are concerned with an input of a small a.c. signal superimposed on a fixed d.c. voltage. Thus our concern is with the effect of small base–emitter voltage changes on the base current about some operating voltage point. For this reason we define the *input resistance* h_{ie} as the $\delta V_{BE}/\delta I_B$, i.e. the reciprocal of the slope of the input characteristic, at the operating point. For the data given in Figure 14.22, at an operating voltage of 0.7 V the current changes by about 50×10^{-6} A for a voltage change of 0.05 V and so the input resistance is about 1000 Ω.

2   *Transfer characteristic*
This shows how the collector current changes with changes in the base current, the collector–emitter voltage being kept constant. Figure 14.23 shows a typical characteristic. The characteristic is almost a straight line passing through the origin. The slope of the characteristic gives the short-circuit current gain h_{fe}. For Figure 14.23, the collector current changes by 1 mA when the base current changes by 4 µA and so $h_{fe} = dI_C/dI_B = 1 \times 10^{-3}/(4 \times 10^{-6}) = 250$.

3   *Output characteristic*
This shows how the collector current changes with changes in the collector–emitter voltage, the base current being kept constant. Figure 14.24 shows a typical characteristic. Each of the curves for the different base currents are basically the same shape with the straight line parts parallel. The *output resistance* is $\delta V_{CE}/\delta I_C$ and is thus the reciprocal of the slope of the output characteristic; the slope dI_C/dV_{CE} is termed the output admittance h_{oe}. For Figure 14.24, the collector current changes by about 0.2 mA when collector–emitter voltage changes by 2 V and so the output resistance is $2/(0.2 \times 10^{-3})$ = 10 000 Ω. Since the collector current changes by about 2 mA for a base current change of 20 µA, then the short-circuit current gain h_{fe} = $\delta I_C/\delta I_b = 2 \times 10^{-3}/(20 \times 10^{-6}) - 100$.

Figure 14.24 shows that there is a small collector current flowing when the base current is zero. This is the *common-emitter leakage current* I_{CEO} and arises from the movement of the minority carriers out of the base and into the emitter.

4   *Mutual characteristic*
This shows how the collector current changes when the base–emitter voltage changes, the collector–emitter voltage being kept constant. Figure 14.25 shows a typical characteristic. The slope of the graph

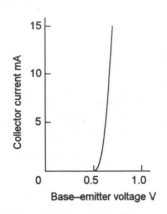

Figure 14.25 *Mutual characteristic*

$\delta I_C/\delta V_{BE}$ is the *mutual conductance* g_m. For the data given in Figure 14.25, the collector current changes from about 7 to 15 mA when the base–emitter voltage changes from 0.6 to 0.65 V and so g_m is about $0.008/0.05 = 0.16$ S.

In summary, for a silicon junction transistor: the collector current is zero until there is a base current, the base current is zero until the base–emitter voltage is about 0.6 V, the base–emitter voltage remains close to 0.6 V for a wide range of values of base current.

In addition to giving the characteristics of transistors, manufacturers' data sheets give maximum values of voltages, currents and power that must not be exceeded. The current gain of a transistor is not the same at all frequencies, being constant at low frequencies and then falling off at high frequencies. The frequency f_t at which the gains falls to 1, the so-termed *transition frequency*, is thus quoted on data sheets.

14.4.3 *h* parameters

The parameters defined as *h* parameters, i.e. hybrid parameters, are so termed because they dp not all have the same units. Some of the parameters are ratios and so without units, some have units of resistance and some have units of conductance. The first subscript used with the *h* is i for input, o for output and f for forward current. The second subscript is *e* for common-emitter and b for common-base connections. For common-emitter connection we have:

$$\text{Short-circuit current gain } h_{fe} = \frac{\delta I_C}{\delta i_B}, \text{ with } V_{CE} \text{ constant}$$

$$\text{Input resistance } h_{ie} = \frac{\delta V_{BE}}{\delta I_B}, \text{ with } I_B \text{ constant}$$

$$\text{Output conductance } h_{oe} = \frac{\delta I_C}{\delta V_{CE}}, \text{ with } I_B \text{ constant}$$

14.4.4 Transistor switching

Figure 14.26 shows a circuit we can use in order to employ a bipolar transistor as a switch. The switching operation is determined by the base current. The output is the collector–emitter voltage and this is $V_{CE} = V_{CC} - I_C R_L$. When the input voltage is zero then the base current is zero and so the transistor is non-conducting (OFF) and $I_C = 0$; the output voltage is then V_{CC}. If now the input voltage is increased then the base current increases and the transistor becomes conducting. There is now a collector current and so a voltage drop across the load resistor. As a consequence of this the collector–emitter voltage drops. When V_{CE} has dropped to a value called the *saturation voltage* $V_{CE(SAT)}$, typically about 0.2 V, it will drop no further and is now said to be ON. The base–emitter voltage required to achieve this condition is about 0.7 V. The output voltage can thus be switched from an OFF, with V_{CC} about 12 V, of 12 V to ON with 0.2 V when the base voltage changes from 0 to about 0.7 V.

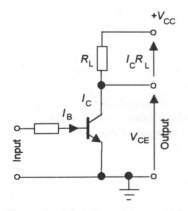

Figure 14.26 *Transistor switch*

We can show the above switching operation on the transistor output characteristic (Figure 14.27). If we have $V_{CC} = 12$ V and $R_L = 2$ kΩ then $V_{CE} = V_{CC} - I_C R_L = 12 - 2000 I_C$. When $V_{CE} = 0$ then the collector current $I_C = 12/2000 = 6$ mA. This load line is shown superimposed on the output characteristic. When the input voltage to the switching circuit is zero there is no base current and so the operating point is A. When the input voltage is such as to give saturation, the operating point moves to B. Any further increase in the base current, above the 120 μA, produces no further increase in collector current. Thus, when the base current is changed from 0 to 120 μA, the collector current changes from its value at A to that at B.

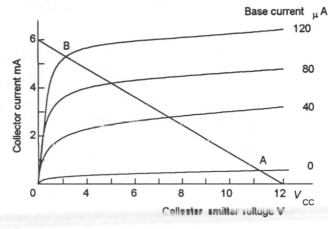

Figure 14.27 *The transistor as a switch*

Such a transistor switch responds very rapidly to changes in input voltage. Figure 14.28 shows the type of response occurring when the input is a rectangular voltage waveform. When this voltage is zero then the transistor is switched OFF and the output is V_{CC}. When the input changes to $+V$, the transistor is ON and the output is $V_{CE(SAT)}$.

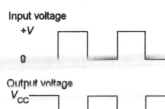

Figure 14.28 *Input and output waveforms*

Example

For the circuit shown in Figure 14.29, with $V_{CE(SAT)} = 0.2$ V, determine the saturated collector current.

$$I_{C(SAT)} = (6 - 0.2)/1000 = 5.8 \text{ mA}$$

14.4.5 Thermal runaway

When the temperature of a semiconductor is increased the number of free charge carriers, electrons and holes, is increased as the increased energy causes more electrons to break free from atoms. As a consequence, an increase in temperature of the collector–base junction of a bipolar transistor results in an increase in the collector leakage current. This increase produces an increase in the power dissipated at the junction and hence to a further increase in temperature. This in turn results in more

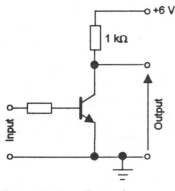

Figure 14.29 *Example*

leakage current and yet further increase in temperature. This can thus lead to thermal runaway and the eventual destruction of the transistor.

The chances of thermal runaway occurring can be reduced by the use of *heat sinks* to conduct heat away from the transistor. These are metal elements in thermal contact with the transistor; they increase the area from which heat can be lost to the surroundings. Heat is removed from the transistor by conduction into the metal and then conducted away from it to the surroundings by convection and radiation. One form of heat sink is a metal clip with the transistor being a push-fit into a hole in the clip. Another form is a corrugated aluminium tube in which the transistor sits. Power transistors are generally supplied on a metal strip which can be mounted in close thermal contact with a metal case.

Activities

1 Use the circuit shown in Figure 14.30 to determine the forward bias characteristic of a junction diode. Then reverse the connections of the diode and obtain the reverse bias characteristic.
2 Determine the static characteristics of a bipolar transistor, e.g. the small signal BC107. Figure 14.31 shows a circuit that can be used and typical values of components.

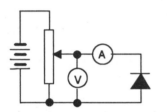

Figure 14.30 *Activity 1*

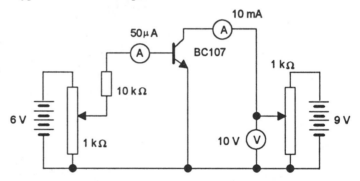

Figure 14.31 *Activity 2*

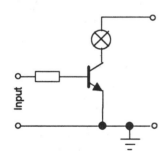

Figure 14.32 *Activity 3*

Adjust the input potentiometer to give a constant base current of about 5 μA, then adjust the output potentiometer to give a collector voltages of 1 V, 2 V, 3 V, etc. and record the collector current at each voltage. Then repeat this with higher base currents and obtain a series of results up to a base current of about 50 μA. Plot the results as graphs of collector current against collector voltage to give the output characteristic. The transfer characteristic can be obtained by using the results to plot base current against collector current.

3 Assemble the lamp switching circuit shown in Figure 14.32, selecting the voltage and resistance values so that the lamp is switched on and off by changing the input voltage.

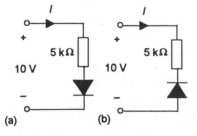

Figure 14.33 *Problem 1*

Problems

1 Determine the current I in the circuits of Figure 14.33 when the voltage is 10 V assuming the diode can be modelled by (i) a switch, (ii) a constant voltage drop of 0.7 V.

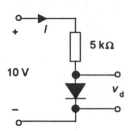

Figure 14.34 *Problem 3*

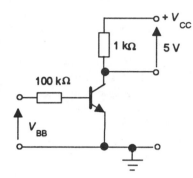

Figure 14.35 *Problem 8*

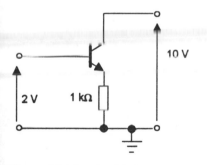

Figure 14.36 *Problem 9*

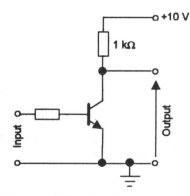

Figure 14.37 *Problem 12*

2 Estimate (a) the d.c. resistance and (b) the a.c. resistance of the diode giving the characteristic shown in Figure 14.13 at a voltage of 0.24 V.

3 For the circuit shown in Figure 14.34, the diode has an a.c. resistance of 10 Ω at the nominal current of 1 mA. What will be the change in the voltage drop across the diode when the supply voltage changes by ±1 V.

4 The following is the current–voltage data for the forward part of the characteristic of a diode. By plotting the characteristic, determine the d.c. and a.c. resistances when the applied voltage is 0.25 V.

Voltage V	0	0.05	0.10	0.15	0.20	0.25	0.30
Current mA	0	0.20	0.40	0.60	4.00	30.00	200

5 Explain the operation of an npn transistor and why the collector current is almost the same as the emitter current.

6 The static collector current for a bipolar transistor is 99.5% of the static emitter current. What is (a) the d.c. common-base gain, (b) the d.c. common-emitter gain?

7 A bipolar transistor has a d.c. current gain of 0.96. What will be the collector current when the emitter current is 9 mA?

8 With the transistor circuit in Figure 14.35, a base current of 50 μA gives a voltage drop of 5 V across the load resistance. What is h_{FB}?

9 Determine the base current, emitter current and collector current for the circuit shown in Figure 14.36 if the transistor has $h_{FB} = 0.98$.

10 A bipolar transistor, connected as common emitter, has a d.c. current gain of 250. What is the emitter current when the collector current is 1 mA?

11 A silicon bipolar transistor has an emitter current of 2.50 mA and a collector current of 2.47 mA. What is the current gain when it is connected as (a) common base, (b) common emitter?

12 For the transistor switching circuit in Figure 14.37, $V_{CE(SAT)}$ is 0.1 V. What is (a) the output voltage when the input voltage is 0 V, (b) the saturated collector current, (c) the output voltage at saturation?

13 A bipolar junction transistor in common-emitter connection gave the following data for a base current of 20 μA. Plot the output characteristic and hence determine the output resistance at this base current for $V_{CE} = 6$ V.

V_{CE} in V	2.0	4.0	6.0	8.0	10.0
I_C in mA	3.5	3.7	3.9	4.1	4.3

14 A bipolar junction transistor in common emitter connection gave the following data for a base current of 80 μA. Plot the output characteristic and hence determine the output resistance at this base current for $V_{CE} = 6$ V.

V_{CE} in V	2.0	4.0	6.0	8.0	10.0
I_C in mA	4.4	4.8	5.2	5.6	6.0

15 Magnetic circuits

15.1 Introduction

This chapter follows on from the discussion of magnetism in Chapter 11 and considers the effect of the materials through which lines of magnetic flux pass. This is important since most devices employing magnetism involve the use of materials such as iron or steel in their construction. In electricity we have electric circuits of conductors through which current can flow; in magnetism we have a similar concept of a circuit of material through which magnetic flux flows. The following is an introduction to the basic principles of magnetic circuits.

15.2 The magnetic circuit

Lines of magnetic flux form closed paths; for example, for the iron ring shown in Figure 15.1 any one line of flux forms a closed path with, in this case, the entire flux path being in iron. The complete closed path followed by magnetic flux is called a *magnetic circuit*. The magnetic flux Φ in a magnetic circuit is analogous to the electric current I in an electric circuit (Figure 15.2), a closed path being necessary for an electric current. The magnetic flux density B is the flux per unit cross-sectional area ($B = \Phi/A$) and is analogous to the current per unit cross-sectional area of conductor (I/A).

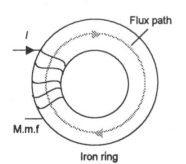

Figure 15.1 *Magnetic circuit*

15.2.1 Magnetomotive force

In an electric circuit an electric current is due to the existence of an electromotive force (e.m.f.) and thus, by analogy, we say that in a magnetic circuit the magnetic flux is due to the existence of a *magnetomotive force* (m.m.f.). A magnetomotive force is produced by a current flowing through one or more turns of wire. When there is more than one turn of wire we consider that the current through each turn generates a m.m.f. and so, for a current I flowing through a coil of N turns, the magnetomotive force (symbol F) is the *total* current linked with the magnetic circuit, namely IN amperes, i.e.

$$\text{m.m.f. } F = IN$$

The unit of m.m.f. is the ampere since the number of turns is dimensionless; however, it is often written as having the unit of ampere-turn.

Example

Calculate the m.m.f. produced by a coil of 400 turns when it carries a current of 4 A.

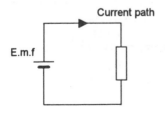

Figure 15.2 *Electrical circuit*

M.m.f. $= IN = 4 \times 400 = 1600$ ampere-turns

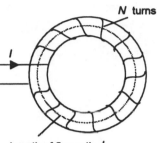

N turns

I

Length of flux path L

Figure 15.3 *Magnetic field strength*

15.2.2 Magnetic field strength

If you connect a length of wire to a source of e.m.f., the current depends on the length of the wire used. Likewise with a magnetic circuit, two magnetic circuits might have the same m.m.f. applied to them but the flux set up in them will be different if the lengths of the flux paths in the two circuits are different. The magnetomotive force per metre length of a magnetic circuit is called the *magnetic field strength H*, unit ampere/metre. Thus, for the magnetic circuit of Figure 15.3, if the length of the flux path is L:

$$H = \text{m.m.f.}/L = IN/L$$

For the electrical circuit, the magnetic field strength is analogous to the p.d. per unit length along a current carrying conductor.

15.2.3 Permeability

With an electric circuit the relationship between the p.d. per unit length of a conductor and the current in the conductor depends on the nature of the material used for the conductor ($R = \rho L/A$ and so $V/I = \rho L/A$ and $V/L = \rho(I/A)$ with ρ being the resistivity). Likewise with a magnetic circuit, the relationship between the magnetic field strength H and the magnetic flux density D depends on the nature of the material used for the magnetic circuit and we write:

$$B = \mu H$$

where μ is the *absolute permeability* of the material and has the unit of tesla/(ampere/metre) which is the same as henry/metre (H/m). It is analogous to the reciprocal of the resistivity of an electrical conductor, i.e. its conductivity.

The *permeability of free space*, i.e. a vacuum μ_0 is the flux density produced in a vacuum divided by the magnetic field strength used to produce it and has the value:

$$\mu_0 = \frac{B}{H} \text{ for a vacuum} = 4\pi \times 10^{-7} \text{ H/m}$$

The value of the permeability in air or in any other non-magnetic material is almost exactly the same as the permeability in a vacuum, i.e. the permeability of free space. Hence, for non-magnetic materials we use:

$$B = \mu_0 H = 4\pi \times 10^{-7} \times H$$

The magnetic flux inside a coil is intensified when an iron core is inserted. The term *relative permeability* μ_r specifies by what factor the flux density in a material is greater than that which would occur in a vacuum for the same magnetic field, i.e. the relative permeability of a material is the ratio of the flux density produced in a material to the flux density produced in a vacuum by the same magnetic field. Hence, for a material having a relative permeability μ_r:

$$B = \mu_r\mu_0 H$$

and so:

absolute permeability $\mu = \mu_r\mu_0$

The relative permeability is the ratio of the absolute permeability to the permeability of free space and is thus a dimensionless quantity. The relative permeability for air is effectively 1. The relative permeability for magnetic materials such as iron is considerably greater than 1. Table 15.1 indicates some typical relative permeability values for magnetic material; it should, however, be noted that the relative permeability of magnetic materials is not constant but depends on the value of the flux density (see later in this chapter).

Table 15.1 *Typical relative permeabilities of magnetic materials*

Magnetic material	Relative permeability
Cast iron	200
Cast steel	700
Mild steel	600
Ferrite (manganese + zinc oxides)	1 500
Silicon iron (iron + 3% silicon)	3 500
Stalloy	4 000
Silicon steel	4 500
Mumetal	6 000
Permalloy (iron + 78.5% nickel)	100 000
Supermalloy (79% nickel, 16% iron, 5% molybdenum)	800 000

Example

A wooden ring, with a mean circumference of 600 mm and a uniform cross-sectional area of 400 mm^2, is wound with a coil of 200 turns. Determine the magnetic field strength, flux density and flux in the ring when there is a current of 4 A in the coil.

The mean circumference can be taken as the length of the flux path and so:

$$H = IN/L = 4 \times 200/0.6 = 1333 \text{ A/m}$$

Because the core is made of wood we can approximate the permeability to be that of free space. Hence:

$$B = \mu_0 H = 4\pi \times 10^{-7} \times 1333 = 0.001\,675\,\text{T} = 1675\,\mu\text{T}$$

This is the flux density over a cross-sectional area of $400 \times 10^{-6}\,\text{m}^2$ and so:

$$\text{flux} = BA = 1675 \times 10^{-6} \times 400 \times 10^{-6} = 0.67\,\mu\text{Wb}$$

Example

An iron ring, with a mean circumference of 600 mm and a uniform cross-sectional area of 400 mm², is wound with a coil of 200 turns. Determine the resulting magnetic field strength, flux density and flux in the ring when there is a current of 4 A in the coil. Take the relative permeability of the iron to be constant at 200.

This is a repeat of the previous example with the wood replaced by iron. The mean circumference can be taken as the length of the flux path and so:

$$H = IN/L = 4 \times 200/0.6 = 1333\,\text{A/m}$$

Because the core is made of iron:

$$B = \mu_r\mu_0 H = 200 \times 4\pi \times 10^{-7} \times 1333 = 0.335\,\text{T}$$

This is the flux density over a cross-sectional area of $400 \times 10^{-6}\,\text{m}^2$ and so:

$$\text{flux} = BA = 0.335 \times 400 \times 10^{-6} = 134\,\mu\text{Wb}$$

15.3 Reluctance

With an electrical circuit (Figure 15.4) having an e.m.f. E giving a current I through a circuit of resistance R, we have the relationship:

$$E = IR$$

We can develop a similar relationship for a magnetic circuit.

Consider a magnetic circuit in the form of an iron ring having a cross-sectional area of A square metres and a mean circumference of L metres (Figure 15.5), wound with N turns carrying a current I amperes:

$$\text{flux } \Phi = \text{flux density} \times \text{area} = BA$$

$$\text{m.m.f. } F = \text{magnetic field strength} \times \text{flux path length} = HL$$

Hence:

$$\frac{\Phi}{\text{m.m.f.}} = \frac{BA}{HL} = \mu_r\mu_0 \times \frac{A}{L}$$

Thus:

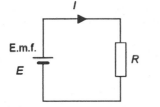

Figure 15.4 *Electrical circuit*

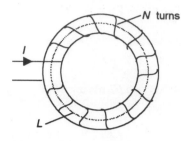

Figure 15.5 *Magnetic circuit*

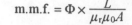

$$\text{m.m.f.} = \Phi \times \frac{L}{\mu_r \mu_0 A}$$

If we write:

$$S = \frac{L}{\mu_r \mu_0 A}$$

then:

$$\text{m.m.f.} = \Phi S$$

This is analogous to $V = IR$ for an electrical circuit and might be termed the 'Ohm's law' for a magnetic circuit. The term ($L/\mu_r\mu_0 A$) is called the *reluctance*, symbol S, and is similar in form to the relationship for the resistance of a conductor in terms of its length and cross-sectional area $R = \rho L/A = L/\sigma A$, where ρ is the resistivity and σ the conductivity. The absolute permeability $\mu_r\mu_0$ thus corresponds to the reciprocal of the resistivity or the conductivity of the conducting material. Reluctance has the unit ampere/weber.

Example

A mild steel ring with a cross-sectional area of 500 mm² and a mean circumference of 400 mm has a coil with 250 turns wound around it. Determine the reluctance of the ring and the current required to produce a flux of 500 μWb in the ring. The relative permeability of mild steel can be assumed to be constant at 600.

$$S = \frac{L}{\mu_r \mu_0 A} = \frac{0.4}{600 \times 4\pi \times 10^{-7} \times 500 \times 10^{-6}}$$

$$= 1.06 \times 10^6 \text{ A/Wb}$$

Hence the required m.m.f. is:

$$\text{m.m.f.} = \Phi S = 500 \times 10^{-6} \times 1.06 \times 10^6 = 530 \text{ A}$$

and so the magnetising current is m.m.f./N = 530/250 = 2.12 A.

15.3.1 Series magnetic circuits

For an electrical circuit with two resistors R_1 and R_2 in series (Figure 15.6), the same current I passes through each resistor and the potential difference across the two is equal to the sum of the potential differences across each resistor, i.e. $V = V_1 + V_2$. Hence we can write $IV = IV_1 + IV_2$ and so the total resistance of the circuit $R = R_1 + R_2$.

Consider the magnetic circuit of Figure 15.7 with the flux path being through an iron core and an air gap. We can consider the iron and the air to be in series if we assume that same flux flows though each, flux leakages being assumed to be negligible. The m.m.f. is the magnetic

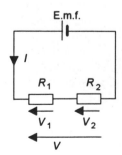

Figure 15.6 *Series circuit*

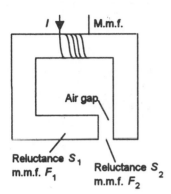

Figure 15.7 *Series circuit*

equivalent of the electrical potential difference and so the m.m.f. F across the two reluctances is the sum of the m.m.f.s across the two, i.e.

$$F = F_1 + F_2$$

The m.m.f. F_1 required for the iron is ΦS_1 and the m.m.f. F_2 required for the air gap is ΦS_2. Thus:

$$F = \Phi S_1 + \Phi S_2 = \Phi(S_1 + S_2)$$

If we replaced the two reluctances by a single reluctance, to give the same flux and total m.m.f., it would need a reluctance $S = F/\Phi$ and so:

reluctance of series composite circuit $S = S_1 + S_2$

The rule to calculate the equivalent reluctances for a series-connected magnetic circuit is the same form as that for series-connected resistors in an electrical circuit.

Consider the effect of introducing an air gap into a magnetic circuit. If, for example, we have an iron circuit with a flux path length in it of 200 mm and it has a relative permeability of 1000, then its reluctance is $L/\mu_r\mu_0 A = 0.0002/\mu_0 A$. Introducing a 1 mm air gap will add a reluctance of $0.001/\mu_0 A$ to give a total reluctance of $0.0012/\mu_0 A$. The effect of introducing the air gap has been to increase the reluctance by a factor of 6. If the air gap had been 2 mm then it would have added a reluctance of $0.002/\mu_0 A$ to give a total reluctance of $0.0022/\mu_0 A$ and so increase the reluctance by a factor of 11. The bigger the reluctance the smaller the flux density produced for a given m.m.f. Thus, air gaps in magnetic circuits have to be kept to the minimum size possible.

Example

A magnetic circuit is shown in Figure 15.8. The cross-sections are rectangular and all sides have the same depth of 100 mm. Three sides of the circuit are of the same uniform width of 150 mm and the fourth side is thinner with a width of 100 mm. A coil of 100 turns is wrapped round the left-hand limb of the circuit and carries a current of 0.5 A. What will be the flux in the circuit if the relative permeability may be assumed to be 2500?

The circuit can be considered to consist of two reluctances S_1 and S_2 in series; one has a flux path length $L_1 = 1300$ mm and cross-sectional area $0.150 \times 0.100 = 0.015$ m² and the other $L_2 = 450$ mm and cross-sectional area $0.100 \times 0.100 = 0.010$ m². The reluctance S_1 is:

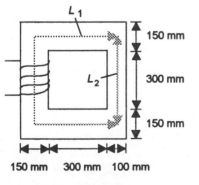

150 mm   300 mm   100 mm

Figure 15.8 *Example*

$$S_1 = \frac{L_1}{\mu_r\mu_0 A_1} = \frac{1.3}{2500 \times 4\pi \times 10^{-7} \times 0.015} = 27\,600 \text{ A/Wb}$$

and the reluctance S_2 is:

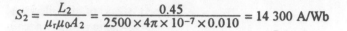

$$S_2 = \frac{L_2}{\mu_r \mu_0 A_2} = \frac{0.45}{2500 \times 4\pi \times 10^{-7} \times 0.010} = 14\,300 \text{ A/Wb}$$

The total reluctance S is thus:

$$S = S_1 + S_2 = 27\,600 + 14\,300 = 41\,900 \text{ A/Wb}$$

The total m.m.f. is $NI = 100 \times 0.5 = 50$ A and so the flux is:

$$\Phi = \frac{F}{S} = \frac{50}{41\,900} = 0.0012 \text{ Wb}$$

15.3.2 Parallel magnetic circuits

For an electrical circuit with two resistors in parallel, the potential difference is the same across each resistor and the circuit current divides so that part goes through one resistor and the remainder through the other. Similar principles apply to parallel magnetic circuits.

Figure 15.9 is an example of a parallel magnetic circuit. The centre limb has a current-carrying coil wound round it and supplies an m.m.f. The flux Φ from the centre limb splits up with part going round the left-hand limb Φ_1 and part round the right-hand limb Φ_2:

$$\Phi = \Phi_1 + \Phi_2$$

The left-hand and the right-hand limbs of the magnetic circuit are in parallel with the same m.m.f. F across them. Thus, for the left-hand limb $F = \Phi_1 S_1$ and for the right-hand limb $F = \Phi_2 S_2$ and so:

$$\Phi = \frac{F}{S_1} + \frac{F}{S_2}$$

Since the total flux Φ divided by the m.m.f. F is the total reluctance S:

$$\frac{1}{S} = \frac{1}{S_1} + \frac{1}{S_2}$$

Reluctances in parallel combine in the same way as resistances in parallel.

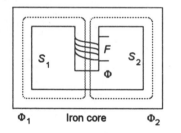

Figure 15.9 *Parallel magnetic circuit*

Example

The magnetic circuit shown in Figure 15.10 has a central limb of cross-sectional area 400 mm² and two equal size outer limbs with cross-sectional area 200 mm². The flux path in the central limb has a length of 60 mm and the length in each outer limb is 150 mm. The air gap has a length of 1 mm. Determine the m.m.f. that must be supplied in the windings on the inner limb to give a flux in the air gap of 0.5 mWb. The permeability of the iron used for the circuit may be assumed to be constant at 800.

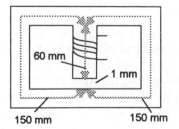

Figure 15.10 *Example*

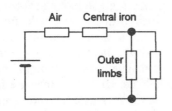

Figure 15.11 *Example*

The two outer limbs are in parallel and in the central limb we have the iron in series with the air. The equivalent electrical circuit is thus as shown in Figure 15.11.

For an outer limb, the reluctance is:

$$S_{\text{outer}} = \frac{0.150}{800 \times 4\pi \times 10^{-7} \times 200 \times 10^{-6}} = 7.46 \times 10^5 \text{ A/Wb}$$

The two outer limbs are in parallel and so the total reactance due to them is:

$$\frac{1}{S_{\text{p}}} = \frac{1}{7.46 \times 10^5} + \frac{1}{7.46 \times 10^5}$$

Hence $S_{\text{p}} = 3.73 \times 10^5$ A/Wb.

The inner limb consists of two reluctances in series. The reluctance of the iron in the central limb is:

$$S_{\text{i}} = \frac{0.060}{800 \times 4\pi \times 10^{-7} \times 400 \times 10^{-6}} = 1.49 \times 10^5 \text{ A/Wb}$$

The reluctance of the air gap, assuming the air can be considered to have the same cross-section for flux as the central limb, is:

$$S_{\text{a}} = \frac{0.001}{4\pi \times 10^{-7} \times 400 \times 10^{-6}} = 19.89 \times 10^5 \text{ A/Wb}$$

The iron and the air are in series and so the total reluctance of the inner part of the circuit is:

$$1.49 \times 10^5 + 19.89 \times 10^5 = 21.38 \times 10^5 \text{ A/Wb}$$

The total reluctance of the circuit is thus:

$$21.38 \times 10^5 + 3.73 \times 10^5 = 25.11 \times 10^5 \text{ A/Wb}$$

The m.m.f. that must be supplied to give a flux of 0.5 mWb is thus $25.11 \times 10^5 \times 0.5 \times 10^{-3} = 1255.5$ ampere turns.

15.3.3 Magnetic flux leakage and fringing

In the above discussion of magnetic circuits it has been assumed that all flux is confined within the magnetic core and that none leaks. This is not quite true, there is some leakage (Figure 15.12). Another assumption that has been made is that the when there are air gaps in the magnetic circuit, the effective cross-sectional area of the air gap is the same as that of the magnetic core. This is not the case, the effective cross-sectional area of the air gap is greater than that of the magnetic material on either side of the gap. This extra area is due, to what is termed, fringing (Figure 15.13).

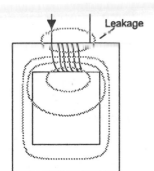

Figure 15.12 *Leakage*

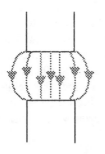

Figure 15.13 *Fringing*

15.4 Magnetisation curves

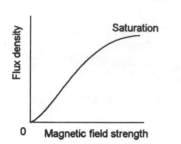

Figure 15.14 *Magnetisation curve*

If a sample of a magnetic material is taken and a graph plotted showing how the flux density B in a material varies with the magnetic field strength H as it is initially magnetised, a graph of the form shown in Figure 15.14 is obtained. such a graph is called a *magnetisation curve*. The flux density is not proportional to the magnetic field strength and there is a tendency for the curve to level off, i.e. no further increase in flux density is obtained by further increasing the magnetic field strength. This is termed *saturation*.

The graphs in Figure 15.15 show typical relationships between the flux density and the magnetic field strength for different types of magnetic materials.

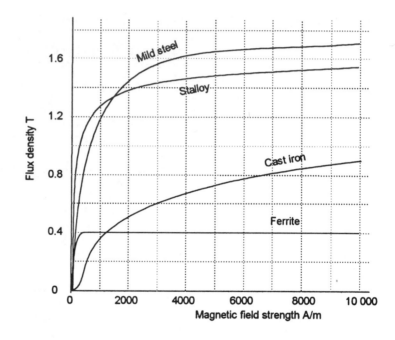

Figure 15.15 *Magnetisation curves*

Because the magnetisation graph is not linear, the permeability of the material, which is B/H, is not constant and its value depends on the magnetic field strength. For example, for the mild steel in Figure 15.15, when H is 500 A/m then B is 0.9 T and so the relative permeability is $B/\mu_0 H = 0.9/(4\pi \times 10^{-7} \times 500) = 1433$. When H is 1000 A/m then B is 1.2 T and so the relative permeability is $B/\mu_0 H = 1.2/(4\pi \times 10^{-7} \times 1000) = 955$. When H is 2000 A/m then B is 1.45 T and the relative permeability $B/\mu_0 H = 1.45/(4\pi \times 10^{-7} \times 2000) = 577$. Figure 15.16 shows how the relative permeability for the mild steel varies with magnetic field strength.

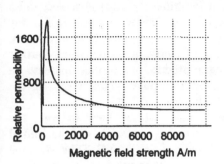

Figure 15.16 *Variation of relative permeability with magnetic field strength for mild steel*

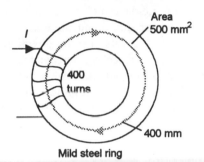

Figure 15.17 *Example*

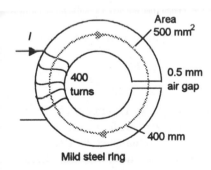

Figure 15.18 *Example*

Example

A mild steel ring has a cross-sectional area of 500 mm² and a mean circumference of 400 mm (Figure 15.17). If a coil of 400 turns is wound around it, calculate the current required to produce a flux of 800 μWb in the ring. Use Figure 15.15 for the data relating flux density and magnetic field strength.

The flux density B in the ring is:

$$B = \frac{\Phi}{A} = \frac{800 \times 10^{-6}}{500 \times 10^{-6}} = 1.6 \text{ T}$$

From Figure 15.15, the magnetic field strength to produce a flux density of 1.6 T in mild steel is approximately 3500 A/m. Therefore:

total m.m.f. required $= HL = 3500 \times 0.400 = 1400$ A

magnetising current $=$ m.m.f.$/N = 1400/400 = 3.5$ A

Example

If the mild steel ring of the previous example has a 0.5 mm wide air gap cut in it (Figure 15.18), calculate the current required to produce a flux density of 1.6 T. Assume that there is no magnetic flux leakage or fringing.

The flux density B in the ring is $800 \times 10^{-6}/(500 \times 10^{-6}) = 1.6$ T. From Figure 15.15, the magnetic field strength to produce a flux density of 1.6 T in mild steel is approximately 3500 A/m. Therefore, for the steel:

m.m.f. required $= HL = 3500 \times 0.3995 = 1398$ A

The magnetic field strength for the air gap is $B/\mu_0 = 1.6/(4\pi \times 10^{-7}) = 1.274 \times 10^6$ A/m. Therefore:

m.m.f. for the air gap $= 1.274 \times 10^6 \times 0.0005 = 637$ A

Hence the total m.m.f. required for the circuit is:

total m.m.f. $= 1398 + 637 = 2035$ A

Hence the current required is 2035/400 = 5.1 A.

15.5 Hysteresis

Figures 15.14 and 15.15 represent the magnetisation of a material when we start off with an initially unmagnetised material. When an initially unmagnetised material (the simplest method of demagnetising so that we start with an unmagnetised specimen is to reverse the magnetising current a large number of times, the maximum value of the current at

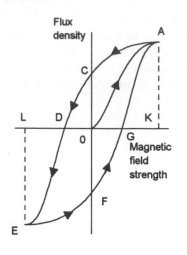

Figure 15.19   *Hysteresis loop*

each reversal being reduced until it is ultimately zero) is subject to an increasing magnetic field strength, the flux density increases in the way shown in Figure 15.14. After a particular magnetising field is reached, the magnetic flux reaches an almost constant value when further increases in magnetic field strength produce no further increase in magnetic flux, this being the saturation point. This stage of the operations is shown by the line 0A in Figure 15.19.

If the magnetic field strength is now reduced back to zero, the material may not simply just retrace its path down the initial magnetisation line from A to 0 but follow the line A to C. Thus when the magnetic field strength is zero, there is still some flux density in the material. The retained flux density 0C is termed the *remanent flux density* or *remanence*. To demagnetise the material, i.e. bring the flux density in it down to a zero value, a reverse direction magnetic field strength 0D has to be applied. This is called the *coercive field* or *coercivity*. If this reverse direction magnetic field strength is still further increased, a reversed direction flux density is produced and saturation in this reverse direction (E) can occur. Reducing this reverse direction magnetic field strength to zero results in reverse direction remanent flux density 0F. Now if the magnetic field strength is increased in its initial direction the material follows the graph line FGA. The resulting closed loop ACDEFGA is termed the *hysteresis loop* (hysteresis is the Greek word for lagging behind). For the loop 0C = 0F, 0G = 0D and 0K = 0L.

15.5.1 Energy loss in hysteresis cycles

The area enclosed by a hysteresis loop is a measure of the energy lost in the material each time the magnetising current goes through a complete cycle. This lost energy appears as heat. Thus a material which has a large area enclosed by its hysteresis loop requires more energy to take it through it magnetising–demagnetising cycle.

15.5.2 Soft and hard magnetic materials

In Figure 15.20 the hysteresis loops are shown for two materials, termed *hard* and *soft magnetic materials*. Compared with a soft magnetic material, a hard magnetic material has high remanence so that a high degree of magnetism is retained in the absence of a magnetic field, a high coercivity so that it is difficult to demagnetise and a large area enclosed by the hysteresis loop and so a large amount of energy is dissipated in the material during each cycle of magnetisation. A soft material is very easily demagnetised, having low coercivity and the hysteresis loop only enclosing a small area.

Hard magnetic materials are used for such applications as permanent magnets while soft magnetic materials are used for transformers where the magnetic material needs to be easily demagnetised and little energy dissipated in magnetising it. A typical soft magnetic material used for a transformer core is an iron–3% silicon alloy. The main materials used for permanent magnets are the iron–cobalt–nickel–aluminium alloys, ferrites and rare earth alloys.

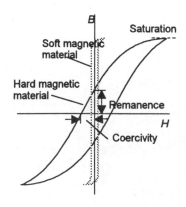

Figure 15.20   *Hysteresis loops for a soft and a hard magnetic material*

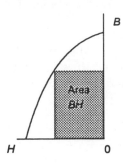

Figure 15.21 *Demagnetisation quadrant*

Table 15.2 gives properties of typical soft magnetic materials and Table 15.3 gives details for hard magnetic materials. For hard magnetic materials an important parameter is the demagnetisation quadrant (Figure 15.21) of the hysteresis loop, it indicates how well a permanent magnet is able to retain its magnetism. The bigger the area the greater the amount of energy needed to demagnetise the material. A measure of this area is given by the largest rectangle which can be drawn in the area, this being the maximum value of the product BH.

Table 15.2 *Soft magnetic materials*

Material	Relative permeability Initial value	Relative permeability Max. value	Remanence T	Coercive field A/m
Silicon steel		90 000		6
Mumetal	60 000		0.5	1
Permendur	300	2 000	1.7	950
Ni–Zn ferrite	20–600			
Mn–Zn ferrite	600–5000			

Table 15.3 *Hard magnetic materials*

Material	Remanence T	Coercive field kA/m	Max. BH T A/m
Alni (Fe,Co,Ni, Al alloy)	0.56	46	10
Alnico (Fe, Co, Ni, Al alloy)	0.72	45	14
6% tungsten steel	1.05	5.2	2.4
6% chromium steel	0.95	5.2	2.4
3% cobalt steel	0.72	10	2.8
Feroba 1 (ferrite)	0.22	135	8

Activities

1 Determine the hysteresis loop for a material. Figure 15.22 shows the traditional circuit that is used to measure the magnetic flux in a ring. The magnetising field is provided by a current through coil A. Another coil B is connected to a ballistic galvanometer; this is a special form of moving coil instrument with virtually no damping. The magnetising current is adjusted to the desired value and then the reversing switch is used to reverse the direction of the current and therefore reverse the flux in the magnetic circuit. An e.m.f. is induced in coil B and the initial galvanometer deflection is proportional to the flux. An alternative to using the ballistic galvanometer is to use a flux meter; this gives a steady reading rather then the need to observe the first swing of the ballistic galvanometer.

If N_A is the number of turns on coil A, L the mean circumference of the ring, I the current through coil A, then the magnetic field strength $H = IN_A/L$. If θ is the initial ballistic galvanometer deflection, or flux meter deflection, when the current through A is reversed, then the change in flux linkage with coil B $= c\theta$, where c

is a constant for the instrument in weber-turns per unit of scale deflection. If the flux changes from Φ to $-\Phi$ when the current through coil P is reversed, and N_B is the number of turns on B, the change of flux linkage with coil B = change of flux × number of turns on B = $2\Phi N_B$. Thus, $\Phi = c\theta/(2N_B)$ and, if A is the cross-sectional area of the ring flux density in ring $B = \Phi/A = c\theta/(2AN_B)$.

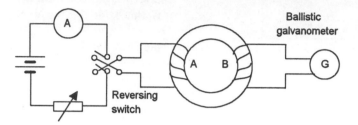

Figure 15.22 *Determination of the hysteresis loop*

Problems

1 Calculate the m.m.f. produced by a coil of 500 turns when it carries a current of 2 A.

2 An iron ring, of constant relative permeability 500, is wound with a coil of 400 turns and a current of 2 A passed through it. If the ring has a mean circumference of 400 mm and a uniform cross-sectional area of 300 mm², determine the resulting magnetic field strength, flux density and flux in the ring.

3 A mild steel ring, of constant relative permeability 600, is wound with a coil of 300 turns. The ring has a mean circumference of 150 mm and a uniform cross-sectional area of 500 mm². Determine the current necessary to produce a flux of 0.5 mWb in the ring.

4 An iron ring has a mean circumference of 250 mm and is wound with 500 turns of wire. A current of 400 mA through the coil produces a flux density of 350 mT in the iron. What is the relative permeability of the iron?

5 A mild steel ring, of relative permeability 500, with a cross-sectional area of 250 mm² and a mean circumference of 200 mm has a coil with 250 turns wound around it. Determine the reluctance of the ring and the current required to produce a flux of 2 mWb in the ring.

6 Determine the reluctance of an air gap in a magnetic circuit if it has a cross-sectional area of 300 mm² and a length of 1 mm.

7 Figure 15.23 shows a magnetic circuit with an iron core, of constant relative permeability 2500, of rectangular cross-section with all sides having the same depth of 100 mm. A coil of 500 turns is wrapped round the left-hand limb of the circuit and carries a current of 1 A. What will be the flux in the circuit?

8 Figure 15.24 shows a magnetic circuit with an iron core, of constant relative permeability 400 and mean length 400 mm, and an air gap of length 0.5 mm. If the circuit has a uniform cross-sectional area of 1200 mm² and the coil has 400 turns, determine the current required to produce a flux density of 0.5 T in the air gap.

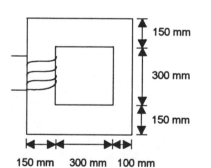

150 mm

300 mm

150 mm

150 mm   300 mm   100 mm

Figure 15.23 *Problem 7*

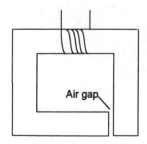

Air gap

Figure 15.24   *Problem 8*

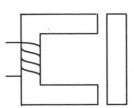

Figure 15.25   *Problem 9*

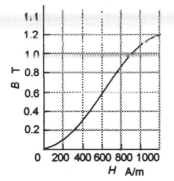

Figure 15.26   *Problem 11*

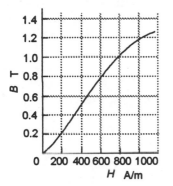

Figure 15.27   *Problem 13*

9 Figure 15.25 shows the magnetic circuit of a relay. The steel has a constant relative permeability of 1250 and a magnetic flux path length of 350 mm. If each of the air gaps has a length of 1.5 mm, determine the flux density produced in the air gap if the coil has 500 turns and a current of 4 A is used. Assume that the magnetic circuit has a uniform cross-sectional area.

10 The following are values of the flux density for various magnetic field intensities for a sample of steel. Calculate the relative permeabilities of the steel at each of the magnetic field intensities.

B in T	0.28	0.72	1.10	1.28
H in A/m	50	100	200	300

11 Figure 15.26 shows the magnetisation curve for a sample of cast steel. What magnetic field intensity will be required to obtain a flux density of 1.0 T in the cast steel?

12 A cast steel ring (magnetisation curve Figure 15.26) has an inner diameter of 120 mm and an external diameter of 200 mm. What m.m.f. is required to produce a flux density of 1.2 T in the core?

13 A cast steel magnetic circuit (magnetisation curve Figure 15.27) has a uniform cross-section with area 1.5×10^{-5} m^2 and a magnetic flux path length of 0.5 m. What current is required in a 200 turn coil wrapped round the core to give a core magnetic flux of 12 µWb?

14 Calculate the relative permeabilities of the steel giving the following data at each of the magnetic field intensities.

B in T	1.1	1.25	1.35	1.45	1.50
H in A/m	500	700	1000	1500	2000

15 A U-shaped electromagnet is made from mild steel (B–H data given in Problem 14) and has a uniform cross-sectional area of 600 mm^2. An armature of the same material, and with the same cross-sectional area, is placed tightly across the ends of the electromagnet. If the total magnetic flux path in the iron is 500 mm, what m.m.f. will need to be applied to generate a flux of 0.75 mWb in the core?

16 An electromagnet has a magnetic flux path in the iron of 300 mm and an air gap of length 2 mm. A coil of 500 turns is wound round it. Determine the current needed to produce a flux density of 0.8 Wb/m^2 in the air gap; neglect fringing. Figure 15.27 gives the B–H data.

16 Inductance

This follows on from Chapter 11 and is concerned with the effect the magnetic flux produced by a changing current in a coil has on the current in the coil itself and also on neighbouring coils. These effects being termed self inductance and mutual inductance.

16.2 Inductance

A current through a coil produces magnetic flux which links the turns of the coil. Thus when the current though the coil changes, the flux linked by that coil changes. Hence an e.m.f. is induced. This phenomenon is known as *self inductance* or just *inductance*.

The induced e.m.f. is proportional to the rate of change of linked flux (Faraday's law). However, the flux produced by a current is proportional to the size of the current. Thus the rate of change of flux will be proportional to the rate of change of the current responsible for it. Hence the induced e.m.f. *e* is proportional to the rate of change of current, i.e. $e \propto$ rate of change of current. Thus we can write $e = L \times$ (rate of change of current), where L is the constant of proportionality, and so, writing dI/dt for the rate of change of current:

General symbol

Inductor with iron core

Figure 16.1 *Inductor*

$$e = L\frac{dI}{dt}$$

where L is called the *inductance* of the circuit. The inductance is said to be 1 henry (H) when the e.m.f. induced is 1 V as a result of the current changing at the rate of 1 A/s. Figure 16.1 shows the circuit symbols used for an inductor, this being a component specifically designed to have inductance.

The effect of inductance on the current in a circuit is that, when the applied voltage is switched on or off, the current does not immediately rise to its maximum value or fall to zero but takes some time. When the voltage is switched on and the current starts to increase from zero, then the changing current results in an induced e.m.f. Thus is in such a direction as to oppose the growing current (Lenz's law) and slow its growth. For this reason, the induced e.m.f. is often referred to as a *back e.m.f.* (Figure 16.2). When the voltage is switched off, then the current starts to fall and so produces an induced e.m.f. This is in such a direction as to oppose the current falling (Lenz's law) and so the current takes longer to fall to zero.

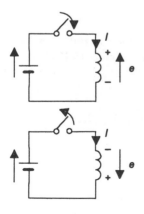

Figure 16.2 *Back e.m.f.*

Example

What is the average back e.m.f. induced in a coil of inductance 500 mH when the current through it is increased from 1.0 A to 3.0 A in 0.05 s?

$$e = L\frac{dI}{dt} = \frac{0.500 \times (3.0 - 1.0)}{0.05} = 20 \text{ V}$$

16.2.1 Inductance of a coil

Suppose we have a coil of N turns for which a current I produces a flux Φ. If the current takes a time t to increase from 0 to I, then the average rate of change of current is I/t and so the average e.m.f. induced in the coil is LI/t. The average rate of change of flux in this time is Φ/t and thus Faraday's law of electromagnetic induction gives the average induced e.m.f. as $N\Phi/t$. Thus $LI/t = N\Phi/t$ and hence:

$$L = \frac{N\Phi}{I}$$

Since the flux density $B = \Phi/A$, where A is the cross-sectional area of the coil, then $L = NBA/I$. Since $B = \mu_0\mu_r H$ (see Section 15.2.3), where μ_0 is the permeability of free space, μ_r is the relative permeability and H the magnetic field strength, and $H = IN/l$ (see Section 15.2.2), where l is the length of the flux path, then we can write:

$$L = N\mu_0\mu_r\left(\frac{IN}{l}\right)\frac{A}{I} = \frac{\mu_0\mu_r N^2 A}{l} = \frac{N^2}{(l/\mu_0\mu_r A)}$$

But the reluctance $S = l/\mu_0\mu_r A$ (see Section 15.3) and so:

$$L = \frac{N^2}{S}$$

Thus an air-cored coil has an inductance which is proportional to the square of the number of turns. The effect of including iron in the core increase the relative permeability and so increases the inductance.

Example

A coil of 1000 turns is wound on a wooden former and a current of 4 A through it produces a magnetic flux of 200 μWb. What is the inductance of the coil?

A wooden former is equivalent to an air-cored coil. Thus:

$$L = \frac{N\Phi}{I} = \frac{1000 \times 200 \times 10^{-6}}{4} = 0.05 \text{ H}$$

Example

What is the inductance of a coil with an air core and 600 turns if the coil has a length of 0.2 m and a cross-sectional area of 600 mm²?

$$L = \mu_0\mu_r N^2 A/l = 4\pi \times 10^{-7} \times 600^2 \times 600 \times 10^{-6}/0.2 = 0.0014 \text{ H}$$

16.2.2 Energy stored in an inductance

When a current is switched through an inductor it grows to a steady value. When the steady current is attained there is a steady magnetic field. Energy is required to set up this magnetic field. During the growth stage there is a rate of change of current di/dt. The back e.m.f. arising from this changing current is $L\ di/dt$ and thus energy is required to overcome this back e.m.f. and maintain the current. The voltage required to overcome the back e.m.f. is $v = L\ di/dt$ and so the power required when the current is i is $vi = i \times L\ di/dt$. Suppose the current increases from 0 to I in a time t. The average value of the current is $\frac{1}{2}I$ and the average rate of change of current with time is I/t. Thus the average power required to obtain a current I is $\frac{1}{2}I \times LI/t = \frac{1}{2}LI^2/t$. The energy stored in the magnetic field in this time is the average power multiplied by the time and so:

energy stored in magnetic field $= \frac{1}{2}LI^2$

Example

Calculate the energy stored in the magnetic field of an inductor with an inductance of 0.2 H when it is carrying a current of 4 A.

Energy stored $= \frac{1}{2}LI^2 = \frac{1}{2} \times 0.2 \times 4^2 = 1.6$ J.

16.3 Mutual inductance

When the current in a coil changes, then changing magnetic flux is produced. If this flux links the turns of a neighbouring coil then an e.m.f. will be induced in it. The two coils are said to possess *mutual inductance*.

If two coils are wound on an iron core (Figure 16.3(a)) then the flux produced by a current through one coil will be concentrated in the core and most of the flux set up by the current in one coil will link with the turns of the other coil. This is the situation that occurs with a transformer. If the coils have air cores (Figure 16.3(b)) then only a small fraction of the flux produced by one coil will link with the turns of the other coil. The fraction of the flux linked between two coils is termed the *coupling coefficient k*. If there is no magnetic coupling of two coils then k is zero; if the coupling is perfect then k is 1. When k is low then coils are said to be loosely coupled, when k is near 1 then tightly coupled.

If the flux set up in the core by a current I_A in coil A is Φ_A, then the flux linking coil B is $k\Phi_A$. Thus if the current in coil A is changing and producing a rate of change of flux of $d\Phi_A/dt$ then the e.m.f. e induced in coil B is:

$$e = -N_B k\frac{d\Phi_A}{dt} = -N_B k\frac{d\Phi_A}{dI_A}\frac{dI_A}{dt} = -M\frac{dI_A}{dt}$$

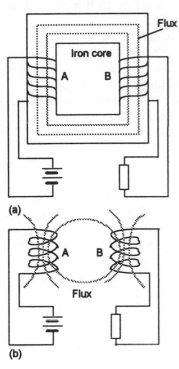

(a)

(b)

Figure 16.3 *Coupling of coils*

where M, termed the *mutual inductance*, is $N_B k \, d\Phi_A/dI_A$. Two coils are said to have a mutual inductance of 1 henry (H) if an e.m.f. of 1 V is induced into one of the coils when the current in the other coil changes at the rate of 1 A/s.

The transformer involves two coils wound on a common core so that, since the two coils are magnetically coupled, a changing current in one coil induces an e.m.f. in the other coil and so an alternating current in one coil gives rise to an alternating current in the other coil. See Section 11.4.1 for a preliminary discussion of the transformer and Chapter 22 for a more detailed discussion.

Example

If the mutual inductance between a pair of coils is 100 mH, what will be the e.m.f. induced in one coil when the rate of change of current in the other coil is 20 A/s?

$$e = -M \, dI_A/dt = -0.100 \times 20 = -2 \text{ V}$$

16.3.1 Relationship between mutual inductance and self inductance

Consider a coil A with N_A turns and self inductance L_A which is coupled with coil B of N_B turns and self inductance L_B. We thus have $L_A = N_A d\Phi_A/dI_A$ and $L_B = N_B d\Phi_B/dI_B$, where Φ_A and Φ_B are the magnetic fluxes in the coils. The mutual inductance M of the coils, when there is a flux change in coil A is $N_B k \, d\Psi_A/dI_A$. Hence we can write:

$$M = N_B k \frac{L_A}{N_A}$$

Similarly, the mutual inductance M of the coils, when there is a flux change in coil B is $N_A k \, d\Phi_B/dI_B$. Hence we can write:

$$M = N_A k \frac{L_B}{N_B}$$

Thus:

$$\frac{N_A}{N_B} = \frac{k L_A}{M} = \frac{M}{k L_B}$$

$$M = k \sqrt{L_A L_B}$$

Example

Determine the value of the coupling coefficient when two coils of inductances 200 mH and 400 mH are coupled and have a mutual inductance of 50 mH.

$$k = \frac{M}{\sqrt{L_A L_B}} = \frac{0.050}{\sqrt{0.200 \times 0.400}} = 0.18$$

Problems

1 What is the inductance of a coil if an e.m.f. of 20 V is induced in it when the current changes at the rate of 10 A/s?

2 What is the e.m.f. induced in a coil of inductance 100 mH when the current through it is changing at the rate of 4 A/s?

3 A coil has 1000 turns and a current of 5 A through it produces a magnetic flux of 50 μWb. What is the inductance of the coil?

4 A coil has 200 turns and a current of 2 A through it produces a magnetic flux of 0.2 mWb. What is the inductance of the coil?

5 An air-cored coil of 500 turns has an inductance of 30 mH. What is the flux produced in the coil by a current of 1.5 A?

6 What is the inductance of an air-cored coil with 1000 turns, cross-sectional area 400 mm² and length 150 mm?

7 A 100 turn coil is wound on an iron rod of diameter 10 mm. If the rod and coil have a length of 40 mm and the relative permeability of the iron is constant at 400, what is its inductance?

8 Calculate the energy stored in the magnetic field of an inductor with an inductance of 400 mH when it is carrying a current of 30 mA.

9 Calculate the energy stored in the magnetic field of an inductor with an inductance of 5 mH when it is carrying a current of 3 A.

10 If the mutual inductance between a pair of coils is 50 mH, what will be the e.m.f. induced in one coil when the rate of change of current in the other coil is 10 A/s?

11 If the mutual inductance between a pair of coils is 100 mH, what will be the e.m.f. induced in one coil when the rate of change of current in the other coil is 5 A/s?

12 What is the mutual inductance of a pair of coils if current changing at the rate of 50 A/s in one coil induces an e.m.f. of 50 mV in the other coil?

13 A pair of coils have a magnetic coupling coefficient of 0.6. What percentage of the flux produced in one coil will link with the other coil?

14 Determine the value of the coupling coefficient when two coils of inductances 20 mH and 60 mH are coupled and have a mutual inductance of 10 mH.

15 The coupling coefficient between two air-cored coils is 0.05, with one having an inductance of 8 mH and the other 12 mH. Determine the voltage induced in the 12 mH coil when the current in the 8 mH coil is changing at the rate of 300 A/s.

16 The coupling coefficient between two air-cored coils is 0.08, with one having an inductance of 8 mH and the other 18 mH. Determine the voltage induced in the 18 mH coil when the current in the 8 mH coil is changing at the rate of 250 A/s.

17 Alternating current

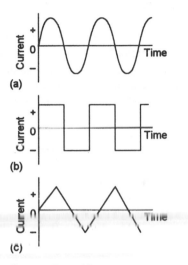

(a)

(b)

(c)

Figure 17.1 *Alternating waveforms*

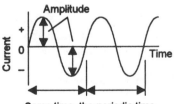

Same time, the periodic time

Figure 17.2 *Terms*

The term *direct* voltage or current is used when the voltage or current is always in the same direction. The term *alternating* voltage or current is used when the polarity or direction of flow of the voltage or current alternates, continually changing with time. Alternating voltages and currents can have many different forms. A particularly important form is one which is in the form of a sine graph, the form shown in Figure 17.1(a). This is because electrical power generation in the entire world is virtually all in the form of such sinusoidal voltages. Consequently the mains electrical supply to houses, offices and factories is sinusoidal. The sinusoidal waveform in the figure has a current which oscillates from positive values to negative values to positive values, to ... and so on.

Figure 17.1(b) and (c) show some further examples of alternating current waveforms: (b) is a rectangular waveform and (c) a triangular waveform. The rectangular waveform in the figure starts with a positive current which then abruptly switches to a negative current, which then abruptly switches to a positive current, which then ... and so on. The triangular waveform shows a similar form of behaviour.

Alternating waveforms oscillate from positive to negative values in a regular, periodic manner. One complete sequence of such an oscillation is called a *cycle* (Figure 17.2). The time T taken for one complete cycle is called the *periodic time* and the number of cycles occurring per second is called the *frequency f*. Thus $f = 1/T$. The unit of frequency is the hertz (Hz), 1 Hz being 1 cycle per second.

This chapter is a consideration of alternating waveforms and the mean values and the root-mean-square values of such alternating waveforms.

Example

The mains voltage supply in Britain has a frequency of 50 Hz. What is the time taken for the voltage to complete one cycle?

Frequency = 1/ (time taken to complete a cycle) and thus the time taken to complete a cycle = 1/frequency = 1/50 = 0.02 s.

17.2 Sinusoidal waveform

We can generate a sinusoidal waveform if we consider the line OA in Figure 17.3 rotating in an anticlockwise direction about O with a constant angular velocity ω so that it rotates through equal angles in equal intervals of time (note that the unit of angular velocity is radians/second).

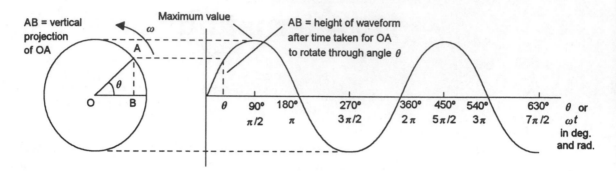

Figure 17.3 *Deriving a sinusoidal waveform*

The line starts from the horizontal position and rotates through an angle θ in a time t. Since AB/OA $= \sin \theta$ we can write:

$$AB = OA \sin \theta$$

where AB is the vertical height of the line at some instant of time, OA being its length. The maximum value of AB will be OA and occur when $\theta = 90°$. But an angular velocity ω means that in a time t the angle θ through which OA has rotated is ωt. Thus the vertical projection AB of the rotating line will vary with time and is described by the equation

$$AB = OA \sin \omega t$$

If we represent an alternating current i by the perpendicular height AB then its maximum value I_m is represented by OA and we can thus write:

$$i = I_m \sin \omega t$$

This is the equation describing the sinusoidal waveform and how its current i at any instant varies with time t. In a similar way we can write for a sinusoidal alternating voltage:

$$v = V_m \sin \omega t$$

One cycle is a rotation of OA through 360° or 2π radians. Rotating at an angular velocity of ω means that the time T taken to complete one cycle is:

$$T = \frac{2\pi}{\omega}$$

The frequency f is $1/T$ and so:

$$\omega = 2\pi f$$

So we can write the equations as:

$$i = I_m \sin 2\pi ft \quad \text{and} \quad v = V_m \sin 2\pi ft$$

Example

A sinusoidal alternating current has a frequency of 50 Hz and a maximum value of 4.0 A. What is the instantaneous value of the current (a) 1 ms, (b) 2 ms after it is at zero current?

We have $i = I_m \sin 2\pi ft = 4.0 \sin 2\pi \times 50t = 4.0 \sin 314t$.
(a) When $t = 1$ ms then $i = 4.0 \sin 314 \times 1 \times 10^{-3} = 4.0 \sin 0.314$. The 0.314 is in units of radians, not degrees. This can be worked out using a calculator operating in the radian mode, the key sequence being change mode to rad, press the keys for 0.314, then for sin, the $\times$ key, the key for 4 and then the = key. Thus $i = 1.24$ A.
(b) When $t = 2$ ms then $i = 4.0 \sin 314 \times 2 \times 10^{-3} = 4.0 \sin 0.628 = 2.35$ A.

Example

A sinusoidal alternating current is represented by $i = 10 \sin 500t$, where i is in mA. What is (a) the size of the maximum current, (b) the angular frequency, (c) the frequency and (d) the current after 1 ms from when it is zero?

(a) The maximum current is 10 mA.
(b) The angular frequency is 500 rad/s.
(c) The frequency is given by the equation $\omega = 2\pi f$ and so $f = \omega/2\pi = 500/2\pi = 79.6$ Hz.
(d) After 0.01 s we have $i = 10 \sin 500 \times 0.01 = 10 \sin 0.50$. This angle is in radians. Thus $i = 4.79$ mA.

17.3 Average value

The *average*, or *mean*, value of a set of numbers is their sum divided by the number of numbers summed. The mean value of some function between specified limits that is described by a graph can be considered to be the mean value of all the ordinates representing the values between these limits. We can give an approximation of this if we divide the area into a number of equal width strips (Figure 17.4); an approximation to the average value is then the sum of all the mid ordinates y_1, y_2, y_3, etc. of the strips divided by the number n of strips considered.

$$\text{average value} = \frac{\text{sum of mid-ordinate values}}{\text{number of mid-ordinates}}$$

Figure 17.5 shows a sinusoidal waveform which has been subdivided into 18 equal width strips. If the instantaneous values of the current for the mid-ordinates of these strips are i_1, i_2, i_3, etc. then the average value over one cycle will be the sum of all these mid-ordinate values divided by the number of values taken. But, because the waveform has a negative

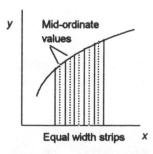

Figure 17.4 *Average value*

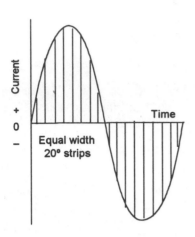

Figure 17.5 *Obtaining the average value*

half cycle which is just the mirror image of the positive half cycle, then the average over a full cycle must be zero. For every positive value there will be a corresponding negative value.

Over one half cycle the average value of the sinusoidal waveform is:

$$I_{av} = \frac{i_1 + i_2 + i_3 + ... + i_9}{9}$$

When the maximum value of the current is 1, readings taken from the graph for the currents (or obtained using a calculator to give the values of the sine at the mid-ordinate values) are:

	i_1	i_2	i_3	i_4	i_5	i_6	i_7	i_8	i_9
Strip	0–20°	20–40°	40–60°	60–80°	80–100°	100–120°	120–140°	140–160°	160–180°
Mid-ordinate	10°	30°	50°	70°	90°	110°	130°	150°	170°
Value	0.17	0.50	0.77	0.94	1.00	0.94	0.77	0.50	0.17

The average of these values is:

$$\frac{0.17 + 0.50 + 0.77 + 0.94 + 1.00 + 0.94 + 0.77 + 0.50 + 0.17}{9}$$

The average is thus 0.64. The accuracy of the average value is improved by taking more values for the average. The average value for a sinusoidal waveform of maximum value I_m over half a cycle is then found to be 0.6371. This is the value of $2/\pi$ and so for a current of maximum value I_m we have:

$$I_{av} = \frac{2I_m}{\pi} = 0.637I_m$$

In a similar way we could have derived the average value for a sinusoidal voltage over half a cycle as:

$$V_{av} = \frac{2V_m}{\pi} = 0.637V_m$$

Example

Determine the average value of the half cycle of the current waveform shown in Figure 17.6.

Dividing the half cycle into segments of width 1 s by the solid lines then the mid-ordinates are indicated by the fainter lines in the figure. The values of these are:

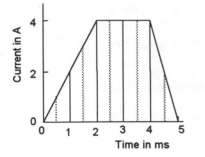

Figure 17.6 *Example*

Mid-ordinate in ms	0.5	1.5	2.5	3.5	4.5
Current in A	1.0	3.0	4.0	4.0	2.0

The average value is thus

$$average = \frac{1.0 + 3.0 + 4.0 + 4.0 + 2.0}{5} = 2.8$$

17.4 Root-mean-square values

Since we are frequently concerned with the power developed by a current passing through a circuit component, a useful measure of an alternating current is in terms of the direct current that would give the same power dissipation in a resistor. For an alternating current, the power at an instant of time is i^2R, where i is the current at that instant and R is the resistance. Thus to obtain the power developed by an alternating current over a cycle, we need to find the average power developed over that time. In terms of mid-ordinates we add together all the values of power given at each mid-ordinate of time in the cycle and divide by the number of mid-ordinates considered. Because we are squaring the current values, negative currents give positive values of power. Thus the powers developed in each half cycle add together.

For the sinusoidal waveform shown in Figure 17.7, the average power P_{av} developed in one complete cycle is the sum of the mid-ordinate powers divided by the number of mid-ordinates used and so:

$$P_{av} = \frac{i_1^2 R + i_2^2 R + i_3^2 R + \dots + i_{18}^2 R}{18} = \frac{i_1^2 + i_2^2 + i_3^2 + \dots + i_{18}^2}{18} R$$

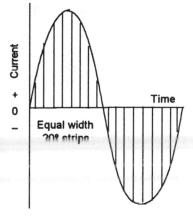

For a direct current I to give the same power as the alternating current, we must have $I^2 R = P_{av}$. Thus the square of the equivalent direct current I^2 is the average (mean) value of the squares of the instantaneous currents during the cycle. This equivalent current is known as the *root-mean-square current* I_{rms}. Hence:

$I_{rms} = \sqrt{}$(mean of the sum of the values of the squares of the alternating current)

Figure 17.7 *Obtaining the average value*

In the above we considered the power due to a current through a resistor. We could have considered the power in terms of the voltage developed across the resistor. The power at an instant when the voltage is v is v^2/R. The equivalent direct voltage V to give the same power is thus when V^2/R is equal to the average power developed during the cycle. Thus, as before, we obtain:

$V_{rms} = \sqrt{}$(mean of the sum of the values of the squares of the alternating voltage)

Consider the use of the of the mid-ordinate rule to obtain values of the root-mean-square current, or voltage, for the sinusoidal waveform given in Figure 17.7. When the maximum value is 1, we have:

	i_1	i_2	i_3	i_4	i_5	i_6	i_7	i_8	i_9
Strip	0–20°	20–40°	40–60°	60–80°	80–100°	100–120°	120–140°	140–160°	160–180°
Mid-ordinate	10°	30°	50°	70°	90°	110°	130°	150°	170°
Value	0.17	0.50	0.77	0.94	1.00	0.94	0.77	0.50	0.17
Square	0.03	0.25	0.59	0.88	1.00	0.88	0.59	0.25	0.03

	i_1	i_2	i_3	i_4	i_5	i_6	i_7	i_8	i_9
Strip	180–200°	200–220°	220–240°	240–260°	260–280°	280–300°	300–320°	320–340°	340–360°
Mid-ordinate	190°	210°	230°	250°	270°	290°	310°	330°	350°
Value	−0.17	−0.50	−0.77	−0.94	−1.00	−0.94	−0.77	−0.50	−0.17
Square	0.03	0.25	0.59	0.88	1.00	0.88	0.59	0.25	0.03

The sum of these values is 9.00 and thus the mean value is 9.00/18 = 0.50. The root-mean-square current is therefore $\sqrt{0.50} = 1/\sqrt{2}$. Thus, with a maximum current of I_m:

$$I_{rms} = \frac{I_m}{\sqrt{2}}$$

Similarly:

$$V_{rms} = \frac{V_m}{\sqrt{2}}$$

where V_m is the maximum voltage.

We could have arrived at the above results for the sinusoidal waveform by considering the form of the graph produced by plotting the square of the current values, or the voltage values. Figure 17.8 shows the graph. The squares of the positive and negative currents are all positive quantities and so the resulting graph oscillates between a maximum value of I_m^2 and 0. The mean value of the i^2 graph is $I_m^2/2$, the i^2 graph being symmetrical about this value. The mean power is thus $R\, I_m^2/2$ and so the root-mean-square current is $I_m/\sqrt{2}$. Thus, the root-mean-square current is the maximum current divided by $\sqrt{2}$.

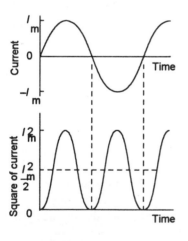

Figure 17.8 *Root-mean-square current*

Example

Determine the root-mean-square value of a current with a peak value of 3 A.

The root-mean-square value = $I_m/\sqrt{2} = 3/\sqrt{2} = 2.1$ A.

17.4.1 Form factor

For alternating currents, or voltages, the relationships between the root-mean-square values and the maximum values depend on the form of the waveform. Figure 17.9 shows some examples of waveforms with their average over half a cycle and root-mean-square values in terms of their maximum values. The ratio of the root-mean-square value to the average value over half a cycle is called the *form factor* and is an indication of the shape of the waveform.

$$\text{form factor} = \frac{\text{rms value}}{\text{average value over half a cycle}}$$

Wave shape	Average	RMS	Form factor
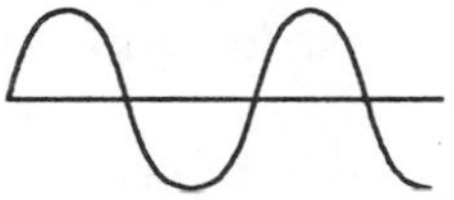 Time	$0.637\ V_m$	$0.707\ V_m$	1.11
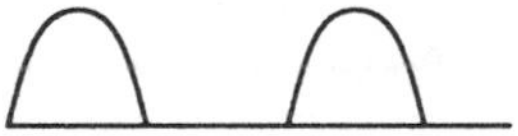 Time	$0.319\ V_m$	$0.500\ V_m$	1.57
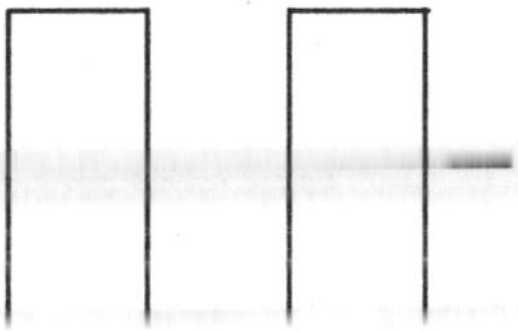 Time	V_m	V_m	1.00
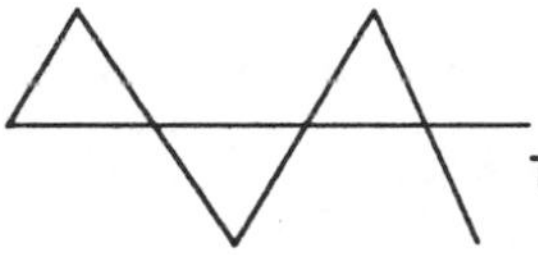 Time	$0.500\ V_m$	$0.577\ V_m$	1.15

Figure 17.9 *Form factors*

Example

Using the mid-ordinate rule the following current values were obtained for the half cycle of a waveform, the other half cycle being just a mirror image. Estimate (a) the average value over half a cycle, (b) the root-mean-square value, (c) the form factor.

Current values in A   2   2   2   2   2   4   4   4

(a) $I_{av} = \dfrac{2+2+2+2+2+4+4+4}{8} = 2.75\ \text{A}$

(b) $I_{rms} = \sqrt{\dfrac{4+4+4+4+4+16+16+16}{8}} = 2.92\ \text{A}$

(c) form factor $= \dfrac{\text{rms value}}{\text{average value}} = \dfrac{2.92}{2.75} = 1.06$

17.5 Basic measurements

For alternating currents and voltages, the moving coil meter can be used with a rectifier circuit. Typically the ranges with alternating currents vary from about 10 mA to 10 A with accuracies up to ± 1% of full scale deflection for frequencies in the region 50 Hz to 10 kHz. With alternating voltages, the ranges are typically from about 3 V to 3 kV with similar accuracies. Multi-meters give a number of direct and alternating current and voltage ranges, together with resistance ranges. A typical meter has full scale deflections for d.c. ranges from 50 µA to 10 A, a.c. from 10 mA to 10 A, direct voltages from 100 mV to 3 kV, alternating voltages 3 V to 3 kV and for resistance 2 kΩ to 20 MΩ. The accuracy for d.c. ranges is about ± 1% of full scale deflection, for a.c. ± 2% of full scale deflection and for resistance ± 3% of the mid-scale reading.

17.5.1 The cathode ray oscilloscope

The *cathode ray oscilloscope* (CRO) is essentially a voltmeter which displays a voltage as the movement of a spot of light on a fluorescent screen and consists of a cathode ray tube with associated circuits. The cathode ray tube (Figure 17.10) consists of an electron gun which produces a beam of electrons which is focused onto a fluorescent screen where it produces a small glowing spot of light. This beam can be deflected in the vertical direction (the Y-direction) and the horizontal direction (the X-direction) by voltages applied to the Y and X inputs, the amount of deflection being proportional to the applied voltage.

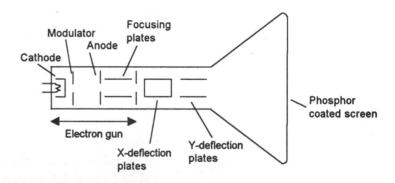

Figure 17.10 *Cathode ray tube*

1   *Brightness*
    The electrons are produced by heating the cathode. The number of electrons passing down the tube per second, and hence the brightness of the spot on the screen, is controlled by a potential applied to the modulator. The more negative it is the more it repels electrons and prevents them passing down the tube.

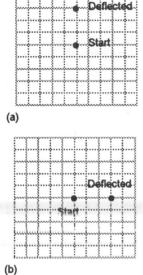

(a)

(b)

Figure 17.11 *Beam deflections in (a) Y and (b) X directions*

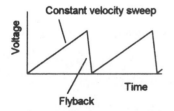

Figure 17.12 *Time-base signal*

2 *Focussing*

The electrons are attracted down the tube as a result of the anode being positive with respect to the cathode. The focusing plates have their potentials adjusted so that the electrons in the beam are made to converge to a spot on the screen. The screen fluoresces when hit by electrons and is marked with a grid so that the position of the spot formed by the beam, and any movement, can be detected.

3 *Deflection*

When a potential difference is applied to the Y-deflection plates it produces an electric field between the plates which causes the beam of electrons to be deflected in the Y-direction. When a potential difference is applied to the X-deflection plates it produces an electric field between the plates which causes the beam of electrons to be deflected in the X-direction. The amount by which the electron beam is deflected is proportional to the potential difference between the plates. A switched attenuator and amplifier enable different ranges of voltage signals to be applied to the Y-input. A general purpose oscilloscope is likely to have calibrated ranges which vary between 5 mV per scale division to 20 V per grid scale division. In order that alternating voltages components of signals can be viewed in the presence of high direct voltages, a blocking capacitor can be switched into the input line. Also, so that the vertical position of the beam can be altered, an internally supplied potential difference can be supplied to the Y-plates.

Thus with a voltage applied to the Y-deflection plates we might move the spot on the screen in the way shown in Figure 17.11(a), with the voltage applied to the X-deflection plates the movement might be in the way shown in Figure 17.11(b). The screen is marked with a grid of centimetre squares to enable such deflections to be easily measured. Thus in (a), if each square corresponds to a voltage of 1 V then the Y-input was 3 V, in (b) if each square corresponded to, say, 2 V then the X-input was 6 V.

4 *Time base*

The purpose of the X-deflection plates is to deflect the beam of electrons in the horizontal direction. They are generally supplied with an internally generated signal which sweeps the beam from left to right across the screen with a constant velocity, then very rapidly returns the beam back (called flyback) to the left side of the screen again (Figure 17.12). The constant velocity means that the distance moved in the X-direction is proportional to the time elapsed, hence the X-direction can be taken as a time axis, i.e. a so-called *time base*. Thus with, say, an alternating voltage applied to the Y-deflection plates and a time base signal to the X-deflection plates, the screen displays a graph of input voltage against time. With a sinusoidal alternating voltage, and a suitable time base signal so that it takes just the right time to move across the screen and get back to the start on the left to always start at the same point on the voltage wave, the screen might look like that shown in Figure 17.13. From measurements of the displacement in the Y-direction to give the

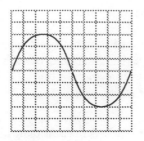

Figure 17.13 *Alternating voltage displayed*

maximum displacement from the zero axis we can determine the value of the maximum voltage. By using a calibrated time base so that we know the time taken to cover each centimetre in the horizontal direction, we can determine the time taken to complete one cycle of the alternating voltage.

A general purpose oscilloscope will have times bases ranging from about 1 s per scale division to 0.2 μs per scale division. Also, so that the horizontal start position of the beam can be altered, an internally supplied potential difference can be supplied to the X-plates.

5 *Trigger*
For a periodic input signal to give rise to a steady trace on the screen it is necessary to synchronise the time base and the input signal using the *trigger circuit* so that the movement of the beam from left to right across the screen always starts at the same point on the signal waveform. The trigger circuit can be set so that its produces a pulse to start the time base sweep across the screen when a particular Y-input voltage is reached and also whether it is increasing or decreasing.

Figure 17.14 shows the controls likely to be found on a very basic cathode ray oscilloscope. The basic procedures to be adopted in order to operate such an oscilloscope are:

1 Brightness control off. Focus control midway. Y-shift control midway. AC/DC switch set to DC. Y volts/cm control set to 1 V/cm. Time base control set to 1 ms/cm.

2 Plug into the mains socket. Switch on using the brightness control. Wait for the oscilloscope to warm up, perhaps a minute. Move the brightness control to full on. A bright trace should appear across the screen. If it does not, the trace may be off the edge of the screen. Try adjusting the Y-shift to see if it can be brought on to the screen.

3 Centre the trace using the Y-shift control. Reduce the brightness to an acceptable level and use the focus to give a sharp trace.

4 Switch the time base control to the required range for the signal concerned. Apply the input voltage. Select a suitable number of volts/cm so that the trace occupies a reasonable portion of the screen and measurements can be made.

Example

A cathode ray oscilloscope is being used to determine the value of a direct voltage. With the Y volts/cm control set at 5 V/cm, connection of the voltage to the Y-input results in the spot on the screen being displaced through 3.5 cm. What is the applied voltage?

Since the spot displaces by 1 cm for every 5 V, a displacement of 3.5 cm means a voltage of $3.5 \times 5 = 17.5$ V.

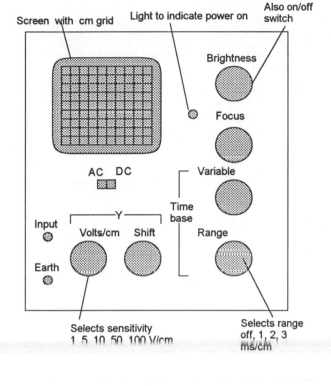

Screen with cm grid   Light to indicate power on   Also on/off switch

Brightness

Focus

AC   DC

Variable

Time base

Input

Earth

—Y—

Volts/cm   Shift

Range

Selects sensitivity
1, 5, 10, 50, 100 V/cm

Selects range
off, 1, 2, 3
ms/cm

1   *Brightness*
This control is also generally used as an off/on switch. It adjusts the brightness of the spot on the screen.

2   *Focus*
This adjusts the electron gun so that is produces a focused, i.e. sharp, spot or trace on the screen.

3   *Time base*
This has two controls. With the variable knob set to off, the calibrated range switch can be used to select the time taken for the spot to move per cm of screen in the horizontal direction from left to right. The variable knob can be used to finely tune the time. Also, when the time base is switched off, the variable control can be used to shift the spot across the screen in the X-direction.

4   *Y-controls*
This has two controls. The Y-shift is to move the spot on the screen in a vertical direction and is used for centring the spot. The volts/cm control is used to adjust the sensitivity of the Y-displacement to the input voltage. With it set, for example, at 1 V/cm then 1 V is needed for each centimetre displacement.

5   *AC/DC switch*
With switch set to d.c. the oscilloscope will respond to both d.c. and a.c. signals. With the switch set to a.c. a blocking capacitor is inserted in the input line to block off all d.c. signals. Thus if the input was a mixture of d.c. and a.c. the deflection on the screen will be only for the a.c. element.

6   *Y-input*
The Y-input is connected to the oscilloscope via two terminals, the lower terminal being earthed.

Figure 17.14 *Controls on a basic cathode ray oscilloscope*

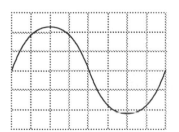

Figure 17.15 *Example*

Example

Figure 17.15 shows the screen seen when an alternating voltage is applied to the Y-input with the Y attenuator set at 1 V/scale division and the time base at 0.1 ms/ scale division. Determine the maximum value of the voltage and its frequency.

The alternating voltage oscillates between $+V_m$ and $-V_m$ about the central axis. The trace on the screen is of a wave which oscillates with a displacement which varies from +2.25 cm to −2.25 cm about the central line through the wave. Thus, since the control is set at 1 V/cm, the maximum voltage is 2.25 V. The number of horizontal divisions in one cycle is 8 and thus corresponds to a time of 0.8 ms. Hence the frequency, which is the reciprocal of this time, is 1/0.8 = 1.25 kHz.

Activities   1   Use a cathode ray oscilloscope to determine the amplitude and frequency of a given alternating voltage.

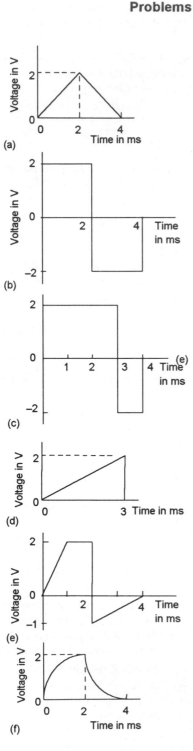

(a)

(b)

(c)

(d)

(e)

(f)

Figure 17.16   *Problem 5*

Problems

1 A sinusoidal voltage has a maximum value of 10 V and a frequency of 50 Hz. (a) Write an equation describing how the voltage varies with time. (b) Determine the voltages after times from $t = 0$ of (i) 0.002 s, (ii) 0.006 s and (iii) 0.012s.

2 A sinusoidal current has a maximum value of 50 mA and a frequency of 2 kHz. (a) Write an equation describing how the current varies with time. (b) Determine the currents after times from $t = 0$ of (i) 0.4 ms, (ii) 0.8 ms, (iii) 1.6 ms.

3 For a sinusoidal voltage described by $v = 10 \sin 1000t$ volts, what will be (a) the value of the voltage at time $t = 0$, (b) the maximum value of the voltage, (c) the voltage after 0.2 ms?

4 Complete the following table for the voltage $v = 1 \sin 100t$ volts.

t in ms	0	2	4	6	8	10	12	14	16
v in V									

5 Determine the average values of the segments of waveforms shown in Figure 17.16.

6 Show that for a triangular waveform, of the form of the full cycle shown in Figure 17.16(a), with a maximum value of V_m that over half the cycle shown the average value is $0.5V_m$.

7 A sinusoidal alternating current has a maximum value of 2 A. What is the average value over (a) half a cycle, (b) a full cycle?

8 A rectangular shaped alternating voltage has a value of 5 V for half a cycle and −5 V for the other half. What is the average value over (a) half a cycle, (b) a full cycle?

9 Show that the average value for a square waveform, as in Figure 4.16(b), over half a cycle is equal to the maximum value.

10 Show that the average value for a sawtooth waveform over half a cycle, as in Figure 17.16(d), is half the maximum value.

11 A sinusoidal alternating current has a maximum value of 4 V. What is the root-mean-square value?

12 Determine the root-mean-square current for a triangular waveform which gave the following mid-ordinate values over a full cycle:

current in mA   5 15 25 15 5 −5 −15 −25 −15 −5

13 Determine the root-mean-square voltage for an irregular waveform which gave the following mid-ordinate values over a full cycle:

voltage in V   5 10 12 8 2 −5 −10 −12 −8 −2

14 Show that the root-mean-square value of a triangular waveform of the form shown in Figure 17.16(a) and having a maximum value of V_m is given by $V_m/\sqrt{3}$.

15 Determine the average value, the root-mean-square value and the form factor of the waveform giving Figure 17.17.

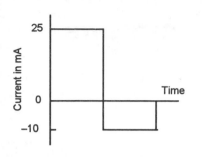

Figure 17.17 *Problem 15*

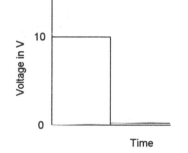

Figure 17.18 *Problem 21*

16 A rectangular shaped alternating current has a maximum value of 4 A for half a cycle and −4 A for the other half. What is the root-mean-square value?

17 A half cycle of a waveform gave the following voltage values for the mid-ordinates, the other half cycle being a mirror image. What are (a) the half cycle average value, (b) the root-mean-square value, and (c) the form factor?

Voltages in V 1 3 5 7 9 9 9 9

18 A half cycle of a waveform gave the following current values for the mid-ordinates, the other half cycle being a mirror image. What are (a) the half cycle average value, (b) the root-mean-square value, and (c) the form factor?

Current in mA   5   15   25   25   15   5

19 Show that the root-mean-square value for a square waveform, as in Figure 17.16(b), over half a cycle is equal to the maximum value.

20 Show that the root-mean-square value for a sawtooth waveform over half a cycle, as in Figure 17.16(d), is the maximum value divided by $\sqrt{3}$.

21 Determine the average value, the root-mean-square value and the form factor for the voltage with the waveform shown in Figure 17.18

22 Determine the root-mean-square value of a square wave voltage waveform, as in Figure 17.9, with a maximum value of 1 V.

23 Determine the root-mean-square value of a sinusoidal voltage waveform with a maximum value of (a) 2 V, (b) 10 V.

24 Determine the root-mean-square value of a sinusoidal current waveform with a maximum value of (a) 10 mA, (b) 3 A.

25 Determine the maximum value of a sinusoidal current waveform with a root-mean-square value of (a) 100 mA, (b) 4 A.

26 Determine the maximum value of a square wave voltage waveform with a root-mean-square value of 10 V.

27 Explain how the following controls on a cathode ray oscilloscope work: (a) brightness, (b) vertical position, (c) time base, (d) trigger.

28 The trace on the screen of a cathode ray oscilloscope is deflected 2.5 scale divisions upwards when a direct voltage signal is applied to the Y-input. If the Y attenuator is set at 0.1 V/scale division, what is the size of the voltage?

29 When a sinusoidal alternating voltage is applied to the Y-input of a cathode ray oscilloscope, the signal on the screen has a peak-to-peak separation of 4.5 screen divisions. If the Y attenuator is set at 5 V/scale division, what is the maximum voltage?

30 When a sinusoidal alternating voltage is applied to the Y-input of a cathode ray oscilloscope, the signal on the screen has a distance of 6.4 scale divisions for one cycle. If the time base is set at 1 ms/scale division, what is the frequency of the voltage?

18 Series a.c. circuits

18.1 Introduction

This chapter follows on from Chapter 17 and considers the characteristics and behaviour of resistors, capacitors and inductors in single phase a.c. series circuits. Phasors are introduced and used to simplify the analysis. The frequency dependent behaviour of capacitors and inductors is determined and series circuits involving resistors, capacitors and inductor circuits analysed. Power dissipation in circuits is analysed and the effects of phase angle, true power, apparent power and power factor considered, together with their significance in electrical engineering.

Half-wave and full-wave rectifier circuits, with the associated smoothing circuits, are also discussed.

18.2 Sine waves and phasors

We can generate a sinusoidal waveform by rotating a line OA in an anticlockwise direction about O with a constant angular velocity ω (Figure 18.1). With the line starting from the horizontal position and rotating through an angle θ in a time t, the vertical height AB = OA sin θ. The maximum value of AB will be OA and occur when θ = 90°. An angular velocity ω means that in a time t the angle θ covered is ωt. Thus AB = OA sin ωt. If we represent an alternating current i by the perpendicular height AB then its maximum value I_m is represented by OA and we can write:

$$i = I_m \sin \omega t$$

In a similar way we can write for a sinusoidal alternating voltage:

$$v = V_m \sin \omega t$$

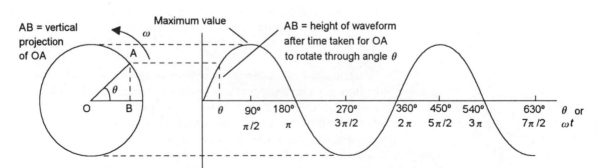

Figure 18.1 *Generating a sinusoidal waveform*

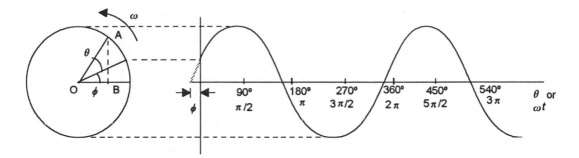

Figure 18.2 *Waveform not starting from a zero value*

The frequency f is $1/T$ and so $\omega = 2\pi f$. Thus the above equations can be written as:

$$i = I_m \sin 2\pi ft \text{ and } v = V_m \sin 2\pi ft$$

In Figure 18.1 the rotating line OA was shown as starting from the horizontal position at time $t = 0$. But we could have an alternating voltage or current starting from some value other than 0 at $t = 0$. Figure 18.2 shows such a situation. At $t = 0$ the line OA is already at some angle ϕ. As the line OA rotates with an angular velocity ω then in a time t the angle swept out is ωt and thus at time t the angle with respect to the horizontal is $\omega t + \phi$. Thus we have:

$$i = I_m \sin(\omega t + \phi) \text{ and } v = V_m \sin(\omega t + \phi)$$

In discussing alternating current circuits we often have to consider the relationship between an alternating current through a component and the alternating voltage across it. If, for a series circuit, we take the alternating current as the reference and consider it to be represented by OA being horizontal at time $t = 0$, then the voltage may have some value at that time and so be represented by another line OB at some angle ϕ at $t = 0$ (Figure 18.3). There is said to be a *phase difference* of ϕ between the current and the voltage. If ϕ has a positive value then the voltage is said to be *leading* the current (as in Figure 18.3), if a negative value then *lagging* the current.

It is thus possible to describe a sinusoidal alternating current by equations of the type given above or by just specifying the rotating line in terms of its length and its initial angle relative to a horizontal reference line. The term *phasor*, being an abbreviation of the term phase vector, is used for such lines. The length of the phasor can represent the maximum value of the sinusoidal waveform or the root-mean-square value, since the maximum value is proportional to the root-mean-square value.

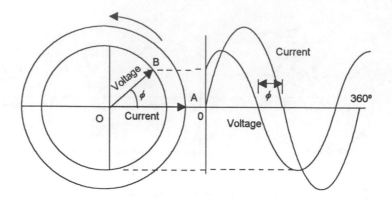

Figure 18.3   *Voltage leading the current by ϕ with $i = I_m \sin \omega t$ and $v = V_m \sin (\omega t + \phi)$*

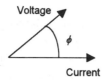

Figure 18.4   *Phasor diagram*

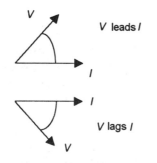

Figure 18.5   *Leading and lagging*

Currents and voltages in the same circuit will have the same frequency and thus the phasors used to represent them will rotate with the same angular velocity and maintain the same phase angles between them at all times; they have zero motion relative to one another. For this reason, we do not need to bother about drawing the effects of their rotation but can draw phasor diagrams giving the relative angular positions of the phasors as though they were stationary. Figure 18.4 shows the phasor diagram for Figure 18.3.

The following summarise the main points about phasors:

1   A phasor has a length that is directly proportional to the maximum value of the sinusoidally alternating quantity or, because the maximum value is proportional to the root-mean-square value, a length proportional to the r.m.s. value.

2   Phasors are taken to rotate anticlockwise and have an arrow-head at the end which rotates.

3   The angle between two phasors shows the phase angle between their waveforms. The phasor which is at a larger anticlockwise angle is said to be leading, the one at the lesser anticlockwise angle lagging (Figure 18.5).

4   The horizontal line is taken as the reference axis and one of the phasors given that direction, the others have their phase angles given relative to this reference axis.

Note that, in textbooks, it is common practice where we are concerned with just the size of a phasor to represent it using italic script, e.g. V, but where we are referring to a phasor quantity with both its size and phase we use bold non-italic text, e.g. **V**. Thus we might say – phasor **V** has size of V and a phase angle of ϕ.

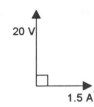

Figure 18.6   *Example*

18.3 *R, L, C* in a.c. circuits

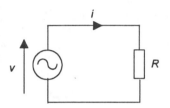

Figure 18.7   *Circuit with only resistance*

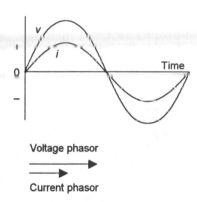

Voltage phasor

Current phasor

Figure 18.8   *Current and voltage with a pure resistor*

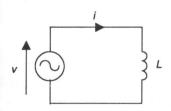

Figure 18.9   *Circuit with only inductance*

Example

Draw the phasor diagram to represent the voltage and current in a circuit where the current is described by $i = 1.5 \sin \omega t$ A and the voltage by $v = 20 \sin (\omega t + \pi/2)$ V.

Figure 18.6 shows the phasors with their lengths proportional to the maximum values of 1.5 A and 20 V.

In the following discussion the behaviour of resistors, inductors and capacitors are considered when each individually is in an a.c. circuit.

18.3.1 Resistance in a.c. circuits

Consider a sinusoidal current:

$$i = I_m \sin \omega t$$

passing through a *pure resistance* (Figure 18.7). A pure resistance is one that has only resistance and no inductance or capacitance. Since we can assume Ohm's law to apply, then the voltage v across the resistance must be $v = Ri$ and so:

$$v = RI_m \sin \omega t$$

The current and the voltage are thus in phase (Figure 18.8). The maximum voltage will be when $\sin \omega t = 1$ and so $V_m = RI_m$.

18.3.2 Inductance in a.c. circuits

Consider a sinusoidal current $i = I_m \sin \omega t$ passing through a *pure inductance* (Figure 18.9). Figure 18.10(a) shows how the current varies with time. A pure inductance is one which has only inductance and no resistance or capacitance. With an inductance a changing current produces a back e.m.f. of $-L \times$ the rate of change of current, i.e. $L \, di/dt$ (see Section 16.2), where L is the inductance.

The gradient of the graph in (a) gives the rate of change of current and thus (b) shows how the rate of change of current varies with time; the rate of change of current is zero when the current is at a maximum and a maximum when the current is zero. The back e.m.f. is $-L \times$ the rate of change of current and so is given by (c). The applied e.m.f. must overcome this back e.m.f. for a current to flow. Thus the voltage v across the inductance is given by (d), i.e. it is $L \, di/dt$. As graphs (a) and (d) show, the current and the voltage are out of phase with the voltage leading the current by 90°. Figure 18.10(e) shows the phasors.

We can thus obtain the graph of the voltage by considering, as above, the gradients of the current–time graph or by differentiating the current equation to give:

$$v = L\frac{di}{dt} = L\frac{d}{dt}(I_m \sin \omega t) = \omega L I_m \cos \omega t$$

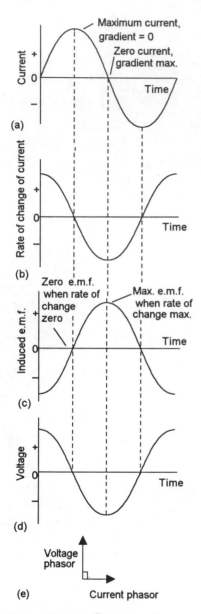

(a)

(b)

(c)

(d)

(e)

Figure 18.10 *Current and voltage with pure inductancee*

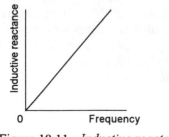

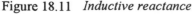

Figure 18.11 *Inductive reactance*

Since $\cos \omega t = \sin (\omega t + 90°)$, the current and the voltage are out of phase with the voltage leading the current by 90°.

The maximum voltage is when $\cos \omega t = 1$ and so we have $V_m = \omega L I_m$. V_m/I_m is called the *inductive reactance* X_L. Thus:

$$X_L = \frac{V_m}{I_m} = \omega L$$

Note that the maximum values do not occur at the same time. The reactance has the unit of ohms. The reactance is a measure of the opposition to the current. The bigger the reactance the greater the voltage has to be to drive the current through it.

Since the root-mean-square current is the maximum current divided by the square root of 2 and the root-mean-square voltage is the maximum voltage divided by the square root of 2, the inductive reactance can also be written as:

$$X_L = \frac{V_m}{I_m} = \frac{\sqrt{2}\,V_{rms}}{\sqrt{2}\,I_{rms}} = \frac{V_{rms}}{I_{rms}} = \omega L$$

Since $\omega = 2\pi f$ then:

$$X_L = 2\pi f L$$

Thus the reactance is proportional to the frequency f. The higher the frequency the greater the opposition to the current (Figure 18.11). With d.c., i.e. $f = 0$, there is zero reactance and so the inductor acts as a short circuit.

Example

The alternating current in milliamperes passing through an inductor which has only inductance of 200 mH is $i = 50 \sin 2000t$. Derive the equation for the voltage across the inductor.

Since the inductive reactance $X_L = \omega L = 2000 \times 0.200 = 400\ \Omega$, the maximum voltage must be $V_m = X_L I_m = 400 \times 50 \times 10^{-3} = 20$ V. Hence, since the voltage will lead the current by 90° we have:

$$v = 20 \sin (2000t + 90°)\ \text{V or } 20 \cos 2000t\ \text{V}$$

18.3.3 Capacitance in a.c. circuits

Consider a circuit having just *pure capacitance* (Figure 18.12) with a sinusoidal voltage $v = V_m \sin \omega t$ being applied across it (Figure 18.13(a)). A pure capacitance is one which has only capacitance and no resistance or inductance.

The charge q on the plates of a capacitor is related to the voltage v by $q = Cv$ (see Section 13.2.1). Thus, since current is the rate of movement of charge dq/dt, we have:

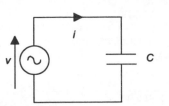

Figure 18.12 *Circuit with only capacitance*

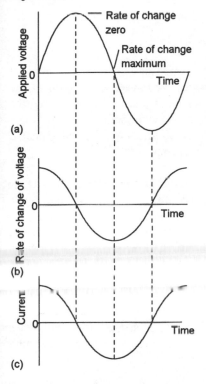

(a)

(b)

(c)

Figure 18.13 *Voltage and current with pure capacitance*

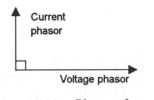

Figure 18.14 *Phasors for pure capacitance*

i = rate of change of q = rate of change of (Cv)

$= C \times$ (rate of change of v)

i.e. $i = C\, dv/dt$. The current is proportional to the rate of change of voltage with time and is thus proportional to the gradient of the voltage–time graph. Figure 18.13(b) how the gradient of voltage–time graph varies with time. Thus Figure 18.13(c) shows how the current varies with time. As the graphs indicate, the current leads the voltage by 90° and Figure 18.14 shows the phasors.

Alternatively, instead of determining the relationship between the voltage and the current graphically, we can use calculus. Since current is the rate of change of charge q:

$$i = \frac{dq}{dt} = \frac{d}{dt}(Cv) = C\frac{d}{dt}(V_m \sin \omega t) = \omega C V_m \cos \omega t$$

Since $\cos \omega t = \sin (\omega t + 90°)$, the current and the voltage are out of phase, the current leading the voltage by 90°.

The maximum current occurs when $\cos \omega t = 1$ and so $I_m = \omega C V_m$. V_m/I_m is called the *capacitive reactance* X_C. Thus:

$$X_C = \frac{V_m}{I_m} = \frac{1}{\omega C}$$

The reactance has the unit of ohms. Since $\omega = 2\pi f$ then $X_C = 1/2\pi f C$ and since $V_m = \sqrt{2}V_{r.m.s.}$ and $I_m = \sqrt{2}I_{r.m.s.}$ then:

$$X_C = \frac{V_m}{I_m} = \frac{V_{r.m.s}}{I_{r.m.s.}} = \frac{1}{\omega C} = \frac{1}{2\pi f C}$$

The reactance is a measure of the opposition to the current. The bigger the reactance the greater the voltage has to be to drive the current through it. The reactance is inversely proportional to the frequency f and so the higher the frequency the smaller the opposition to the current. Figure 18.15 shows how the reactance varies with frequency. With d.c., i.e. $f = 0$, the reactance is infinite and so no current flows. A capacitor can thus be used to block d.c. current.

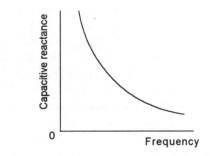

Figure 18.15 *Effect of frequency on capacitive reactance*

Example

Determine the reactance of a 220 pF capacitor at a frequency of 5 kHz and hence the root-mean-square current through it when the root-mean-square voltage across it is 4 V.

$$X_C = \frac{1}{\omega C} = \frac{1}{2\pi f C} = \frac{1}{2\pi \times 5 \times 10^3 \times 220 \times 10^{-12}} = 145 \text{ k}\Omega$$

$V_{rms} = V_m/\sqrt{2}$ and $I_{rms} = I_m/\sqrt{2}$, hence $V_{rms}/I_{rms} = V_m/I_m = X_C$. Thus:

$$I_{rms} = \frac{V_{rms}}{X_C} = \frac{4}{145 \times 10^3} = 0.028 \text{ mA}$$

18.4 Components in series

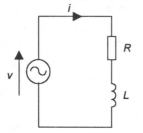

Figure 18.16 *Resistance and inductance in series*

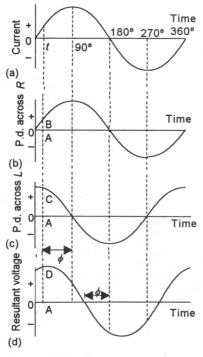

Figure 18.17 *Resistance and inductance in series*

Suppose we want to add the potential differences across two components in series. If they are alternating voltages we must take account of the possibility that the two voltages may not be in phase, despite having the same frequency since they are supplied by the same source.

Consider a circuit having a resistance R in series with an inductance L (Figure 18.16) and supplied with a current $i = I_m \sin \omega t$ (Figure 18.17(a)).

The p.d. across R will be in phase with the current (Section 18.3.1), as in Figures. 18.17(a) and (b). The voltage applied to an inductance leads the current by 90° (Section 18.3.2), as in Figures 18.17(a) and (c).

Because R and L are in series, at any instant the resultant voltage is the sum of the p.d.s across R and L. Thus at time t when the p.d. across R is AB and that across L is AC, the total applied voltage is AB + AC = AD. By adding together the two curves for the p.d.s in this way, we can derive the curve representing the resultant voltage across R and L. As the result of such addition (Figure 18.17(d)) shows, the resultant voltage attains its maximum positive value ϕ° before the current does and passes through zero ϕ° before the current passes through zero in the same direction; the resultant voltage leads current by ϕ.

18.4.1 Adding phasors

For a series circuit, the total voltage is the sum of the p.d.s across the series components, though the p.d.s may differ in phase. This means that if we consider the phasors, they will rotate with the same angular velocity but may have different lengths and start with a phase angle between them. As illustrated above, we can obtain the sum of two series voltages by adding the two voltage graphs, point-by-point, to obtain the resulting voltage. However, exactly the same result is obtained by using the *parallelogram law* of vectors the add the two phasors.

If two phasors are represented in size and direction by adjacent sides of a parallelogram, then the diagonal of that parallelogram is the sum of the two.

The procedure for adding two phasors is thus:

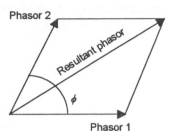

Figure 18.18   *Adding phasors*

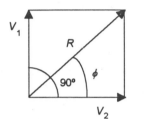

Figure 18.19   *Adding phasors*

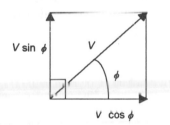

Figure 18.20   *Resolving a phasor*

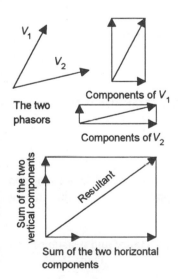

Figure 18.21   *Adding phasors*

1   Draw the phasors as the two adjacent sides of a parallelogram (Figure 18.18).

2   Complete the parallelogram and draw the diagonal from the same origin as the phasors being added. The diagonal represents in magnitude and direction the resultant phasor.

If the phase angle between the two phasors of sizes V_1 and V_2 is 90° (Figure 18.19), then the resultant can be calculated by the use of the Pythagoras theorem as having a size V given by $V^2 = V_1^2 + V_2^2$ and is at a phase angle ϕ relative to the phasor for V_2 given by $\tan \phi = V_2/V_1$.

When phasors are not at right angles to each other, the simplest procedure is generally to first resolve the phasors into their *horizontal and vertical components*. Resolving a phasor into its components (Figure 18.20) is the reverse procedure of adding two phasors to give a single resultant; two phasors at right angles to each other are determined which, when added, gives the original phasor. The horizontal component of the phasor V is $V \cos \phi$ and the vertical component is $V \sin \phi$. Thus, for the addition of two phasors V_1 at phase angle ϕ_1 and V_2 at phase angle ϕ_2, the two horizontal components are $V_1 \cos \phi_1$ and $V_2 \cos \phi_2$ and the two vertical components are $V_1 \sin \phi_1$ and $V_2 \sin \phi_2$ (Figure 18.21). The sum of the horizontal components is $V_1 \cos \phi_1 + V_2 \cos \phi_2$ and the sum of the vertical components is $V_1 \sin \phi_1 + V_2 \sin \phi_2$. We have now replaced the two original phasors by these two phasors at right angles to each other and can obtain the resultant phasor using Pythagoras.

18.4.2 Resistance and inductance in series

Consider a circuit having resistance and inductance in series (Figure 18.22(a)). For such a circuit, the voltage for the resistance is in phase with the current and the voltage for the inductor leads the current by 90° (Figure 18.22(b)). Thus the phasor for the sum of the voltage drops across the two series components is given by Figure 18.22(c) as a voltage phasor with a phase angle ϕ. We can use the Pythagoras theorem to give the magnitude V of the voltage:

$$V^2 = V_R^2 + V_L^2$$

and trigonometry to give the phase angle ϕ, i.e. the angle by which the voltage leads the current (this is in the direction of $\mathbf{V_R}$):

$$\tan \phi = \frac{V_L}{V_R}$$

$$\cos \phi = \frac{V_R}{V}$$

Since $V_R = IR$ and $V_L = IX_L$:

$$V^2 = (IR)^2 + (IX_L)^2 = I^2(R^2 + X_L^2)$$

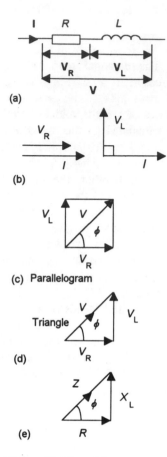

(a)

(b)

(c) Parallelogram

(d)

(e)

Figure 18.22 *RL series circuit*

The term *impedance Z* is used for the opposition of a circuit to the flow of current, being defined as $Z = V/I$ with the unit of ohms. Thus, for the resistance and inductance in series, the circuit impedance is given by:

$$Z = \sqrt{R^2 + X_L^2} = \sqrt{R^2 + (\omega L)^2}$$

If we consider half of the parallelogram, i.e. the voltage triangle shown in Figure 18.22(d), then multiplying each of the sides by I gives the *impedance triangle* with sides of lengths IR, IX_L and IZ, as shown in Figure 18.22(e). The values of the impedance and the phase angle can be determined from the impedance triangle by the use of the Pythagoras theorem and trigonometry:

$$Z^2 = R^2 + X_L^2$$

$\tan \phi = X_L/R$, $\sin \phi = X_L/Z$, and $\cos \phi = R/Z$.

Example

In a series *RL* circuit, the resistance is 10 Ω and the inductance 50 mH. Determine the value of the current and its phase angle with respect to the voltage if a 10 V r.m.s., 50 Hz supply is connected to the circuit.

The inductive reactance is $X_L = 2\pi f L = 2\pi \times 50 \times 0.05 = 15.7$ Ω. Thus:

$$Z = \sqrt{R^2 + X_L^2} = \sqrt{10^2 + 15.7^2} = 18.6 \ \Omega$$

$$\phi = \tan^{-1}\frac{X_L}{R} = \tan^{-1}\frac{15.7}{10} = 57.5°$$

Hence the magnitude of the current I is given by:

$$I = \frac{V}{Z} = \frac{10}{18.6} = 0.54 \ A$$

and, since the current is in the direction of the phasor V_R, its phase angle is 57.5° lagging behind the applied voltage V.

Example

A coil of resistance 30 Ω and inductive reactance 40 Ω is connected across a 240 V r.m.s. alternating supply. Determine the current taken from the supply:

A single coil will generally have both inductance and resistance and we can consider these two elements to be in series. Thus:

$$Z = \sqrt{R^2 + X_L^2} = \sqrt{30^2 + 40^2} = 50 \ \Omega$$

Thus the current $I = V/Z = 240/50 = 4.8$ A.

18.4.4 Resistance and capacitance in series

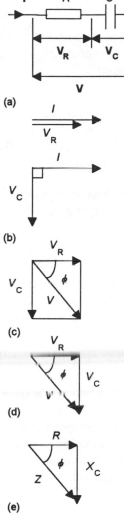

(a)

(b)

(c)

(d)

(e)

Figure 18.23 *Series RC circuit*

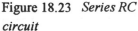

Consider a circuit having resistance and capacitance in series (Figure 18.23(a)). For such a circuit, the voltage across the resistance is in phase with the current and the voltage across the capacitor lags the current by 90° (Figure 18.23(b)). Thus the phasor for the sum of the voltage drops across the two series components is given by Figure 18.23(c) as a voltage phasor with a phase angle ϕ. We can use the Pythagoras theorem to give the magnitude V of the voltage:

$$V^2 = V_R^2 + V_C^2$$

and trigonometry to give the phase angle ϕ, i.e. the angle by which the current leads the voltage:

$$\tan \phi = \frac{V_C}{V_R}$$

$$\cos \phi = \frac{V_R}{V}$$

Since $V_R = IR$ and $V_C = IX_C$:

$$V^2 = (IR)^2 + (IX_C)^2 = I^2(R^2 + X_C^2)$$

The impedance Z is V/I and thus:

$$Z = \sqrt{R^2 + X_C^2} = \sqrt{R^2 + (1/\omega C)^2}$$

If we consider half of the parallelogram, i.e. the voltage triangle shown in Figure 18.23(d), then multiplying each of the sides by I gives the *impedance triangle* with sides of lengths IR, IX_L and IZ, as shown in Figure 18.23(e). This gives:

$$Z^2 = R^2 + X_C^2$$

$\tan \phi = X_C/R$, $\sin \phi = X_C/Z$ and $\cos \phi = R/Z$.

Example

In a series RC circuit the resistance is 20 Ω and the capacitance 50 μF. Determine the value of the current and its phase angle with respect to the voltage if a 240 V r.m.s., 50 Hz supply is connected to the circuit.

The capacitive reactance is $1/2\pi fC = 1/(2\pi \times 50 \times 50 \times 10^{-6}) = 63.7$ Ω. Thus the impedance triangle gives:

$$Z = \sqrt{R^2 + X_C^2} = \sqrt{20^2 + 63.7^2} = 66.8 \ \Omega$$

$$\phi = \tan^{-1} \frac{X_C}{R} = \tan^{-1} \frac{63.7}{20} = 72.6°$$

Hence the magnitude of the current I is given by $I = V/Z = 240/66.8$ = 3.59 A and, since the current is in the direction of the phasor $\mathbf{V_R}$, its phase angle is 72.6° leading the applied voltage $\mathbf{V}$.

Example

A capacitor of reactance 3 kΩ is connected in series with a resistor of 4 kΩ. An alternating voltage of 20 V r.m.s. is applied to the circuit. Determine the circuit current magnitude and phase.

The impedance $Z = \sqrt{(3^2 + 4^2)} = 5$ kΩ and the phase angle ϕ is given by $\tan \phi = X_C/R = 3/4$ and so $\phi = 36.9°$. The current is thus 20/5 = 4 mA and lags the voltage by 36.9°.

18.4.5 *RCL* series circuit

For such a circuit (Figure 18.24), the voltage across the resistance is in phase with the current, the voltage across the capacitor lags the current by 90° and the voltage across the inductor leads the current by 90°. There are three different operating conditions: when $V_L > V_C$, i.e. $X_L > X_C$; when $V_L < V_C$, i.e. $X_L < X_C$; and when $V_L = V_C$, i.e. $X_L = X_C$.

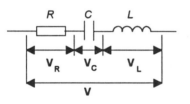

Figure 18.24   *Series RCL*

1   For $V_L > V_C$, i.e. $X_L > X_C$ (Figure 18.25)
Because the voltage phasors for the inductor and capacitor are in opposite directions (Figure 18.25(a)) we can subtract them to give a phasor for the voltage drop across the inductor and capacitor of size $V_L - V_C$. We can then use the parallelogram relationship of phasors to obtain the resultant voltage (Figure 18.25(b)). The circuit behaves like a resistance in series with an inductance, the voltage across the series arrangement leading the current by ϕ:

$$Z^2 = R^2 + (X_L - X_C)^2$$

$$\tan \phi = \frac{X_L - X_C}{R}$$

$$\cos \phi = \frac{R}{Z}$$

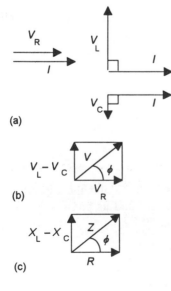

Figure 18.25 $V_L > V_C$

2   For $V_L < V_C$, i.e. $X_L < X_C$ (Figure 18.26)
Because the voltage phasors for the inductor and capacitor are in opposite directions (Figure 18.26(a)) we can subtract them to give a phasor for the voltage drop across the inductor and capacitor of size $V_C - V_L$. We can then use the parallelogram relationship of phasors to obtain the resultant voltage (Figure 18.26(b)). As a result the circuit behaves like a resistance in series with capacitance, the

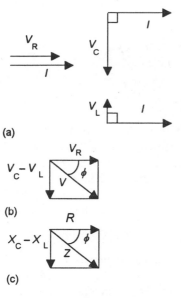

(a)

(b)

(c)

Figure 18.26 $V_L < V_C$

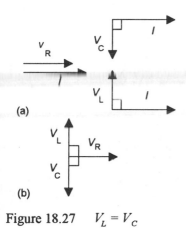

(a)

(b)

Figure 18.27 $V_L = V_C$

voltage across the series arrangement lagging behind the current by phase ϕ:

$$Z^2 = R^2 + (X_C - X_L)^2$$

$$\tan \phi = \frac{X_C - X_L}{R}$$

$$\cos \phi = \frac{R}{Z}$$

3   For $V_L = V_C$, i.e. $X_L = X_C$ (Figure 18.27)
The voltage phasor for the capacitor is equal in magnitude to the voltage phasor for the inductor but in exactly the opposite direction (Figure 18.27(a)). The two voltage phasors when added cancel each (Figure 18.27(b)) other out. The result is that the circuit behaves as though it was just the resistance with the impedance $Z = R$ and phase $\phi = 0°$.

Example

In a series RLC circuit the resistance is 10 Ω, the inductance 60 mH and the capacitance 300 µF. Determine the value of the current and its phase angle with respect to the voltage if a 24 V r.m.s., 50 Hz supply is connected to the circuit.

The inductive reactance is $X_L = 2\pi fL = 2\pi \times 50 \times 0.060 = 18.8 \, \Omega$ and the capacitive reactance is $X_C = 1/(2\pi fC) = 1/(2\pi \times 50 \times 300 \times 10^{-6}) = 10.6 \, \Omega$. Since the inductive reactance is greater than the capacitive reactance the situation is like that shown in Figure 18.25. The circuit impedance Z is:

$$Z = \sqrt{R^2 + (X_L - X_C)^2} = \sqrt{10^2 + (18.8 - 10.6)^2} = 12.9 \, \Omega$$

$$\phi = \tan^{-1} \frac{X_L - X_C}{R} = \tan^{-1} \frac{18.8 - 10.6}{10} = 39.4°$$

The circuit current has a magnitude $I = V/Z = 24/12.9 = 1.86$ A and the current, in phase with V_R, lags the voltage by 39.4°.

18.5 Series resonance Inductive reactance is given by $X_L = 2\pi fL$ and so increases as the frequency f increases (Figure 18.28(a)). Capacitive reactance $X_C = 1/2\pi fC$ and so decreases as the frequency f increases (Figure 18.28(b)). Resistance does not vary with frequency (Figure 18.28(c)). Thus for a series RLC circuit connected to a variable frequency voltage supply, there will be a frequency f_0 at which we have $X_L = X_C$ and so:

$$2\pi f_0 L = \frac{1}{2\pi f_0 C}$$

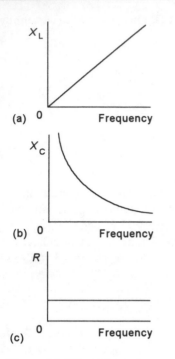

(a)

(b)

(c)

Figure 18.28 *Effect of changes in frequency*

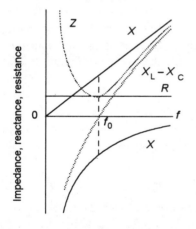

Figure 18.29 *Resonance*

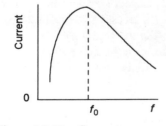

Figure 18.30 *Current*

$$f_0 = \frac{1}{2\pi\sqrt{LC}}$$

This frequency is known as the *resonant frequency*.

The impedance Z of a RCL circuit is given by:

$$Z = \sqrt{R^2 + (X_L \sim X_C)^2}$$

The 'curly minus' sign is used to indicate that it is the difference between the two reactances we are concerned with and it is either $X_L - X_C$ or $X_C - X_L$ depending which reactance is the greater. Figure 18.29 shows how the impedance varies with frequency. When $X_L = X_C$, i.e. at the resonant frequency, the impedance is just R. At all other frequencies the impedance is greater than R. The circuit thus has a minimum impedance at the resonant frequency.

Since the circuit current $I = V/Z$, the consequence of the circuit impedance being a minimum at the resonant frequency f_0 is that the circuit current is at its maximum value, being V/R (Figure 18.30). At frequencies much lower than the resonant frequency and much higher, the impedance is high and so the current is very low.

Example

A circuit, of a resistance of 4 Ω, an inductance of 0.5 H and a variable capacitance in series, is connected across a 10 V, 50 Hz supply. Determine (a) the capacitance to give resonance and (b) the voltages across the inductance and the capacitance.

(a) For resonance the inductive reactance equals the capacitive reactance and thus:

$$2\pi f L = \frac{1}{2\pi f C}$$

$$C = \frac{1}{4\pi^2 f^2 L} = \frac{1}{4\pi^2 \times 50^2 \times 0.5^2} = 20.3 \times 10^{-6} \text{ F} = 20.3 \ \mu\text{F}$$

(b) At resonance the impedance equals the resistance of 4 Ω. Thus:

$$I = \frac{V}{R} = \frac{10}{4} = 2.5 \text{ A}$$

$$V_L = IX_L = 2.5 \times 2\pi \times 50 \times 0.5 = 392.5 \text{ V}$$

Since we have $X_C = X_L$, the p.d. across the capacitance equals the p.d. across the inductance and so is also 392.5 V.

18.5.1 Q-factor

The voltage across the inductance at resonance is IX_L and thus can be very large, likewise the voltage across the capacitor IX_C. A factor known as the *Q-factor* or 'quality factor' is used to indicate the voltage magnification across either the inductor or capacitor at resonance compared with the voltage across the resistance:

$$Q\text{-factor} = \frac{\text{voltage across } L \text{ or } C \text{ at resonance}}{\text{voltage across } R \text{ at resonance}}$$

$$= \frac{IX_L}{IR} = \frac{2\pi f_0 L}{R}$$

Since $f_0 = 1/2\pi\sqrt{(LC)}$ we can write:

$$Q\text{-factor} = \frac{1}{R}\sqrt{\frac{L}{C}}$$

Example

A series RLC circuit has a resonant frequency of 50 Hz, a resistance of 20 Ω, an inductance of 300 mH, capacitance and a supply voltage of 24 V. Calculate (a) the value of the capacitance, (b) the circuit current at resonance, (c) the values of the voltages across each component and (d) the Q-factor of the circuit.

(a) The resonant frequency $f_0 = 1/2\pi\sqrt{(LC)}$ and so:

$$C = \frac{1}{4\pi^2 f_0^2 L} = \frac{1}{4\pi^2 \times 50^2 \times 0.300} = 33.8 \ \mu\text{F}$$

(b) At resonance $I = V/R = 24/20 = 1.2$ A.

(c) At resonance, the voltage across the resistor will be the entire supply voltage of 24 V. The inductive reactance $X_L = 2\pi f L = 2\pi \times 50 \times 0.300 = 94.2 \ \Omega$. At resonance $X_L = X_C$ and so the capacitive reactance is also 94.2 Ω. Hence the value of the voltage across the inductor and across the capacitor is $IX = 1.2 \times 94.2 = 113.0$ V.

(d) The Q-factor is voltage across L or C divided by the voltage across R and so is $113.0/24 = 4.7$.

18.6 Subtracting phasors

There can be situations where we need to subtract phasors, e.g. we might know the voltage across a series combination of two elements and the voltage across one of them and need to find the voltage across the other. The procedure for subtracting one phasor (phasor 2) from another (phasor 1) is:

1   For the phasor to be subtracted (phasor 2), reverse its direction to give –phasor 2.

2   Draw the parallelogram with the reversed direction phasor (–phasor 2) and the other phasor (phasor 1) as the two adjacent sides of a parallelogram (Figure 18.31).

3   Add phasor 1 and –phasor 2 by completing the parallelogram and drawing the diagonal from the same origin as the phasors being

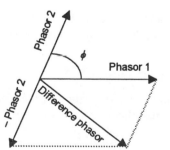

Figure 18.31 *Subtracting phasors*

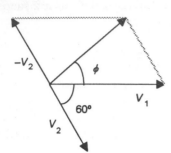

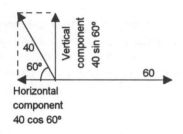

Figure 18.32   *Example*

Figure 18.33   *Example*

added. The diagonal represents in magnitude and direction the difference phasor.

Example

The instantaneous values of two alternating voltages are represented by $v_1 = 60 \sin \theta$ volts and $v_2 = 40 \sin (\theta - \pi/3)$ volts. Determine the difference of these voltages.

Figure 18.32 shows the phasors $\mathbf{V_1}$ and $\mathbf{V_2}$ for the two voltages. To subtract $\mathbf{V_2}$ from $\mathbf{V_1}$ we reverse the direction of $\mathbf{V_2}$ and then add $-\mathbf{V_2}$ to $\mathbf{V_1}$. We can add these by the use of a scale diagram of the phasors. Alternatively, we can use calculation if we resolve the phasors into their horizontal and vertical components and determine the resultants in these two directions (Figure 18.33). The resultant horizontal component is $60 - 20 = 40$ V and the resultant vertical component is 34.64 V. Therefore:

$$\text{maximum value of the difference voltage}$$
$$= \sqrt{40^2 + 34.64^2} = 52.9 \text{ V}$$

$$\tan \phi = \frac{DC}{OD} = \frac{34.64}{40} = 0.866$$

Thus $\phi = 40.9° = 0.714$ radian and so the instantaneous value of the difference voltage $= 52.9 \sin (\theta + 9.714)$ V.

18.7 Power in a.c. circuits

The power developed in a d.c. circuit is the product of the current and the resistance. With an a.c. circuit the instantaneous power p is the product of the instantaneous current i and instantaneous voltage v:

$$p = iv$$

Since the current and voltage are changing with time, the product varies. For this reason we quote the average power dissipated over a cycle.

18.7.1 Purely resistive circuit

For a purely resistive circuit the current and voltage are in phase and we have $i = I_m \sin \omega t$ and $v = V_m \sin \omega t$. Hence the instantaneous power is:

$$p = I_m V_m \sin^2 \omega t$$

Figure 18.34 shows the graphs of i, v and p. By inspection of the graph of p varying with time we might deduce that the average value of p over one cycle is:

$$\text{average power } P = \tfrac{1}{2} I_m V_m$$

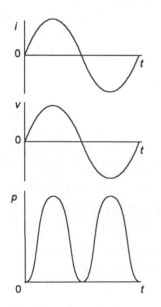

Figure 18.34   *Power in a resistive circuit*

Alternatively, we can use the trigonometric relation $\cos 2\theta = 1 - 2\sin^2\theta$ to write this as:

$$p = I_m V_m \tfrac{1}{2}(1 - \cos 2\omega t) = \tfrac{1}{2}I_m V_m - \tfrac{1}{2}I_m V_m \cos 2\omega t$$

The average value of a cosine function over one cycle is zero. Hence the average power $P = \tfrac{1}{2}I_m V_m$.

The r.m.s. current I is $I_m/\sqrt{2}$ and the r.m.s. voltage V is $V_m/\sqrt{2}$. Hence we can write the equation for the average power dissipated with a purely resistive circuit as:

$$\text{average power } P = IV$$

18.7.2 Purely inductive circuit

Consider the power dissipated in a purely inductive circuit. At any instant the power is iv. But the voltage leads the current by 90°. Thus we have $i = I_m \sin \omega t$ and $v = V_m \sin(\omega t - 90°)$. The power p is thus $p = I_m \sin \omega t \, V_m \sin(\omega t - 90°)$. Figure 18.35 shows the current and voltage graphs. We can obtain the power at an instant of time by multiplying the current and voltage values at that time. Repeating these for each value of time gives the power graph shown.

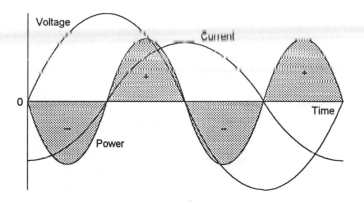

Figure 18.35 *Power with a purely inductive circuit*

An alternative to graphically determining the power is to use the trigonometric relationship $2\sin A \sin B = \cos(A - B) - \cos(A + B)$ then:

$$2\sin \omega t \sin(\omega t - 90°) = \cos(90°) - \cos(2\omega t - 90°) = 0 - \sin 2\omega t$$

Hence:

$$\text{power} = vi = V_m \sin \omega t \times I_m \sin(\omega t - 90°) = -\tfrac{1}{2}V_m I_m \sin 2\omega t$$

The power alternates about the zero axis with a frequency which is twice that of the current or voltage. Over one cycle the average value is thus zero. This is because in that part of the cycle where the current is positive, energy is being stored in the magnetic field of the inductor, while in that part of the cycle where the current is negative the magnetic field is releasing its energy into the circuit.

18.7.3 Purely capacitive circuit

For a pure capacitance, the current leads the voltage by 90°, i.e. we have $i = I_m \sin (\omega t + 90°)$ and $v = V_m \sin \omega t$ and so the power is given by $p = I_m \sin (\omega t + 90°) V_m \sin \omega t$. Figure 18.36 shows the current and voltage graphs. We can obtain the power at an instant of time by multiplying the current and voltage values at that time. Repeating these for each value of time gives the power graph shown.

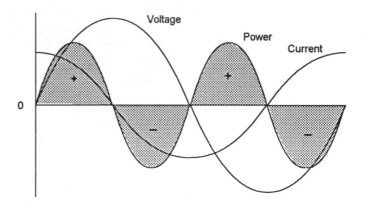

Figure 18.36 *Power with purely capacitive circuit*

An alternative to graphically determining the power is to use the trigonometric relationship $2 \sin A \sin B = \cos (A - B) - \cos (A + B)$:

$$\text{power} = vi = V_m \sin \omega t \times I_m \sin (\omega t + 90°) = -\tfrac{1}{2}V_m I_m \cos (2\omega t + 90°)$$

and so:

$$\text{power} = \tfrac{1}{2}V_m I_m \sin 2\omega t$$

The power alternates about the zero axis with a frequency which is twice that of the current or voltage. Thus the average power over one cycle is zero. This is because in that part of the cycle where the voltage is positive, energy is being stored in the electric field of the capacitor as it becomes charged, while in that part of the cycle where the voltage is negative the electric field is releasing its energy into the circuit.

18.7.4 Resistance in series with inductance or capacitance

When a circuit consists of a resistance in series with an inductance, or a resistance in series with a capacitance, then no power is dissipated in either the inductance or the capacitance but only in the resistance.

Example

A coil, equivalent to a resistance of 500 Ω in series with an inductance of 500 mH, is connected across a 50 V r.m.s., 50 Hz, supply. What is the power dissipated in the coil?

The power dissipated will just be that in the resistance, i.e. I^2R. The impedance of the circuit is

$$Z = \sqrt{R^2 + X_L^2} = \sqrt{500^2 + (2\pi \times 50 \times 0.5)^2} = 524\ \Omega$$

Thus the current $I = V/Z = 50/524 = 0.095$ A. Hence the power dissipated is $0.095^2 \times 500 = 4.6$ W.

Example

Determine the resistance and inductance of a series RL circuit if, when connected to a 240 V r.m.s., 50 Hz supply, it takes a power of 0.8 kW and a current of 4 A?

The circuit has a resistive component of R and a reactive component of X_L. The true power is the power dissipated in just the resistive component and thus $P = I^2R$ and so $R = 800/4^2 = 50\ \Omega$. The circuit impedance $Z = V/I = 240/4 = 60\ \Omega$. Since $Z = \sqrt{(R^2 + X_L^2)}$ then $X_L = \sqrt{(Z^2 - R^2)} = \sqrt{(60^2 - 50^2)} = 33.2\ \Omega$. Since the reactance $X_L = 2\pi fL$ then $L = 33.2/(2\pi \times 50) = 0.11$ H.

18.8 Power factor

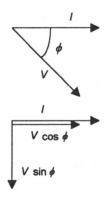

Figure 18.37 *Current and voltage phasors*

Consider a component or circuit where there is a phase difference between the current and the voltage (Figure 18.37), i.e. we have $i = I_m \sin \omega t$ and $v = V_m \sin (\omega t - \phi)$ with ϕ being the phase difference between the current and voltage. If we resolve the voltage phasor into horizontal and vertical components of $V \cos \phi$ and $V \sin \phi$, then the component at 90° to $\mathbf{I}$ is the situation we would get with a purely reactive circuit and so the average power over a cycle is zero for that component. Thus the average power P over a cycle is given by the in-phase components as:

$$P = IV \cos \phi$$

where I and V are r.m.s. values, the average power being in watts.

We can obtain the above relationship by using trigonometric relation $2 \sin A \sin B = \cos (A - B) - \cos (A + B)$. Thus:

$$p = I_m \sin \omega t \times V_m \sin (\omega t - \phi)$$

$$= \tfrac{1}{2} I_m V_m \cos \phi - \tfrac{1}{2} I_m V_m \cos (2\omega t - \phi)$$

The average value of $\cos (2\omega t - \phi)$ over one cycle is zero and thus:

$$\text{average power } P = \tfrac{1}{2} I_m V_m \cos \phi = IV \cos \phi$$

The product of the r.m.s. values of the current and voltage is called the *apparent power* S, the unit of apparent power being volt amperes (VA). Multiplying the apparent power by $\cos \phi$ gives the real power dissipated. For this reason, $\cos \phi$ is called the *power factor*. It has no units. Thus:

$$\text{average power} = \text{apparent power} \times \text{power factor}$$

For a purely reactive circuit the power factor is 0, for a purely resistive factor it is 1.

Example

A coil has a resistance of 6 Ω and an inductance of 30 mH and is connected across a 50 V, 50 Hz supply. Calculate (a) the current, (b) the phase angle between the current and the applied voltage, (c) the power factor, (d) the apparent power and (e) the true power.

(a) The reactance of coil $X_L = 2\pi f L = 2\pi \times 50 \times 0.03 = 9.42$ Ω. Hence, the impedance is:

$$Z = \sqrt{R^2 + X_L^2} = \sqrt{6^2 + 9.42^2} = 11.17\ \Omega$$

and so:

$$\text{current} = \frac{V}{Z} = \frac{50}{11.17} = 4.48 \text{ A}$$

(b) The phase is given by:

$$\tan \phi = \frac{X_L}{R} = \frac{9.42}{6} = 1.57$$

Hence, $\phi = 57.5°$.
(c) The power factor $= \cos \phi = \cos 57.5° = 0.537$.
(d) The apparent power $= VI = 50 \times 4.48 = 224$ VA.
(e) The true power = apparent power × power factor $= 224 \times 0.537 = 120.3$ W. Alternatively, we can consider the true power as just that dissipated in the resistance element and so true power $= I^2 R = 4.48^2 \times 6 = 120.4$ W.

18.8.1 Power triangle

If we consider an alternating current circuit such as an RL series circuit, the voltage will be leading the current by some phase angle ϕ. We can resolve the voltage phasor into two components, one in phase with the current of magnitude $V \cos \phi$ and one at a phase of 90° to the current of magnitude $V \sin \phi$ (Figure 18.38(a)). Multiplying the magnitude of each of these by the magnitude of the current I transforms the phasor diagram into a power triangle (Figure 18.38(b)). For this triangle, the horizontal is the true power $P = IV \cos \phi$, the hypotenuse is the apparent power $S = IV$ and the vertical is $VI \sin \phi$, this being termed the *reactive power Q* and given the unit of volt amperes reactive (V Ar).

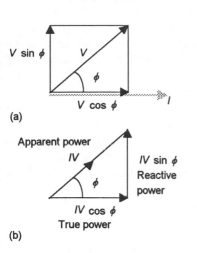

(a)

(b)

Figure 18.38 *(a) Voltage components, (b) power diagram*

Example

A load takes 120 kW at a power factor of 0.6 lagging. Determine the apparent power and the reactive power.

The true power $P = 120$ kW. Thus $120 = VI \cos \phi = VI \times 0.6$ and so the apparent power $S = IV = 120/0.6 = 200$ kV A. The reactive power $Q = IV \sin \phi$. Since $\cos \phi = 0.6$ then $\phi = 53.1°$ and so $\sin \phi = 0.8$ and $Q = 200 \times 0.8 = 160$ kV Ar.

Example

A motor takes a current of 15 A from a 100 V supply at a power factor of 0.8 lagging. Calculate (a) the apparent power, (b) the power taken from the supply and (c) the reactive power and (d) the output power from the motor, assuming it to have an efficiency to be 85 per cent.

(a) The apparent power is $IV = 400 \times 45 = 18\ 000$ VA = 18 kVA.
(b) The true power = apparent power × power factor = $18 \times 0.8 = 14.4$ kW.
(c) The reactive power = $VI \sin \phi = 18 \times 0.6 = 10.8$ kVAr.
(e) The output power = input power × efficiency = $14.4 \times 0.85 = 12.24$ kW.

18.8.2 The significance of the power factor in electrical engineering

Consider a generator which is rated to give a current of 2000 A at a voltage of 400 V. The apparent power rating of the generator is thus $400 \times 2000/1000 = 800$ kVA. Now consider this generator being used to supply a load. The true power delivered to the load depends on the power factor, i.e. the phase difference between the voltage and the current, and this is determined by the nature of the load; generally industrial loads include inductive motor coils and so have lagging power factors. If the power factor of the load is unity then the true power delivered to the load is the same as the apparent power and thus 800 kW. But if the power factor of the load were to be 0.5, then the true power = apparent power × power factor = $800 \times 0.5 = 400$ kW. Thus, though the generator is supplying its rated output of 800 kVA, it is only delivering a true power of 400 kW. It is therefore evident that the higher the power factor of the load, the greater is the power that can be delivered by a given generator.

The conductors connecting the generator to the load have to be capable of carrying the 2000 A without excessive temperature rise and consequently they can transmit 800 kW if the power factor is unity, but only 400 kW at 0.5 power factor for the same rise of temperature. The higher the power factor the greater is the power that can be transmitted by a given conductor.

Thus, the lower the power factor the greater is the cost of generation and transmission of electrical energy. For this reason electrical supply authorities do all they can to improve the power factor of their loads. A common method of improving the power factor of a load is to connect a capacitor in parallel with it.

Example

An a.c. generator is supplying a load of 300 kW at a power factor of 0.6. If the power factor is raised to unity, how many more kilowatts can the alternator supply for the same apparent power?

Apparent power = true power/power factor = 300/0.6 = 500 kVA. When the power factor is raised to unity the true power = apparent power × power factor = 500 × 1 = 500 kW. Hence the increase in power supplied by the alternator is 500 – 300 = 200 kW.

18.9 Rectification

The mains supply is, in Britain, 240 V a.c. and generally electronic equipment requires relatively low d.c. voltages. To provide such voltages, power supplies are used which, when connected to the a.c. mains, converts the 240 V into the required voltage and rectifies it to give d.c. Transformers are used to step-down the mains voltage to the required level. *Rectification* is the conversion of an a.c. voltage into a d.c. voltage. Because of their ability to conduct current in only one direction, diodes (see Section 14.3) are the basis of rectifier circuits.

18.9.1 Half-wave rectification

Figure 18.39 shows the basic half-wave rectifier circuit. When the input sine wave goes positive, the diode is forward biased and conducts current to the load resistor. When the input sine wave goes negative, the diode is reverse biased and there is no current. The result is that only the positive half-cycles of the input voltage appear across the load and so a pulsating d.c. voltage is produced, hence the term half-wave rectification.

In order to smooth out the fluctuations, a *reservoir capacitor C* is connected in parallel with the load (Figure 18.40). When the diode conducts and the voltage across the load rises, the capacitor charges. When the voltage begins to drop, the capacitor begins to discharge. This discharge continues until the applied voltage rises to a value greater than the discharge voltage. The resulting voltage across the load is thus

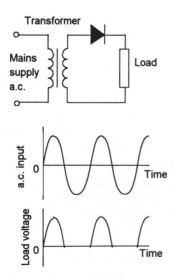

Figure 18.39   *Half-wave rectifier*

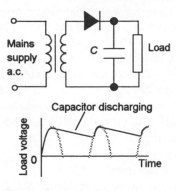

Figure 18.40 *Half-wave rectifier with reservoir capacitor*

smoother but still shows some variations. It can be considered to be effectively the sum of a perfectly smooth d.c. voltage with an a.c. voltage, called the *ripple*, superimposed on it.

18.9.2 Full-wave rectification

Smoother d.c. can be obtained by the use of full-wave recfication with two diodes (Figure 18.41). The secondary winding of the transformer is centre-tapped so that equal voltages are applied to each diode. When the half cycle applied to diode D1 is positive then the half cycle applied to diode D2 is negative. Thus D1 conducts and D2 does not. When the half cycle applied to diode D1 is negative then the half cycle applied to diode D2 is positive. Thus D1 does not conduct and D2 does. The results are as shown in Figure 18.41.

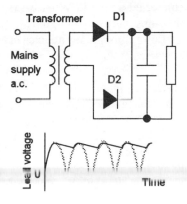

Figure 18.42 *Full-wave rectifier with reservoir capacitor*

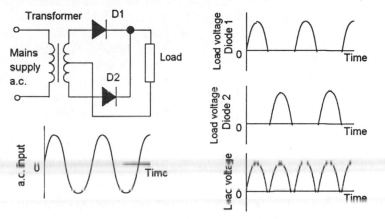

Figure 18.41 *Full-wave rectifier*

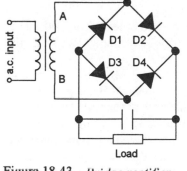
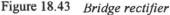

Figure 18.43 *Bridge rectifier*

As with the half-wave rectifier, described above, a reservoir capacitor can be connected across the load to give some smoothing of the output (Figure 18.42).

An alternative to the full-wave rectifier with its two diodes, described above, is the full-wave *bridge rectifier*. This employs four diodes (Figure 18.43). During the half-cycles when A is positive and B negative, then D2 and D3 conduct. During the half-cycles when A is negative and B positive, then D1 and D3 conduct. The result of this is that during each half-cycle the current flows the same way through the load. As with the rectifiers described above, a reservoir capacitor can be used to smooth the full-wave output.

Activities

1 Connect a signal generator in series with an inductor (e.g. a 240 turn coil on a magnetic core) and a lamp (1.25 V, 0.25 A). With the voltage maintained constant, vary the frequency of the generator and explain how the brightness of the lamp changes. Repeat the experiment with a capacitor (e.g. 8 μF) replacing the inductor. Then repeat the experiment with the inductor in series with the capacitor.

Problems

1 Draw the phasors to represent the following sinusoidal alternating quantities when each has the same frequency: (a) Two currents with root-mean-square values of 3 A and 5 A, with the second current leading the first by a phase difference of 60°, (b) Two voltages with root-mean-square values of 1 V and 2 V, with the second voltage lagging the first by a phase difference of 30°.

2 An alternating voltage of frequency 1 kHz and of root-mean-square value 10 V is applied across a 100 mH with pure inductance. What is (a) the reactance, (b) the current flowing through the inductor?

3 If the voltage across a 1 H inductor is $v = 10 \sin 200t$ volts, what is the equation for the current through the inductor?

4 A coil has an inductance of 0.5 mH. What is its reactance at (a) 50 Hz, (b) 1 kHz?

5 A pure inductor of inductance 10 mH has an alternating voltage of 25 V r.m.s. at a frequency of 5 kHz connected across it. What is (a) the reactance of the inductor and (b) the r.m.s. current through it?

6 An alternating current of $10 \sin 150t$ A flows through an inductor of inductance 0.5 H and negligible resistance. What is the voltage across the inductor?

7 An 1 kHz alternating voltage of root-mean-square value 10 V is applied across a capacitor with pure capacitance of 2 μF. What is (a) the reactance, (b) the current flowing through the capacitor?

8 A 10 μF capacitor has a 1.5 kHz alternating voltage of root-mean-square value 15 V applied across it. What is the current?

9 What is its reactance of a 0.1 μF capacitor at (a) 50 Hz, (b) 1 kHz?

10 A pure capacitor of capacitance 10 μF has an alternating voltage of 20 V r.m.s. at a frequency of 3 kHz connected across it. What is (a) the reactance of the capacitor and (b) the r.m.s. current through it?

11 An alternating current of 1 A r.m.s. at a frequency of 100 kHz passes through a coil of resistance 20 Ω and inductance 0.4 mH. Determine the voltage across the coil.

12 A coil having a resistance of 50 Ω and an inductance of 500 mH is connected across a 240 V r.m.s., 50 Hz, supply. What is the magnitude and phase angle of the current?

13 A coil when connected across a 10 V d.c. supply takes a current of 0.4 A. When connected across a 240 V r.m.s., 50 Hz, supply it takes 2 A. What are the resistance and inductance of the coil?

14 A coil of inductance 50 mH and resistance 15 Ω is connected across a 240 V r.m.s., 50 Hz supply. Determine the circuit impedance, the current taken and the phase angle between the current and voltage.

15 A coil has a resistance of 15 Ω and an inductive reactance of 8 Ω. Determine the impedance of the coil and the phase angle between the voltage across the coil and the current through it.

16 A voltage of 100 V r.m.s., 50 Hz is connected to a series *RL* circuit, the resistance being 45 Ω and the inductance having an inductive reactance of 60 Ω. Determine (a) the circuit impedance, (b) the current through the circuit, (c) the voltages across each component.

17 A voltage of 100 V r.m.s., 50 Hz is connected to a series *RL* circuit, the resistance being 15 Ω and the inductance being 60 mH.

Determine (a) the circuit impedance, (b) the magnitude and phase angle of the current through the circuit.

18 When a capacitor is in series with a resistance of 30 Ω, the current taken from a 240 V r.m.s., 50 Hz, supply is 4.8 A r.m.s. What is the (a) the impedance of the circuit, (b) the capacitance?

19 A voltage of 50 V r.m.s., 50 Hz is applied to a series *RC* circuit. If the current taken is 1.7 A when the voltage drop across the resistor is 34 V, what is the capacitance of the capacitor?

20 A voltage of 240 V r.m.s., 50 Hz is applied to a series *RC* circuit. If the resistance is 40 Ω and the capacitance 50 μF, determine (a) the circuit impedance, (b) the magnitude and phase angle of the current.

21 A voltage of 24 V r.m.s., 400 Hz is applied to a series *RC* circuit. If the resistance is 40 Ω and the capacitance 15 μF, determine (a) the circuit impedance, (b) the magnitude and phase angle of the current.

22 A series *RLC* circuit takes a current of 5 A r.m.s. If the resistance is 10 Ω, the inductive reactance 20 Ω and the capacitive reactance 10 Ω, what will be (a) the magnitude and phase of the voltage applied to the circuit, (b) the voltages across each component?

23 A *RLC* circuit has a coil having a resistance of 75 Ω and an inductance of 0.15 H in series with a capacitance of 8 μF. If a voltage of 100 V r.m.s., 200 Hz is applied to it, determine (a) the magnitude and phase angle of the current, (b) the voltage across the coil, (c) the voltage across the capacitor.

24 A series *RLC* circuit has a resistance of 10 Ω, an inductance of 100 mH and a capacitance of 2 μF. What is the resonant frequency?

25 A series *RLC* circuit resonates at a frequency of 100 Hz. If the resistance is 20 Ω and the inductance 300 mH, what is (a) the circuit capacitance, (b) the *Q*-factor of the circuit?

26 A series *RLC* circuit has a resistance of 4 Ω, an inductance of 60 mH and a capacitance of 30 μF. Determine (a) the resonant frequency and (b) the *Q*-factor of the circuit.

27 A coil has a resistance of 10 Ω and an inductive reactance of 25 Ω when connected across a 250 V r.m.s., 50 Hz, supply. Determine the current taken from the supply and the power dissipated.

28 A circuit consists of a resistance of 3 kΩ in series with a capacitor of 0.22 μF. What will be the power dissipated in the circuit when an alternating voltage of 10 V r.m.s., 200 Hz, is applied to it?

29 A series *RC* circuit has a resistance of 2 kΩ and a capacitance of 1 μF. A 240 V r.m.s., 50 Hz supply is applied to the circuit. Determine (a) the power factor, (b) the true power, (c) the apparent power, (d) the reactive power.

30 A load takes 80 kW at a power factor of 0.6 lagging. Determine the apparent power and the reactive power.

31 A load takes a current of 5 A from the 240 V a.c. mains supply and develops a power of 960 W. What is the power factor?

32 A motor takes a current of 8 A from the 240 V a.c. mains supply. What will be the power consumed if it has a power factor of 0.8?

33 What is the power input and power output for a single-phase motor which takes 8 A from the a.c. supply of 250 V if it has a power factor of 0.75 lagging and has an efficiency of 80%?

19 Parallel a.c. circuits

19.1 Introduction

For a parallel d.c. circuit there is the same voltage drop across each component and the current entering the parallel arrangement is the sum of the currents through each component. For a parallel a.c. circuit there is the same voltage phasor for the voltage across each parallel component and the phasor for the current entering the parallel arrangement is the sum of the phasors for the currents through each parallel component. This chapter is a discussion of parallel a.c. circuits and the conditions required for resonance. It follows on from Chapter 18.

19.2 Parallel circuits

The following are illustrations of the analysis of parallel circuits when supplied with alternating current.

19.2.1 Parallel resistance and inductance

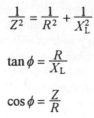

(a)

(b)

(c)

Figure 19.1   *Parallel R and L*

Figure 19.1(a) shows the parallel circuit. The voltage across each component is the same, being the supply voltage **V**. The current **I**$_R$ through the resistance is in phase with the supply voltage; the current **I**$_L$ through the inductor lags behind the supply voltage by 90° (Figure 19.1(b)). The phasor sum of the currents **I**$_R$ and **I**$_L$ must be the supply current **I** entering the parallel arrangement (Figure 19.1(c)). Thus:

$$I^2 = I_R^2 + I_L^2$$

$$\tan \phi = \frac{I_L}{I_R}$$

$$\cos \phi = \frac{I_R}{I}$$

But $V = IZ$, where Z is the impedance of the parallel arrangement and $V = I_R R$ and $V = I_L X_L$, thus the above equations can be written as:

$$\frac{1}{Z^2} = \frac{1}{R^2} + \frac{1}{X_L^2}$$

$$\tan \phi = \frac{R}{X_L}$$

$$\cos \phi = \frac{Z}{R}$$

Example

A resistance of 10 Ω is in parallel with an inductance of 50 mH. What will be the current drawn from the supply and the impedance of the circuit when a voltage of 24 V r.m.s., 50 Hz is applied to it?

The inductive reactance is $X_L = 2\pi f L = 2\pi \times 50 \times 0.050 = 15.7 \, \Omega$. Thus $I_L = V/X_L = 24/15.7 = 1.53$ A. The current through the resistor $I_R = V/R = 24/10 = 2.4$ A. The magnitude of the current drawn from the supply is thus:

$$I = \sqrt{I_R^2 + I_L^2} = \sqrt{2.4^2 + 1.53^2} = 2.85 \text{ A}$$

The phase angle ϕ by which the supply current lags the supply voltage is $\phi = \tan^{-1}(R/X_L) = \tan^{-1}(10/15.7) = 32.5°$. The circuit impedance $Z = V/I = 24/2.85 = 8.42 \, \Omega$.

19.2.2 Parallel resistance and capacitance

Figure 19.2(a) shows the parallel circuit. The voltage across each component is the same, being the supply voltage **V**. The current **I**$_R$ flowing through the resistance is in phase with the supply voltage, the current **I**$_C$ through the inductor leads the supply voltage by 90° (Figure 19.2(b)). The phasor sum of the currents **I**$_R$ and **I**$_C$ must be the supply current **I** entering the parallel arrangement (Figure 19.2(c)). Thus:

$$I^2 = I_R^2 + I_C^2$$

$$\tan \phi = \frac{I_C}{I_R}$$

$$\cos \phi = \frac{I_R}{I}$$

But $V = IZ$, where Z is the impedance of the parallel arrangement and $V = I_R R$ and $V = I_C X_C$, thus the above equations can be written as:

$$\frac{1}{Z^2} = \frac{1}{R^2} + \frac{1}{X_C^2}$$

$$\tan \phi = \frac{R}{X_C}$$

$$\cos \phi = \frac{Z}{R}$$

Example

A resistance of 10 Ω is in parallel with a capacitance of 10 μF. What will be the current drawn from the supply and the impedance of the circuit when a voltage of 20 V r.m.s., 1 kHz is applied to it?

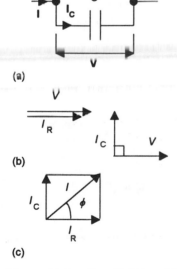

(a)

(b)

(c)

Figure 19.2 *Parallel R and C*

The inductive capacitance is $X_C = 1/2\pi fC = 1/2\pi \times 1000 \times 10 \times 10^{-6}$ $= 15.9\ \Omega$. Thus $I_C = V/X_C = 20/15.9 = 1.26$ A. The current through the resistor $I_R = V/R = 20/10 = 2.0$ A. The magnitude of the current drawn from the supply is thus:

$$I = \sqrt{I_R^2 + I_C^2} = \sqrt{2.0^2 + 1.26^2} = 2.36\ \text{A}$$

The phase angle ϕ by which the supply current leads the voltage is ϕ $= \tan^{-1}(R/X_C) = \tan^{-1}(10/15.9) = 32.2°$. The circuit impedance $Z = V/I = 20/2.36 = 8.47\ \Omega$.

19.2.3 Parallel R-L and C

A coil will have both resistance and inductance and can be regarded as a resistance in series with an inductance. Consider such a coil in parallel with capacitance (Figure 19.3(a)). Figure 19.3(b) shows the phasor diagram for the two series items in the coil branch. For this branch we thus have a voltage V which is leading I_{RL}, the current in phase with V_R, by a phase angle ϕ_1. We can rotate these phasors to give the phasors for the coil branch as that shown in Figure 19.3(c), together with the phasors for the capacitor branch. The total circuit current is the phasor sum of I_{RL} and I_C (Figure 19.3(d)).

Figure 19.3(d) shows the situation when the resolved component of I_{RL} in the vertical direction has a magnitude less than that of I_C (Figure 19.3(a)). There are two other possible conditions that we might have: the resolved component of I_{RL} in the vertical direction has a magnitude greater than that of I_C (Figure 19.3(b)) and the resolved component of I_{RL} in the vertical direction has a magnitude equal to that of I_C (Figure 19.3(c)). The three conditions give:

1    $I_C > I_{RL} \sin \phi_1$, **I** leads **V** by ϕ (Figure 19.4(a))

2    $I_C < I_{RL} \sin \phi_1$, **I** lags **V** by ϕ (Figure 19.4(b))

3    $I_C = I_{RL} \sin \phi_1$, **I** and **V** are in phase (Figure 19.4(c))

We can determine the total circuit current **I** by a scale drawing of the phasors to give one of the phasor diagrams in Figure 19.4.

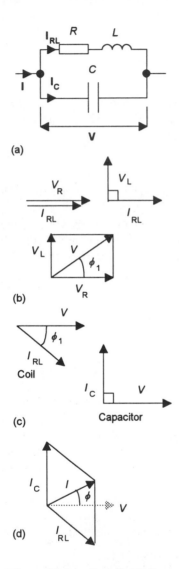

(a)

(b)

(c)

(d)

Figure 19.3   *Parallel R-L and C*

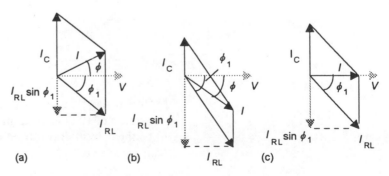

Figure 19.4 *The three possible conditions*

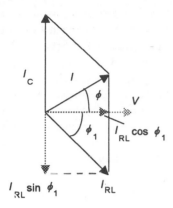

Figure 19.5 *Resolving I_{RL}*

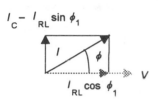

Figure 19.6 *Circuit current*

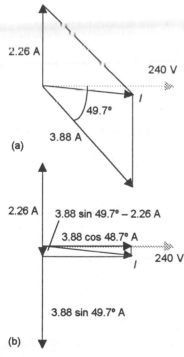

Figure 19.7 *Example*

Alternatively, we can resolve the current phasor I_{RL} into horizontal and vertical component phasors: $I_{RL} \cos \phi_1$ horizontally and $I_{RL} \sin \phi_1$ vertically. The horizontal component is in phase with V and is termed the 'in phase' component, the other component is at 90° and is termed the 'quadrature' component. For Figure 19.4(a) we thus have the situation shown in Figure 19.5. The vertical current has the magnitude of $I_C - I_{RL} \sin \phi_1$ and the horizontal current is $I_{RL} \cos \phi_1$. We can thus draw the phasor diagram shown in Figure 19.6 and so have:

$$I^2 = (I_{RL} \cos \phi_1)^2 + (I_C - I_{RL} \sin \phi_1)^2$$

$$\tan \phi = \frac{I_C - I_{RL} \sin \phi_1}{I_{RL} \cos \phi_1}$$

$$\cos \phi = \frac{I_{RL} \cos \phi_1}{I}$$

Example

A circuit consists of a coil with inductance 150 mH and resistance 40 Ω in parallel with a capacitor of capacitance 30 μF. A 240 V r.m.s., 50 Hz voltage supply is connected across the circuit. Determine the current in (a) the coil, (b) the capacitor and (c) drawn from the supply.

(a) Inductive reactance $X_L = 2\pi f L = 2\pi \times 60 \times 0.150 = 47.1 \, \Omega$. The impedance Z_{RL} of the coil is thus:

$$Z_{RL} = \sqrt{R^2 + X_L^2} = \sqrt{40^2 + 47.1^2} = 61.8 \, \Omega$$

The current through the coil has thus the magnitude $I_{RL} = V/Z_{RL} = 240/61.8 = 3.88$ A with the phase angle $\phi_1 = \tan^{-1}(X_L/R) = 47.1/40 = 49.7°$. This is the angle by which the current lags the supply voltage.

(b) The capacitive reactance is $X_C = 1/2\pi f C = 1/(2\pi \times 50 \times 30 \times 10^{-6}) = 106.1 \, \Omega$. The current through the capacitor has thus the magnitude $X_C = V/X_C = 240/106.1 = 2.26$ A. The current leads the supply voltage by 90°.

(c) Figure 19.7(a) shows the phasor diagram for the currents in the circuit and Figure 19.7(b) the phasor diagram when I_{RL} has been resolved to the in phase and quadrature components. Hence:

$$I = \sqrt{(3.88 \sin 49.7° - 2.26)^2 + (3.88 \cos 48.7°)^2} = 2.65 \text{ A}$$

Its phase angle by which it lags the supply voltage is:

$$\phi = \tan^{-1} \frac{3.88 \sin 49.7° - 2.26}{3.88 \cos 48.7°} = 15.2°$$

19.3 Parallel resonance

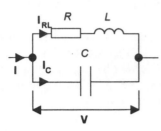

Figure 19.8 *RL in parallel with C*

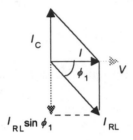

Figure 19.9 *Resonance*

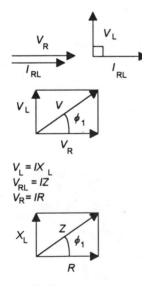

$V_L = IX_L$
$V_{RL} = IZ$
$V_R = IR$

Figure 19.10 *RL arm of circuit*

For the parallel circuit shown in Figure 19.8, resonance occurs when I and V are in phase. This is when the magnitude of the current through the capacitor $I_C = I_{RL} \sin \phi_1$, where ϕ_1 is the phase angle between the current I_{RL} through the RL arm of the circuit and the voltage supply (Figure 19.9). One way of describing this equality is that the reactive current through the capacitor equals the reactive current through the RL arm, the current through the RL arm having a reactive component which is at 90° to the voltage and a resistive component which is in phase with the voltage. Since $I_C = V/X_C$, $I_{RL} = V/Z_{RL}$ and $\sin \phi_1 = X_L/Z_{RL}$ (Figure 19.10), where X_C is the reactance of the capacitor and Z_{RL} is the impedance of the RL combination, then at resonance:

$$I_C = I_{RL} \sin \phi_1$$

gives:

$$\frac{V}{X_C} = \frac{V}{Z_{RL}} \frac{X_L}{Z_{RL}}$$

and hence:

$$Z_{RL}^2 = X_L X_C$$

Since $X_L = 2\pi f_0 L$, $X_C = 1/2\pi f_0 C$ and $Z_{RL} = \sqrt{[R^2 + (2\pi f_0 L)^2]}$, where f_0 is the resonant frequency, then the above equation can be written as:

$$R^2 + (2\pi f_0 L)^2 = 2\pi f_0 L \times \frac{1}{2\pi f_0 C} = \frac{L}{C}$$

$$f_0 = \frac{1}{2\pi} \sqrt{\frac{1}{LC} - \frac{R^2}{L^2}}$$

Generally R^2/L^2 is much smaller than $1/LC$ and so the equation can be approximated to:

$$f_0 \approx \frac{1}{2\pi \sqrt{LC}}$$

This is the same as the equation for the resonance of a series RLC circuit.

At resonance, the current drawn from the supply is in phase with the supply voltage and thus the circuit presents a purely resistive load. The value of this resistance is known as the *dynamic resistance* R_D:

$$R_D = \frac{V}{I}$$

Since $I = I_{RL} \cos \phi_1$ and $I_C = I_{RL} \sin \phi_1$ (see Figure 19.9), we can write:

$$I = \frac{I_C}{\sin \phi_1} \cos \phi_1 = \frac{I_C}{\tan \phi_1}$$

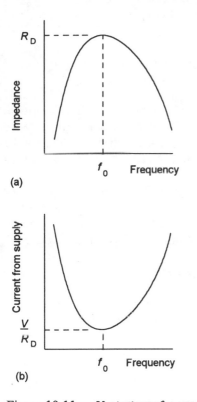

(a)

(b)

Figure 19.11 *Variation of (a) impedance, (b) current with frequency*

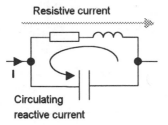

Resistive current

Circulating reactive current

Figure 19.12 *RL in parallel with C at resonance*

and hence:

$$R_D = \frac{V \tan \phi_1}{I_C} = \frac{V \tan \phi_1}{V/X_C} = X_C \tan \phi_1$$

But $X_L/R = \tan \phi_1$ (see Figure 19.9), thus:

$$R_D = \frac{1}{2\pi f_0 C} \frac{2\pi f_0 L}{R} = \frac{L}{CR}$$

A consequence of the above equation is that if $R = 0$ then R_D equals infinity. This implies zero current at resonance when $R = 0$.

Figure 19.11 shows how (a) the impedance and (b) the current drawn from the supply by the circuit varies with frequency. The impedance is at a maximum value of R_D and the current at a minimum value of V/R_D at the resonant frequency. For this reason the parallel resonant circuit is known as the *rejecter circuit* since it 'rejects current' at the resonant frequency.

At resonance the current I_C can be much larger than the current I drawn from the supply (in Figure 19.9 it is shown as only being a little bit larger) We can think of this reactive current as circulating round the RL–C circuit with only the resistive component of the current through RL passing through the system (Figure 19.12). The *Q-factor* of the parallel circuit is defined as being:

$$Q \text{ factor} = \frac{I_C}{I_R}$$

Hence, using the phasor diagram in Figure 19.10:

$$Q\text{-factor} = \tan \phi_1 = \frac{X_L}{R} = \frac{2\pi f_0 L}{R}$$

and since at resonance $X_L = X_C$, we also have:

$$Q\text{-factor} = \frac{1}{2\pi f_0 CR}$$

Example

A circuit has a coil of resistance 100 Ω and inductance 50 mH in parallel with a capacitor of 0.01 µF. Determine the resonant frequency, the dynamic resistance and the Q-factor.

$$f_0 = \frac{1}{2\pi} \sqrt{\frac{1}{LC} - \frac{R^2}{L^2}} = \frac{1}{2\pi} \sqrt{\frac{1}{0.050 \times 0.01 \times 10^{-6}} - \frac{100^2}{0.050^2}}$$

Hence the resonant frequency is 7111 Hz. If, instead of the above equation, we had used the approximate version of the equation the answer would have been 7118 Hz.

$$R_D = \frac{L}{CR} = \frac{0.050}{0.01 \times 10^{-6} \times 100} = 50 \text{ k}\Omega$$

$$Q\text{-factor} = \frac{1}{2\pi f_0 CR} = \frac{1}{2\pi \times 7111 \times 0.01 \times 10^{-6} \times 100} = 22.4$$

19.4 Power in a parallel circuit

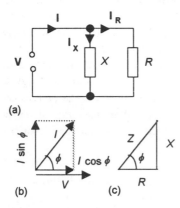

(a)

(b)   (c)

Figure 19.13 *Parallel circuit*

For a parallel arrangement of reactance, i.e. a pure inductance or a pure capacitance, and resistance (Figure 19.13(a)), power will only be dissipated in the resistive branch. The current $\mathbf{I}$ from the supply can be resolved into two current components, one of magnitude $I \cos \phi$ in phase with the supply voltage and the other of magnitude $I \sin \phi$ at 90° to it (Figure 19.13(b)). The current through the resistive branch is the component in phase with the voltage; the current through the reactive branch is the component at 90° to the voltage. Thus the true power dissipation is $VI_R = VI \cos \phi$. The apparent power dissipated is VI and so:

true power = $VI \cos \phi$ = apparent power × $\cos \phi$

As for the series circuit, $\cos \phi$ is known as the power factor.
    Since $R = V_R/I_R = V_R/I \cos \phi$ and $Z = V/I$ then we also have

$\cos \phi = R/Z$

and this can be represented by the impedance triangle shown in Figure 19.13(c).

Example

A circuit consists of a coil with inductance 150 mH and resistance 40 Ω in parallel with a capacitor of capacitance 30 μF. A 240 V r.m.s., 50 Hz voltage supply is connected across the circuit. Determine the power factor of the circuit, the true power and the apparent power.

This is the example for which the current in the coil, the capacitor and drawn from the supply was calculated (four pages earlier with Figure 19.7). The supply current was found to lag the supply voltage by 15.2° and so the power factor of the circuit = cos 15.2° = 0.97. The true power = $IV \cos \phi$ = 2.65 × 240 × 0.97 = 617 W. The apparent power = IV = 2.65 × 240 = 636 VA.

19.4.1 Power factor correction

Power factor correction (see Section 18.8.2), i.e. making the power factor closer to the value 1, can be achieved by connecting a suitable capacitor in parallel with an inductive load. As a consequence of the load being inductive, the current I taken from the supply lags the supply voltage by ϕ (Figure 19.14(a)). Connecting a capacitor in parallel reduces the reactive component of the current and so reduces ϕ and the overall power factor (Figure 19.14(b)).

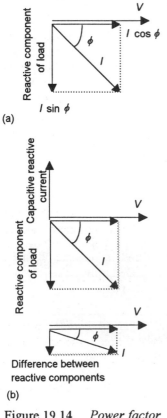

(a)

Difference between
reactive components

(b)

Figure 19.14   *Power factor correction*

Example

A 240 V, 50 Hz motor running at full load has a power of 500 W and a lagging power factor of 0.7. What value of capacitance should be connected in parallel with it to raise the overall power factor to 1?

Since $VI \cos \phi = 500$ W then $I = 500/(240 \times 0.7) = 2.98$ A. This current lags the supply voltage by ϕ (Figure 19.14(a)); since $\cos \phi = 0.7$ then $\phi = 45.6°$. The component of this current in phase with the voltage $= I \cos \phi = 2.98 \times 0.7 = 2.09$ A and the component at 90° to it $= I \sin \phi = 2.98 \sin 45.6° = 2.13$ A. To bring the overall power factor to 1 means reducing the $VI \sin \phi$ component to 0 means taking an additional current from the supply which is equal and opposite to the $I \sin \phi$ component of the supply current (Figure 19.14(b)). This can be supplied by connecting a capacitor in parallel. The current taken by the capacitor must thus be 2.13 A and so its reactance $X_C = 240/2.13 = 112.7$ Ω. Thus $1/2\pi fC = 112.7$ Ω and so $C = 28.2$ μF.

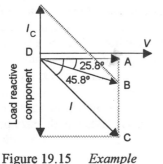

Figure 19.15   *Example*

Example

A 240 V, 50 Hz load takes a current of 5 A at a lagging power factor of 0.7. What capacitor should be connected in parallel with the load to increase the power factor to 0.9 lagging?

The initial current lags the supply voltage by ϕ and, since $\cos \phi = 0.7$, then $\phi = 45.6°$. The corrected power factor is to be 0.9 and, since $\cos \phi = 0.9$, we require ϕ to be 25.8°. The phasor diagram for the currents is as shown in Figure 19.15. Since $I = 5$ A then $AC/5 = \sin 45.6°$ and $AC = 3.57$ A. Since $AB/AD = \tan 25.8°$ and $AD = 5 \cos 45.6°$, then $AB = 1.69$ A. The current $I_C = BC = AC - AB = 3.57 - 1.69 = 1.88$ A. The capacitive reactance $X_C = V/I_C = 240/1.88 = 128$ Ω. Since $X_C = 1/2\pi fC$ then $C = 1/(2\pi \times 50 \times 128) = 24.9$ μF.

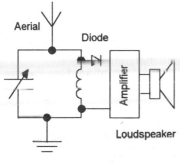

Figure 19.16   *Activity 1*

Activities

1 Assemble a simple radio receiver and explain how adjustment of the capacitance of the tuning capacitor enables different radio stations to be received. Figure 19.16 shows the basic circuit. The aerial can be a length of PVC-covered wire, the tuning capacitor a variable capacitor of about 0–500 pF and, for medium wave radio signals, the inductor a coil of about 50 turns of wire of diameter about 70 mm.

Problems

1 A circuit consists of a resistance of 10 Ω in parallel with an inductance of 50 mH. What is the magnitude and phase angle of the current taken from the supply when a voltage of 240 V r.m.s., 50 Hz is applied?

2 A circuit consists of a resistance of 40 Ω in parallel with an inductance of reactance 30 Ω. What is the magnitude and phase angle of the current taken from the supply when a voltage of 120 V r.m.s., 50 Hz is applied?

3 A circuit consists of a resistance of 150 Ω in parallel with an inductance of 2 mH. What is the magnitude and phase angle of the current taken from the supply of 2 V r.m.s., 20 kHz?

4 A circuit consists of a resistance in parallel with a capacitance. What are the values of the resistance and capacitance if, when a voltage of 240 V r.m.s., 200 Hz is applied, the current taken from the supply is 2 A with a phase angle of 53.1° leading?

5 A resistance is in parallel with a capacitance. If, when a voltage of 100 V r.m.s. is applied to the circuit, the current drawn from the supply is 2 A at a phase angle of 30° leading, what are (a) the magnitudes of the currents through the resistance and the capacitance, and (b) the values of the resistance and the capacitive reactance?

6 A circuit consists of a resistance of 60 Ω in parallel with an 8 μF capacitor and a supply of 10 V r.m.s., 200 Hz. What is the size of the current taken by (a) the resistor, (b) the capacitor, and (c) the overall current taken from the supply?

7 A coil with a resistance of 10 Ω and an inductance of 50 mH is in parallel with a capacitance of 0.01 μF. Determine the resonant frequency.

8 A capacitor which can be varied between 50 pF and 350 pF is connected in parallel with an inductance of 100 μH. Over what range of frequencies will the circuit give resonance?

9 A coil with a resistance of 5 Ω and an inductance of 50 mH is in parallel with a capacitance of 0.1 μF and a voltage supply of 100 V r.m.s., variable frequency, is applied. Determine the resonant frequency, the dynamic resistance, the current drawn at resonance and the Q-factor.

10 A coil with a resistance of 10 Ω and an inductance of 120 mH is in parallel with a capacitance of 60 μF and a voltage supply of 100 V r.m.s., variable frequency, is applied. Determine the resonant frequency, the dynamic resistance, the current drawn at resonance and the Q-factor.

11 A circuit consists of a coil, resistance 200 Ω and inductance 10 H, in parallel with a capacitance of 5 μF. A voltage of 240 V r.m.s., 50 Hz is applied to the circuit. Determine (a) the current drawn from the supply, (b) the power factor of the circuit, (c) the true power, (d) the apparent power, (e) the reactive power.

12 An inductor has a reactance of 100 Ω and an impedance of 200 Ω and is connected in parallel with a resistance of 200 Ω to a 100 V, 50 Hz supply. What is the total power dissipated in the circuit?

13 A 240 V, 50 Hz load has a power of 100 W and a lagging power factor of 0.6. What value of capacitance should be connected in parallel with it to raise the overall power factor to unity?

14 A 240 V, 50 Hz load has a power of 500 W and takes a current of 3 A from the supply. What value of capacitance should be connected in parallel with it to raise the overall power factor to 0.9 lagging?

15 A 3 kW motor draws a current of 8 A from a 240 V, 50 Hz supply. What is (a) the power factor and (b) the capacitor which when connected in parallel will improve the power factor to 0.8?

20 Transients

20.1 Introduction

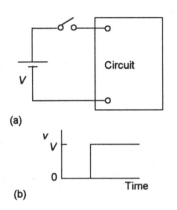

(a)

(b)

Figure 20.1   *Step voltage input*

This chapter is concerned with the current and voltage changes that occur in a circuit when there is a change in applied voltage. This might be when the d.c. voltage V applied to some circuit is switched on (Figure 20.1(a)). Before the switch is closed the voltage input to the circuit is 0 and it rises to V when the switch is closed and then remains at that value. Such an input voltage is termed a *step voltage* and is represented by Figure 20.1(b). The current and voltages that occur in the circuit while they are reacting to the change in input voltage from 0 to V and changing to their steady state values are termed *transients*. Such transients occur whenever the applied voltage changes and thus can occur, for example, when we apply a constantly changing input voltage or switch on a sinusoidal alternating voltage. In this chapter we restrict the consideration to the transients produced in circuits when subject to a step input voltage.

The circuits considered are those involving just resistance, capacitance plus resistance, and inductance plus resistance. The aim is to be able to predict the transient response of such circuits.

20.2 Purely resistive circuit

For a purely resistive circuit (Figure 20.2(a)), the circuit current will be proportional to the applied voltage. Thus at any instant of time the current $i = v/R$, where v is the value of the voltage at that instant and R the circuit resistance. Thus when we have a step input voltage (Figure 20.2(b)) and v abruptly changes from 0 to V, since the circuit current is always proportional to v the current will change abruptly from 0 to I, where $I = V/R$ (Figure 20.2(c)). The current changes at the same time as the applied voltage. Since no circuit is purely resistive, there being always some inductance and capacitance, the above represents the ideal scenario which is only roughly followed by a circuit in which no inductance or capacitance has deliberately been introduced.

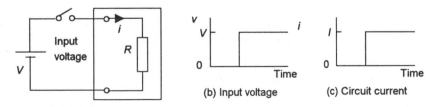

(a) Resistive circuit

(b) Input voltage

(c) Circuit current

Figure 20.2 *Step voltage applied to a purely resistive circuit*

20.3 *RC* circuit: charging

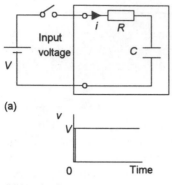

Input voltage

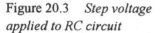

(a)

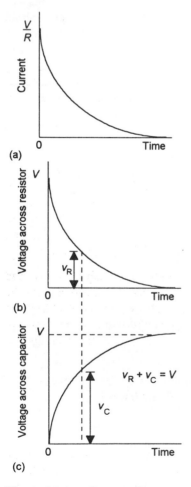

(b) Input voltage

Figure 20.3 *Step voltage applied to RC circuit*

(a)

(b)

(c)

Figure 20.4   *Series RC circuit*

Consider a circuit with series resistance and capacitance and there is a step voltage input to the circuit at time $t = 0$ (Figure 20.3(a)). The voltage is applied across two series components and so we must have:

$$V = \text{voltage across resistance } v_R + \text{voltage across capacitor } v_C$$

The charge q on a capacitor is related to the voltage v_C between its plates by $q = Cv_C$, where C is the capacitance. Initially there is no charge on the capacitor because there has been no current to move charge onto and off its plates. Thus, at the instant we close the switch to start the step voltage, there is no voltage across the capacitor. Therefore the voltage V must be entirely across the resistance. The current i through the resistor, and hence the initial circuit current, is v_R/R and so initially is V/R. When the capacitor begins to acquire charge then the voltage across it increases. Since $V = v_R + v_C$, this must result in a decrease in the voltage across the resistor. Thus, since $i = v_R/R$, the circuit current i must decrease. When the capacitor is fully charged the circuit current has dropped to zero. Thus there is then no voltage across the resistor and so the entire input voltage is across the capacitor. Figure 20.4(a) shows how the circuit current changes with time, Figures 20.4(b) and (c) showing how the voltages across the resistance and capacitance change with time. The graphs are exponentials.

20.3.1 Rate of change of v_C

At any instant of time $V = v_C + v_R$. But $v_R = iR$ and so we can write:

$$V = v_C + iR$$

The circuit current i is the rate of movement of charge through the circuit. But each bit of charge moved through the circuit changes the voltage across the capacitor; $q = Cv_C$ and so:

$$i = \text{rate of movement of charge} = \text{rate of change of } Cv_C$$

$$= C \times \text{rate of change of } v_C$$

We can write this, in calculus notation, as:

$$i = \frac{dq}{dt} = \frac{d(Cv_C)}{dt} = C\frac{dv_C}{dt}$$

The current is thus proportional to the rate of change of the voltage across the capacitor and so:

$$V = v_C + iR = v_C + RC\frac{dv_C}{dt}$$

$$V - v_C = RC\frac{dv_C}{dt}$$

Initially, $v_C = 0$ and so:

initial rate of change of voltage across the capacitor $= V/RC = V/\tau$

The product RC has the unit of time and we define *time constant* $\tau = RC$. The significance of the time constant is that, the bigger the time constant the smaller the initial rate of charging of the capacitor and the longer it will take to reach the voltage V.

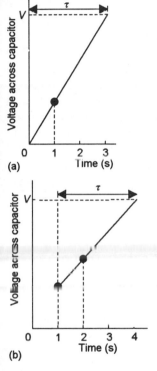

(a)

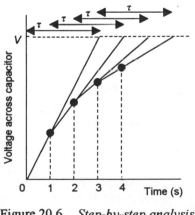
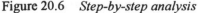

(b)

Figure 20.5 *Steps: (a) first, (b) second*

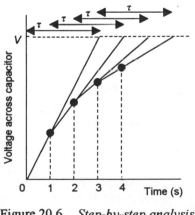

Figure 20.6 *Step-by-step analysis*

Example

What is the initial rate at which the voltage across a 100 µF capacitor will change with time when it is charged through a resistance of 5 kΩ by a voltage being switched from 0 to 6 V?

$$\text{Initial rate} = \frac{V}{RC} = \frac{6}{5 \times 10^3 \times 100 \times 10^{-6}} = 12 \text{ V/s}$$

Example

What is the time constant for a series RC circuit which has a resistance of 2 MΩ and capacitance 10 µF?

The time constant $\tau = RC = 2 \times 10^6 \times 10 \times 10^{-6} = 20$ s.

20.3.2 Step-by-step determination of current and voltage values

The following is an illustration of a step-by-step method that can be used when there is a step voltage input to a RC circuit.

First step: The initial rate of change of v_C with time is $V/RC = V/\tau$, where τ is the time constant. Thus we can draw the initial slope of the graph line on a graph of v_C against time in the way shown in Figure 20.5(a), i.e. we mark out the point from time 0 along the horizontal line for voltage V for a time equal to the time constant. Suppose we consider this to represent the slope for the first 1 s.

Second step: At 1 s we consider what the slope would be if we had started off with the voltage at this point and obtain a new slope by repeating the marking out a point from that time along the horizontal line for voltage V for a time equal to the time constant (Figure 20.5(b)). We now consider this line to give the slope for the time between 1 s and 2 s.

Following steps: We then repeat the above step for successive time intervals and so obtain the graph shown in Figure 20.6.

Example

Use a graphical step-by-step method to plot the graph of the voltage across a 15 µF capacitor in a series RC circuit, the resistance being 40 kΩ, and determine its value 1 s after if it is connected to a step voltage of 12 V.

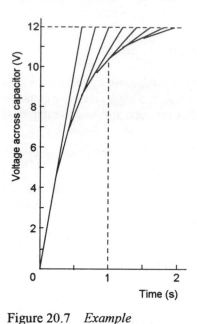

Figure 20.7 *Example*

The time constant $\tau = RC = 40 \times 10^3 \times 15 \times 10^{-6} = 0.6$ s. Initially the voltage across the capacitor is 0. To obtain reasonable accuracy with the voltage value for 1 s we need to get a reasonable number of points in the interval from 0 to 1 s. Suppose we determine the slope of the graph every 0.2 s. Figure 20.7 shows the resulting graph. After 1 s the voltage is estimated as about 10.4 V.

20.3.3 Calculus derivation of current and voltage values

We can obtain the equation of the graph of v_C with time by integrating the equation:

$$V - v_C = RC\frac{dv_C}{dt}$$

Separation of variables gives:

$$\frac{dv_C}{V - v_C} = \frac{dt}{RC}$$

Integration then gives:

$$\int_0^V \frac{dv_C}{V - v_C} = \int_0^t \frac{dt}{RC}$$

$$-\ln(V - v_C) + \ln V = \frac{t}{RC}$$

$$\frac{V}{V - v_C} = e^{t/RC}$$

$$\frac{V - v_C}{V} = e^{-t/RC}$$

and so:

$$v_C = V(1 - e^{-t/RC})$$

Since $V = v_R + v_C$ then:

$$v_R = V - V(1 - e^{-t/RC}) = V e^{-t/RC}$$

Since $i = v_R/R$ then:

$$i = \frac{V}{R} e^{-t/RC}$$

The above three equations describe the graphs in Figure 20.4.

Example

Determine the equation describing the voltage across a 15 μF capacitor in a series RC circuit if the resistance is 40 kΩ and the voltage 1 s after if it is connected to a step voltage of 12 V.

$RC = 40 \times 10^3 \times 15 \times 10^{-6} = 0.6$ s and so the voltage is described by the equation:

$$v_C = V(1 - e^{-t/RC}) = 12(1 - e^{-t/0.6})$$

After 1 s, $v_C = 12(1 - e^{-1/0.6}) = 9.7$ V.

Example

A series RC circuit has a resistance of 100 kΩ and a capacitance of 1 μF. Determine the current (a) 0.05 s, (b) 0.1 s after a steady voltage of 20 V is connected to the circuit.

$RC = 100 \times 10^3 \times 1 \times 10^{-6} = 0.1$ s. Thus:

$$i = \frac{V}{R} e^{-t/RC} = \frac{20}{100 \times 10^{-3}} e^{-t/0.1} = 0.2 \, e^{-t/0.1} \text{ mA}$$

(a) After 0.05 s, $i = 0.2 \, e^{-0.05/0.1} = 0.121$ mA.
(b) After 0.1 s, $i = 0.2 \, e^{-0.1/0.1} = 0.074$ mA.

20.3.4 Time constant

The time constant τ is RC. Thus:

$$v_C = V(1 - e^{-t/RC}) = V(1 - e^{-t/\tau})$$

What time will be required for v_C to reach $0.5V$?

$$0.5V = V(1 - e^{-t/\tau})$$

$$e^{-t/\tau} = 0.5$$

$$-\frac{t}{\tau} = \ln 0.5 = -0.693$$

Thus in a time of 0.693τ the voltage will reach half its steady state voltage. The time taken to reach $0.75V$ is given by:

$$0.75V = V(1 - e^{-t/\tau})$$

$$e^{-t/\tau} = 0.25$$

$$-\frac{t}{\tau} = \ln 0.25 = -1.386$$

Thus in a time of 1.386τ the voltage will reach three-quarters of its steady state value. This is twice the time taken to reach half the steady state voltage. This is a characteristic of exponential graphs: if t is the time taken to reach half the steady state value, then in $2t$ it will reach three-quarters, in $3t$ it will reach seven-eighths, etc. *In each successive time interval of 0.7τ the p.d. across the capacitor reduces its value by a half* (Figure 20.8) (Table 20.1).

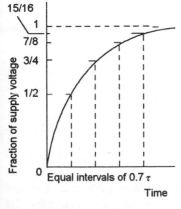

Figure 20.8 *Voltage across the capacitor*

Table 20.1 *Growth of the p.d. across the capacitor*

Time	v_C
0	0
0.7T	0.5V
1.4T	0.75V
2.1T	0.875V
2.8T	0.938V
3.5T	0.969V

When $t = 1\tau$ then $v_C = V(1 - e^{-1}) = 0.632V$. Thus in a time equal to the time constant the voltage across the capacitor rises to 63.2% of the steady state voltage. When $t = 2\tau$ then $v_C = V(1 - e^{-2}) = 0.865V$. Thus the voltage across the capacitor rises to 86.5% of the steady state voltage. When $t = 3\tau$ then $v_C = V(1 - e^{-3}) = 0.950V$. Thus the voltage across the capacitor rises to 95.0% of the steady state voltage.

Example

What is the time constant for a circuit having a capacitance of 8 μF in series with a resistance of 1 MΩ?

Time constant $= RC = 1 \times 10^6 \times 8 \times 10^{-6} = 8$ s

20.4 *RC* circuit: discharging

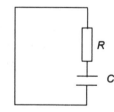

Figure 20.9 *Discharge of a capacitor*

When a voltage is applied to a capacitor and current flows to one of its plates and from the other, it becomes charged. This is what has been considered earlier in this chapter as a consequence of the application of a step voltage input to a *RC* circuit. Now consider what happens if a capacitor that has been charged by a voltage *V* being applied to the *RC* circuit now has the voltage removed and a current path from one terminal to the other provided through the resistor (Figure 20.9). A charged capacitor has a voltage between its terminals and this will result in a current flowing through the resistor and a voltage developing across it. Because there are no other sources of voltage in the circuit, if v_C is the voltage across the capacitor at some instant of time and v_R the voltage across the resistor, we must have:

$$v_C + v_R = 0$$

and so $v_R = -v_C$. The circuit current *i* is thus, at any instant:

$$i = \frac{v_R}{R} = -\frac{v_C}{R}$$

As the charge flows from one plate of the capacitor through the circuit to the other plate, so the charge on the capacitor decreases with time and hence the voltage across the capacitor decreases. Consequently the voltage across the resistor and the circuit current decreases with time.

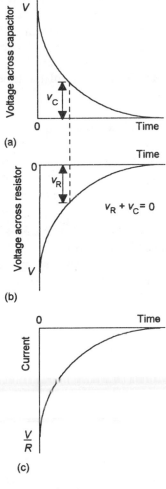

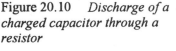

Figure 20.10   *Discharge of a charged capacitor through a resistor*

Figure 20.10 shows how they vary with time. Note that the current flows in the opposite direction to that occurring during charging.

20.4.1 Rate of change of v_C

As with the charging of a capacitor (Section 20.3.1), the circuit current i is the rate of movement of charge through the circuit. But each bit of charge moved through the circuit changes the voltage across the capacitor; $q = Cv_C$ and so:

i = rate of movement of charge = rate of change of Cv_C

$= C \times$ rate of change of v_C

We can write this, in calculus notation, as:

$$i = \frac{dq}{dt} = \frac{d(Cv_C)}{dt} = C\frac{dv_C}{dt}$$

The current is thus proportional to the rate of change of the voltage across the capacitor. At any instant of time $V = v_C + v_R = 0$. But $v_R = iR$ and so we can write:

$$0 = v_C + v_R = v_C + iR = v_C + RC\frac{dv_C}{dt}$$

and so:

$$-v_C = RC\frac{dv_C}{dt}$$

Initially, $v_C = V$ and so the initial rate of change with time of the voltage across the capacitor is $-V/RC$.

The product RC has the unit of time and we define *time constant* $\tau = RC$. The significance of the time constant is that, the bigger the time constant the smaller the initial rate of discharging of the capacitor and the longer it will take to completely discharge.

Example

A capacitor is charged to a voltage of 12 V. It is then allowed to discharge through a resistance of 2 kΩ. What will be (a) the initial circuit current, (b) the circuit current when the voltage across the capacitor has dropped to 2 V?

(a) Initially the voltage across the capacitor is 12 V and so the voltage across the resistor is –12 V. The initial current is thus –12/1000 = –12 mA.

(b) When the voltage across the capacitor has dropped to 2 V the voltage across the resistor will be –2 V and so the circuit current is –2/1000 = –2 mA.

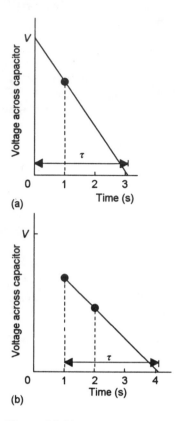

(a)

(b)

Figure 20.11   *Steps: (a) first, (b) second*

20.4.2 Step-by-step determination of current and voltage values

In the same way as earlier in this chapter step-by-step analyses were carried out of the charging of a capacitor in a series RC circuit connected to a voltage source, so such analyses can be carried out for the discharge. For example, for the step-by-step method based on the drawing of rates of change of the voltage across the capacitor with the aid of the time constant, the steps would be:

First step: The initial rate of change of v_C with time is $-V/RC = -V/\tau$, where τ is the time constant. Thus we can draw the initial slope of the graph line on a graph of v_C against time in the way shown in Figure 20.11(a), i.e. we mark out the point from time 0 along the horizontal axis for a time equal to the time constant. Suppose we consider this to represent the slope for the first 1 s.

Second step: At 1 s we consider what the slope would be if we had started off with the voltage at this point and obtain a new slope by repeating the marking out of a point from that time along the horizontal axis for a time equal to the time constant (Figure 20.11(b)). We now consider this line to give the slope for the time between 1 s and 2 s.

Following steps: We then repeat the above step for successive time intervals and so obtain the graph shown in Figure 20.12.

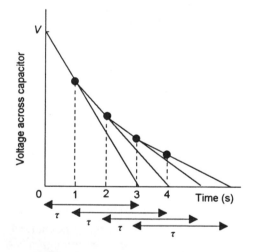

Figure 20.12 *Step-by-step analysis of a series RC circuit*

20.4.3 Calculus derivation of current and voltage values

We can obtain the equation of the graph of v_C with time by integrating the equation:

$$-v_C = RC\frac{dv_C}{dt}$$

Separation of variables gives:

$$\frac{dv_C}{-v_C} = \frac{dt}{RC}$$

Integration then gives:

$$-\int_V^{v_C} \frac{dv_C}{v_C} = \int_0^t \frac{dt}{RC}$$

$$-\ln v_C + \ln V = \frac{t}{RC}$$

$$\frac{V}{v_C} = e^{t/RC}$$

$$\frac{v_C}{V} = e^{-t/RC}$$

and so:

$$v_C = V e^{-t/RC}$$

Since $v_R = -v_C$ then:

$$v_R = -V e^{-t/RC}$$

Since $i = v_R/R$ then:

$$i = -\frac{V}{R} e^{-t/RC}$$

The above three equations are those describing the graphs in Figure 20.10.

Example

Determine after 25 ms the circuit current and voltage across the capacitor of a circuit in which the capacitor of capacitance 0.01 µF, which has been charged to a voltage of 10 V, discharges through a resistance of 5 MΩ.

$RC = 0.01 \times 10^{-6} \times 5 \times 10^6 = 0.05$ s. Thus:

$$i = -\frac{V}{R} e^{-t/RC} = -\frac{10}{5 \times 10^6} e^{-0.025/0.05} = -1.2 \ \mu A$$

$$v_C = V e^{-t/RC} = 10 \ e^{-0.025/0.05} = 6.1 \ V$$

20.4.4 Time constant

As before we can write the time constant $\tau = RC$. Thus, $v_C = V e^{-t/\tau}$. The time taken for v_C to drop from V to $0.5V$ is thus given by:

$$0.5V = V e^{-t/\tau}$$

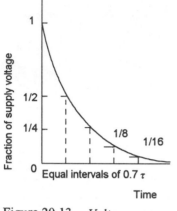

Figure 20.13   *Voltage across the capacitor*

$$e^{-t/\tau} = 0.5$$

$$-\frac{t}{\tau} = \ln 0.5 = -0.693$$

Thus in a time of 0.693τ the voltage will drop to half its initial voltage. The time taken to drop to $0.25V$ is given by:

$$0.25V = V\,e^{-t/\tau}$$

$$e^{-t/\tau} = 0.25$$

$$-\frac{t}{\tau} = \ln 0.25 = -1.386$$

Thus in a time of 1.386τ the voltage will drop to one-quarter of its initial voltage. This is twice the time taken to drop to half the voltage. This is a characteristic of a decaying exponential graph: if t is the time taken to reach half the steady state value, then in $2t$ it will reach one-quarter, in $3t$ it will reach one-eighth, etc. In each of these time intervals it reduces its value by a half (Figure 20.13) (Table 20.2).

Table 20.2 *Discharge of a capacitor*

Time	v_C
0	V
$0.7T$	$0.5V$
$1.4T$	$0.25V$
$2.1T$	$0.125V$
$2.8T$	$0.0625V$
$3.5T$	$0.03125V$

When $t = 1\tau$ then $v_C = V\,e^{-1} = 0.632V$. Thus in a time equal to the time constant the voltage across the capacitor drops to 63.2% of the initial voltage. When $t = 2\tau$ then $v_C = V\,e^{-2} = 0.135V$. Thus the voltage across the capacitor drops to 13.5% of the initial voltage. When $t = 3\tau$ then $v_C = V\,e^{-3} = 0.050V$. Thus the voltage across the capacitor drops to 5.0% of the initial voltage.

20.5 Rectangular waveforms and *RC* circuits

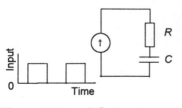

Figure 20.14   *RC circuit*

Consider a series *RC* circuit with the input of a rectangular pulse (Figure 20.14. Such an input is comparable to a step input to the circuit which is followed some time later by the voltage supply effectively short-circuiting and so discharging the capacitor through the resistance. The waveform of the voltage across the capacitor, the voltage across the resistor and the circuit current will depend on the time constant of the circuit and so the extent to which the capacitor can become fully charged and fully discharged in the time given by the input waveform. Figure 20.15(a), (b) and (c) shows some typical waveforms.

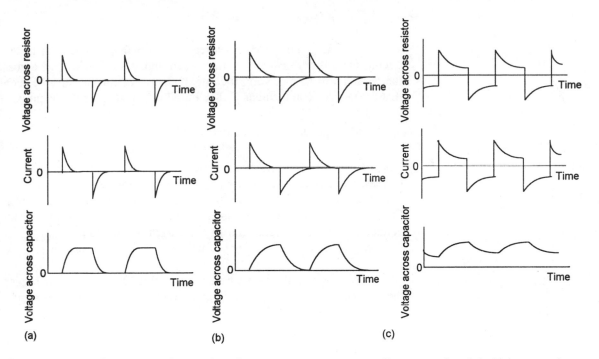

Figure 20.15 *(a) Output waveforms when the time constant is (a) small compared with half the periodic time of the input pulse, (b) the same as half the periodic time, (c) longer than half the periodic time*

20.6 *RL* circuit: current growth

Consider a circuit of an inductance in series with resistance and to which a step voltage is applied (Figure 20.16). When the switch is closed the current in the circuit starts to grow. The changing current in the inductance generates a changing magnetic field in the inductance coil and generates a back e.m.f. which slows down the changing current. The back e.m.f. *e* depends on the rate of increase of the current d*i*/d*t* through the inductor:

$$e = -L\frac{di}{dt}$$

where *L* is the inductance. To maintain a current through the inductor, and hence the circuit, the source must supply a voltage across the inductor *v* to cancel out the induced e.m.f. Thus the voltage drop across the inductor when there is a current *i* is

$$\text{voltage across inductor} = L\frac{di}{dt}$$

Thus for the circuit we have *V* = voltage across resistor + voltage across inductor and so:

$$V = iR + L\frac{di}{dt}$$

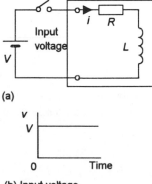

(a)

(b) Input voltage

Figure 20.16 *Step voltage applied to RL circuit*

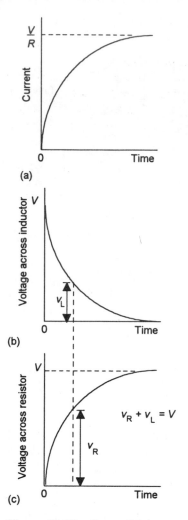

(a)

(b)

(c)

Figure 20.17 *Series RL*

$$\frac{di}{di} = \frac{V}{L} - \frac{R}{L}i$$

At the instant the switch is closed and the voltage is applied to the circuit, there is zero current in the circuit and, since the voltage across the resistor will then be zero, the rate of change of current must be such that the induced e.m.f. equals V. The initial rate of change of current is:

$$\text{initial rate of change of current with time} = \frac{V}{L}$$

As the current in the circuit increases, so the voltage across the resistor increases and hence the voltage across the inductor decreases. This can only mean that the rate of change of current with time is decreasing. Eventually the entire voltage V is across the resistor, there then being no voltage across the inductor and so the current ceases to change. When this occurs the current has reached its steady state value of V/R (Figure 20.17).

For such a circuit, the *time constant* τ is defined as $\tau = L/R$. The bigger the time constant the longer it will take for the current to reach its steady value.

Example

A coil has a resistance of 20 Ω and an inductance of 0.5 H. What will be the maximum rate of change of current with time when a d.c. voltage of 100 V is connected across the coil?

The coil is effectively a resistance in series with an inductance with a step voltage applied to the circuit. The maximum rate of change of current with time occurs at the instant the voltage is applied and is $V/L = 100/0.5 = 200$ V/s.

Example

A coil has a resistance of 20 Ω and an inductance of 0.5 H. What is its time constant?

Time constant $= L/R = 0.5/20 = 0.025$ s

20.6.1 Step-by-step determination of current and voltage

In the same way as earlier in this chapter step-by-step analyses were carried out of the charging of a capacitor in a series RC circuit connected to a voltage source, so such analyses can be carried out for the growth of current in a series RL circuit.

First step: The initial rate of change of current with time is $V/L = IR/L = I/\tau$, where I is the final steady state current when all the voltage is across the resistance and τ is the time constant. Thus we can draw the initial slope of the graph line on a graph of i against time in the way shown in Figure 20.18(a), i.e. we mark out the point from time

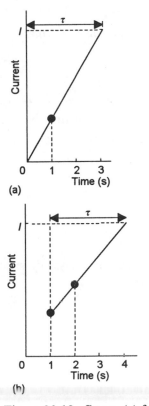

(a)

(b)

Figure 20.18 *Steps: (a) first, (b) second*

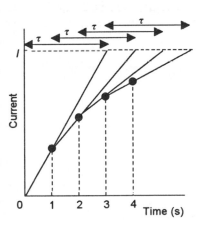

Figure 20.19 *Step-by-step analysis*

0 along the horizontal line at current I for a time equal to the time constant. Suppose we consider this to be the slope for the first 1 s.

Second step: At 1 s we consider what the slope would be if we had started off with the current at this point and obtain a new slope by repeating the marking out of a point from that time along the horizontal line at I for a time equal to the time constant (Figure 20.18(b)). We now consider this line to give the slope for the time between 1 s and 2 s.

Following steps: We then repeat the above step for successive time intervals and so obtain the graph shown in Figure 20.19.

20.6.2 Calculus determination of current and voltage

We can obtain the equation of the graph of the current i with time by integrating the equation:

$$V = iR + L\frac{di}{dt}$$

Separation of variables gives:

$$\frac{di}{(V/R) - i} = \frac{dt}{R/L}$$

Integration then gives:

$$\int_0^t \frac{di}{(V/R) - i} = \int_0^t \frac{dt}{R/L}$$

$$-\ln[(V/R) - i] + \ln(V/R) = \frac{Rt}{L}$$

$$\ln\left[\frac{(V/R) - i}{(V/R)}\right] = -\frac{Rt}{L}$$

$$\frac{(V/R) - i}{(V/R)} = e^{-Rt/L}$$

$$i = \frac{V}{R}(1 - e^{-Rt/L})$$

The voltage across the resistor $v_R = iR$ and so:

$$v_R = V(1 - e^{-Rt/L})$$

The voltage across the inductor $v_L = V - v_R$ and so is:

$$v_L = V e^{-Rt/L}$$

The above three equations describe the graphs in Figure 20.17.

20.6.3 Time constant

For a series LR circuit we can write the time constant τ as L/R. Thus:

$$i = \frac{V}{R}(1 - e^{-t/\tau})$$

When t is very large then the exponential term becomes 0 and so the current becomes the steady state current $I = V/R$. What time will be required for i to reach $0.5I$?

$$0.5I = I(1 - e^{-t/\tau})$$

$$e^{-t/\tau} = 0.5$$

$$-\frac{t}{\tau} = \ln 0.5 = -0.693$$

Thus in a time of 0.693τ the current will reach half its steady state current. The time taken to reach $0.75I$ is given by:

$$0.75I = I(1 - e^{-t/\tau})$$

$$e^{-t/\tau} = 0.25$$

$$-\frac{t}{\tau} = \ln 0.25 = -1.386$$

Thus in a time of 1.386τ the current will reach three-quarters of its steady state value. This is twice the time taken to reach half the steady state current. This is a characteristic of exponential graphs: if t is the time taken to reach half the steady state value, then in $2t$ it will reach three-quarters, in $3t$ it will reach seven-eighths, etc. In each of these time intervals it reduces its value by a half (Figure 20.20) (Table 20.3).

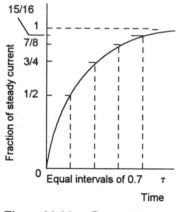

Figure 20.20   *Current in series LR circuit*

Table 20.3 *Growth of current in LR circuit*

Time	i_L
0	0
0.7T	0.5I
1.4T	0.75I
2.1T	0.875I
2.8T	0.938I
3.5T	0.969I

When $t = 1\tau$ then $i = I(1 - e^{-1}) = 0.632I$. Thus in a time equal to the time constant the current rises to 63.2% of the steady state current. When $t = 2\tau$ then $i = I(1 - e^{-2}) = 0.865I$. Thus the current rises to 86.5% of the steady state current. When $t = 3\tau$ then $i = I(1 - e^{-3}) = 0.950I$. Thus the current rises to 95.0% of the steady state current.

Example

A series RL circuit with a d.c. voltage supply of 24 V has a resistance of 100 Ω and an inductance of 50 mH. Determine (a) the time constant, (b) the voltage across the inductor and (c) the circuit current 0.4 ms after the supply is switched on, (d) the steady state current.

(a) The time constant = $L/R = 50 \times 10^{-3}/100 = 0.5$ ms.
(b) The voltage across the inductor after 0.4 ms is:

$$v_L = V\,e^{-Rt/L} = V\,e^{-t/\tau} = 24\;e^{-0.4/0.5} = 10.8 \text{ V}$$

(c) The current after 0.4 ms is:

$$i = \frac{V}{R}(1 - e^{-Rt/L}) = \frac{24}{100}(1 - e^{-0.4/0.5}) = 0.13 \text{ A}$$

(d) The steady state current is $I = V/R = 24/100 = 0.24$ A.

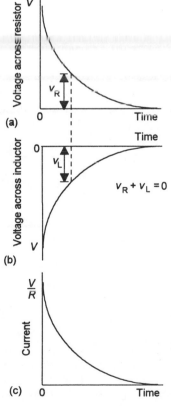

Figure 20.21 *RL circuit*

20.7 *RL* circuit: current decay

When the current through a series RL circuit is switched off (Figure 20.21), the magnetic field of the inductance changes and induces an e.m.f. in the inductance which opposes the decreasing current. As a result the current does not abruptly drop to a zero value when the current is switched off but decays exponentially to a zero value. Figure 20.22 shows how the circuit current and the voltages across the resistor and inductor vary with time.

The voltage across resistor + voltage across inductor = 0, and so:

$$Ri + L\frac{di}{dt} = 0$$

$$\frac{di}{dt} = -\frac{R}{L}i$$

Initially the current in the circuit is $I = V/R$ and so immediately the current is switched off the rate of change of current is:

$$\text{initial rate of change of current} = -\frac{R}{L}I$$

For such a circuit the *time constant* τ is defined as $\tau = L/R$. The larger the time constant the longer it takes for the current to decay to zero.

Example

A coil with a resistance of 15 Ω and an inductance of 2 H has been connected to a 24 V supply for some time. Determine the current through the coil and the initial rate of change of the current with time when the supply is removed and replaced by a shorting link.

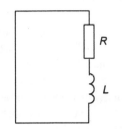

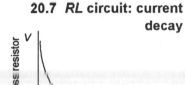

Figure 20.22 *(a) Voltage across R, (b) voltage across L, (c) current*

The current through the coil is $V/R = 24/15 = 1.6$ A. When the coil is shorted the initial rate of change of current with time is $-RI/L = -15 \times 1.6/2 = 12$ A/s.

20.7.1 Step-by-step determination of current and voltage

In the same way as earlier in this chapter step-by-step analyses were carried out of the current growth in a series RL circuit connected to a voltage source, so such analyses can be carried out for the decay of current.

First step: The initial rate of change of i with time is $-RI/L = -I/\tau$, where τ is the time constant. Thus we can draw the initial slope of the graph line on a graph of current against time in the way shown in Figure 20.23(a), i.e. we mark out the point from time 0 along the horizontal axis for a time equal to the time constant. Suppose we consider this to represent the slope for the first 1 s.

Second step: At 1 s we consider what the slope would be if we had started off with the current at this point and obtain a new slope by repeating the marking out of a point from that time along the horizontal axis for a time equal to the time constant (Figure 20.23(b)). We now consider this line to give the slope for the time between 1 s and 2 s.

Following steps: We then repeat the above step for successive time intervals and so obtain the graph shown in Figure 20.24.

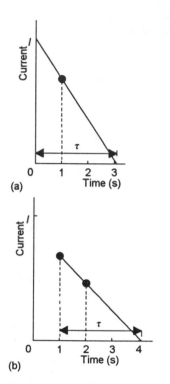

Figure 20.23   *Steps: (a) first, (b) second*

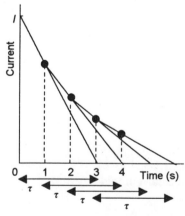

Figure 20.24   *Step-by-step analysis*

20.7.2 Calculus determination of current and voltage

We can obtain the equation of the graph of the current i with time by integrating the equation:

$$0 = iR + L\frac{di}{dt}$$

Separation of variables gives:

$$-\frac{di}{i} = \frac{dt}{R/L}$$

Integration then gives:

$$-\int_I^i \frac{di}{i} = \int_0^t \frac{dt}{R/L}$$

$$-\ln i + \ln I = \frac{Rt}{L}$$

$$\ln\left(\frac{i}{I}\right) = -\frac{Rt}{L}$$

$$i = I\,e^{-Rt/L}$$

The voltage across the resistor $v_R = iR$ and so:

$$v_R = IR\,e^{-Rt/L}$$

The voltage across the inductor $v_L = -v_R$ and so is:

$$v_L = -IR\,e^{-Rt/L}$$

The above three equations describe the graphs in Figure 20.22.

20.7.3 Time constant

For such a LR circuit, we can write the time constant $\tau = L/R$. Thus:

$$i = I\,e^{-t/\tau}$$

The time taken for the current to drop from I to $0.5I$ is given by $0.5I = I\,e^{-t/\tau}$ and so $e^{-t/\tau} = 0.5$ and we have:

$$-\frac{t}{\tau} = \ln 0.5 = -0.693$$

Thus in a time of 0.693τ the current will drop to half its initial voltage. The time taken to drop to $0.25I$ is given by $0.25I = I\,e^{-t/\tau}$ and so $e^{-t/\tau} = 0.25$ and we have:

$$-\frac{t}{\tau} = \ln 0.25 = -1.386$$

Thus in a time of 1.386τ the current will drop to one-quarter of its initial voltage. This is twice the time taken to drop to half the voltage. This is a characteristic of a decaying exponential graph: if t is the time taken to reach half the steady state value, then in $2t$ it will reach one-quarter, in $3t$ it will reach one-eighth, etc. In each of these time intervals it reduces its value by a half (Figure 20.25) (Table 20.4).

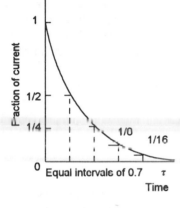

Figure 20.25   *Current decay*

Table 20.4   *Decay of current in series LR circuit*

Time	i_C
0	I
0.7T	0.5I
1.4T	0.25I
2.1T	0.125I
2.8T	0.0625I
3.5T	0.03125I

When $t = 1\tau$ then $i = I\,e^{-1} = 0.632I$. Thus in a time equal to the time constant the current drops to 63.2% of the initial voltage. When $t = 2\tau$ then $i = I\,e^{-2} = 0.135I$. Thus the current drops to 13.5% of the initial current. When $t = 3\tau$ then $i = I\,e^{-3} = 0.050I$. Thus the current drops to 5.0% of the initial current.

Example

A coil of inductance 200 mH and resistance 8 kΩ is connected to a d.c. voltage source of 16 V. Some time after a steady current exists, the voltage source is short-circuited. What will be (a) the current at the time the short-circuit occurs, (b) the current 10 μs later, (c) the time taken for the current to fall to 10% of its initial value?

(a) The initial current is $I = V/R = 16/8000 = 2$ mA.
(b) The time constant of the circuit $= L/R = 0.200/8000 = 25$ μs. Thus the current after 10 μs is $i = I e^{-t/\tau} = 2 e^{-10/25} = 1.3$ mA.
(c) The time taken for the i to equal $0.1I$, is given by $i = I e^{-t/\tau}$ and so $0.1I = I e^{-t/25}$. Thus $\ln 0.1 = -t/25$ and hence $t = 57.6$ μs.

20.8 Rectangular waveforms and RL circuits

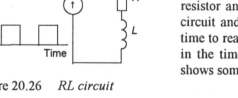

Figure 20.26   *RL circuit*

Consider a series *RL* circuit with the input of a rectangular pulse (Figure 20.26). Such an input is comparable to a step input to the circuit which is followed, some time later by the voltage supply effectively short-circuiting and so causing the decay of the current through the resistance. The waveform of the voltage across the inductor, the voltage across the resistor and the circuit current will depend on the time constant of the circuit and so the extent to which the current through the inductor has time to reach the steady state value that would otherwise occur and decay in the time given by the input waveform. Figure 20.27(a), (b) and (c) shows some typical waveforms.

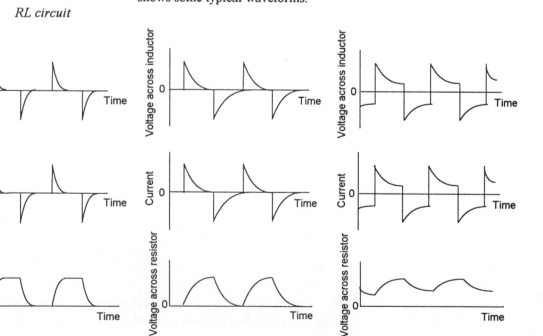

(a)   (b)   (c)

Figure 20.27 *Output waveforms when the time constant is (a) small compared with half the periodic time of the input pulse, (b) the same as half the periodic time, (c) longer than half the periodic time*

Activities

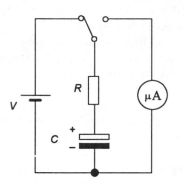

Figure 20.28 *Activity 1*

1 Determine the graph for the variation of current with time for the discharge of a charged capacitor. Figure 20.28 shows a possible circuit; possible values to give a discharge graph which will take a reasonable amount of time are $C = 500$ μF, $R = 100$ kΩ, 100 μA meter, $V = 10$ V or $C = 10\ 000$ μF, $R = 5$ kΩ, 1 mA meter, $V = 10$ V.

2 Determine the inductance L of an inductor by measuring the initial rate of change of current through it when a voltage V is first applied to a series circuit of it and a resistor; the initial rate of change of current is V/L. An oscilloscope can be used to monitor the way the voltage across the resistor varies with time and hence, since $i = v_R/R$ give the circuit current. To give a large inductance, the inductor can be a coil mounted on an iron magnetic circuit with the resistance being perhaps 100 Ω and the voltage $V = 1.5$ V.

Problems

1 What is the initial rate at which the voltage across a 0.1 μF capacitor will change with time when it is charged through a resistance of 5 kΩ by a voltage being switched from 0 to 10 V?

2 What is the initial rate at which the voltage across a 10 μF capacitor will change with time when it is charged through a resistance of 20 kΩ by a voltage being switched from 0 to 6 V?

3 A series RC circuit has a resistance of 1 MΩ and a capacitance of 0.01 μF. Using a graphical step-by-step method, estimate the voltage across the capacitor 20 ms after a steady voltage of 20 V is applied to the circuit.

4 A series RC circuit has a resistance of 50 kΩ and a capacitance of 20 μF. Determine, (i) by a graphical step-by-step method and (ii) algebraically, the current (a) initially and (b) 1 s, after a steady voltage of 20 V is connected to the circuit.

5 A series RC circuit has a resistance of 15 kΩ and a capacitance of 0.02 μF and a steady voltage of 30 V is connected to the circuit. Determine (a) the time constant τ of the circuit and, (i) by a graphical step-by-step method and (ii) algebraically, (b) the current after 1τ, (c) the current after 1.5τ, (d) the steady state current.

6 A series RC circuit has a resistance of 15 kΩ and a capacitance of 0.02 μF and a steady voltage of 30 V is connected to the circuit. Determine, (i) by a graphical step-by-step method and (ii) algebraically, the voltage across the resistor (a) initially, (b) after 0.3 ms, (c) after 1 ms.

7 A series RC circuit has a resistance of 0.5 kΩ and a capacitance of 500 μF and a steady voltage of 10 V is connected to the circuit. Determine (a) the time constant τ of the circuit and, (i) by a graphical step-by-step method and (ii) algebraically, (b) the voltage across the capacitor after 0.25 s, (c) the voltage across the capacitor after 0.5 s.

8 A 16 μF capacitor is charged to 10 V and then discharged through a resistance of 50 kΩ. Determine (a) the time taken for the voltage across the capacitor to drop to 2 V, (b) the circuit current after 0.5 s.

9 A 5 μF capacitor is charged to 10 V and then discharged through a resistance of 2 MΩ. Determine (a) the initial current, (b) the time taken for the voltage across the capacitor to drop to 6.3 V.

10 A 0.2 μF capacitor is charged to 5 V and then discharged through a resistance of 40 kΩ. Determine (a) the voltage across the capacitor after 15 ms, (b) the voltage across the resistor after 15 ms.

11 A coil with a resistance of 50 Ω and an inductance of 2.5 H has a d.c. voltage of 100 V connected across it. Determine (a) the time constant of the circuit, (b) the initial rate of change of current, (c) the circuit current 0.15 s after the voltage is applied to the coil, (d) the final steady state current.

12 Determine by a graphical step-by-step method the current after 0.4 s when a voltage of 10 V is applied to a coil of resistance 25 Ω and inductance 5 H.

13 A coil with a resistance of 25 Ω and an inductance of 2.5 H has a d.c. voltage of 100 V connected across it. Determine (a) the time constant of the circuit, (b) the initial rate of change of current, (c) the circuit current 0.15 s after the voltage is applied to the coil, (d) the final steady state current.

14 A series RL circuit has a d.c. voltage of 10 V applied to a resistance of 125 Ω and an inductance of 0.25 H. Determine (i) graphically by a step-by-step method, (ii) algebraically, the circuit current after 5 ms.

15 A coil of inductance 5 H and resistance 20 Ω is connected to a d.c. voltage source of 100 V. Some time after a steady current exists, the voltage source is short-circuited. What will be the current (a) at the time the short-circuit occurs, (b) 1 s later?

16 A relay coil has a resistance of 100 Ω and an inductance of 100 mH. A rectangular voltage pulse of amplitude 5 V and duration 5 ms is applied to the coil. If the relay contacts close when the current in the coil reaches 40 mA and open when it drops to 15 mA, determine the length of time for which the contacts remain open.

17 A coil of inductance 20 H and resistance 5 Ω is connected to a d.c. voltage source. When the voltage is applied, the initial rate of increase of current is 4 A/s. What is (a) the value of the applied voltage, (b) the rate of growth of the current when the circuit current is 5 A?

18 A coil of inductance 10 H and resistance 10 Ω is connected to a d.c. voltage source of 100 V. What is (a) the current 0.1 s after switching the voltage on, (b) the time taken for the current to decrease to half its initial value?

19 If the time constant of a coil is measured and found to be 50 ms, what will be its inductance if it has a resistance of 2 Ω?

20 When a d.c. voltage of 100 V is applied to the field coil of a machine, after 2 s the current has risen to 6.32 A and eventually it reaches 10 A. What is the inductance of the coil?

21 Three-phase a.c.

21.1 Introduction

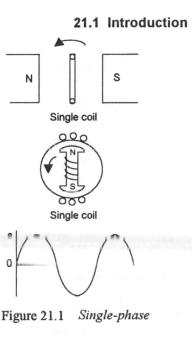

Single coil

Single coil

Figure 21.1 *Single-phase*

The mains electricity supply you obtain in Britain in your home from a power point, between the live and neutral connections, is alternating current which can be described by a simple sinusoidal waveform. It might be thought that such alternating current is generated and transmitted in the same form, i.e. a coil rotating in a magnetic field and producing alternating current which is transmitted out along the live wire and returned back along the neutral wire. Figure 21.1 illustrates this with either a coil rotating in a magnetic field, or a magnetic field rotating in a coil. Such a form of output is termed *single-phase*. However, the generation, transmission and distribution of alternating current in Britain is by means of three phases. This involves three separate voltages of equal amplitude and frequency being generated by using three coils rotating in a magnetic field (or the more practical form of a magnetic field rotating in three coils) and giving voltages which are separated in phase by 120° (Figure 21.2). Each such voltage is termed a *phase* of the supply and the supply termed *three-phase*. A particular house in Britain will utilise just one of the phases, different houses being connected to different phases. This chapter is a discussion of the characteristics of three-phase supply.

21.2 The three phases

For the three coils rotating in Figure 21.2, coil a is shown at the point in its rotation when the generated e.m.f. is zero. At the same instant of time, coil b is at the point in its rotation when it is 120° round from coil 1 and coil c is 240° round from coil 1 (note that I have put the coil letters at the ends of the coils so that the same direction of voltages is obtained in each when it is in the same angular position). The three voltages can be described by:

$$e_a = E_m \sin \omega t, \quad e_b = E_m \sin (\omega t - 2\pi/3), \quad e_c = E_m \sin (\omega t - 4\pi/3)$$

This sequence of e.m.f.s is known as the *positive phase sequence*; this is the sequence in which the e.m.f.s reach their peak voltages in the sequence a, b, c. The sequence a, c, b gives, what is termed, a *negative phase sequence*.

If the three phases have the same peak value and the same frequency (as in Figure 21.2) then:

$$e_a + e_b + e_c = 0$$

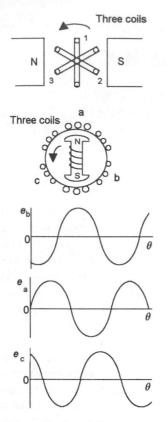

Figure 21.2   *Three-phase*

Look at the result of the addition of the three graphs in Figure 21.2. When the instantaneous sum of the e.m.f.s is zero, the system is said to be *balanced*.

21.2.1 Advantages of three-phase supply

The generation and transmission of three-phase alternating current has the advantages of:

1   With a single-phase alternator we have just a single coil rotating in the magnetic field; by using three coils we can make more efficient use of the magnetic field and obtain a greater power output without increasing the size of the alternator.

2   The same supply can be used for both industrial and domestic use since domestic users can obtain a single-phase alternating current and industrial users three-phase. Three-phase motors are able to generate a greater power output than single phase for the same size motor. This is because, rather than just having one coil which is made to rotate in a magnetic field, they have three coils at 120° spacing. There is also the advantage that the power delivered to a load is constant at all times and not pulsing as with single-phase; the instantaneous value of the power is a constant with three-phase and not fluctuating, as occurs with single-phase.

3   A single-phase supply requires the use of two wires for its transmission; a three phase supply can be transmitted by three wires. Thus, effectively, three single-phase alternating currents can be transmitted by just three wires rather than the six they would have taken if transmitted as just single-phase. The amount of copper required for the cables used for the transmission as three-phase is thus less than that which would be required for single-phase.

21.3 Connection of phases

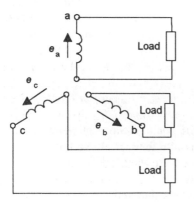

Figure 21.3   *Independent connection of loads*

Each of the three outputs from a three-phase alternator may be independently connected to its own particular load (Figure 21.3). This would require two wires (the term *lines* is generally used for these conductors) for each phase and so a total of six wires between the alternator and the loads. However, it is possible to connect the phases in such a way that less than six wires are used. The two methods are *star connection* and *delta connection*.

21.3.1 Star connection

With star connection, the three alternator coils inputs are connected to a common point, called the *neutral* or *star point* (Figure 21.4). The line from this point is called the *neutral line*. Thus this system involves four lines to connect to the loads. The voltages of each of the output lines with reference to the neutral line, i.e. V_{aN}, V_{bN} and V_{cN}, are the *phase voltages*. The term *line-to-line voltages* is used for the voltages between output lines, e.g. the voltage V_{ab} between the a and b lines.

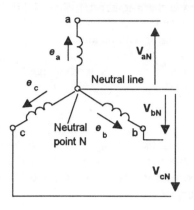

Figure 21.4 *Star connection*

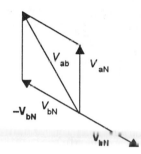

Figure 21.5 $\mathbf{V_{ab}}$

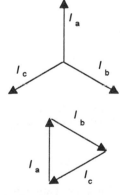

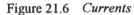

Figure 21.6 *Currents*

The line-to-line voltage between the a and b lines is the phasor difference between the a and b phase voltages:

$$\mathbf{V_{aY}} = \mathbf{V_{aN}} - \mathbf{V_{bN}}$$

Figure 21.5 shows the phasor diagram. The phase difference between $\mathbf{V_{aN}}$ and $-\mathbf{V_{bN}}$ is 60°. If we have a balanced system then $V_{aN} = V_{bN}$ and so $\mathbf{V_{ab}}$ must split the 60° angle in two equal parts. Thus, if we resolve $\mathbf{V_{aN}}$ and $-\mathbf{V_{bN}}$ into directions along $\mathbf{V_{ab}}$ and at right angles to it, then:

$$V_{ab} = V_{aN} \cos 30° + V_{bN} \cos 30°$$

Since $\cos 30° = \sqrt{3}/2$ and $V_{aN} = V_{bN}$ = phase voltage V_p:

$$V_{ab} = \sqrt{3}\, V_p$$

We can draw similar diagrams for the line-to-line voltages $\mathbf{V_{ac}}$ and $\mathbf{V_{bc}}$ and obtain the same basic result:

$$\text{line-to-line voltage } V_L = \sqrt{3}\, V_p$$

The line-to-line voltage is generally just referred to as the line voltage. The three line voltages are 120° apart and 30° displaced from the phase voltages. For star connection, the current along the red line is the same as the phase current through the alternator coil, likewise for the other lines. Thus, in general, the relationship between the line current I_L and the phase current I_p is:

$$\text{line current } I_L = I_p$$

With loads connected across the alternator coils, since the neutral line is the 'return' wire for each of the three circuits, the current along the neutral line $\mathbf{I_N}$ will be the phasor sum of the currents along the a, b and c lines.

$$\mathbf{I_N} = \mathbf{I_a} + \mathbf{I_b} + \mathbf{I_c}$$

If the loads across each alternator coil are balanced, i.e. the same, then the three currents along the a, b and c lines will have the same size but will differ in phase by 120°. The current phasor diagram is thus as shown in Figure 21.6. If we place the phasors arrow-to-tail then we obtain a closed equilateral triangle. The phasor sum is zero and so the neutral current is zero. Under such conditions, the neutral line is not necessary and so we can have a three line connection to the loads.

The three-wire system requires balanced loads. A four-wire system is used when there are three separate loads and the loads might not be equal, e.g. for the supply to houses when some houses in a road will obtain their supply from one phase of the supply and other houses from other phases. Thus one house might have the a-to-neutral phase while another has the b-to-neutral phase.

Thus for a balanced star-connected alternator with balanced loads:

Line current	Phase current	Line voltage	Phase voltage
$I_L = I_p$	$I_p = I_L$	$V_L = \sqrt{3}\, V_p$	$V_p = V_L/\sqrt{3}$

Example

For the balanced three-phase star-connected system shown in Figure 21.7, if the line voltage is 440 V and the three loads each have a resistance of 100 Ω, what are the sizes of (a) the phase voltage, (b) the phase current and (c) the line current?

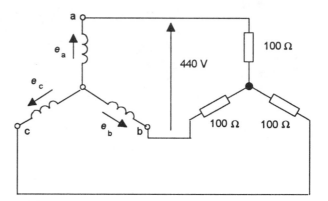

Figure 21.7 *Example*

(a) $V_p = V_L/\sqrt{3} = 440/\sqrt{3} = 254$ V.
(b) Phase current $= V_p/R = 254/100 = 2.54$ A.
(c) The line current = phase current = 2.54 A.

Example

A balanced star-connected three-phase system, similar to that shown in Figure 21.7, has three identical loads of coils with each have a reactance of 30 Ω and a resistance of 40 Ω. If the line voltage is 440 V, what is the size of the line current?

For each load, the impedance $Z = \sqrt{(R^2 + X_C^2)} = \sqrt{(40^2 + 30^2)} = 50\ \Omega$. Since $V_p = V_L/\sqrt{3}$ then $V_p = 440/\sqrt{3} = 254$ V. For a star connection $I_L = I_p = V_p/Z = 254/50 = 5.1$ A.

Example

A three-phase four-wire star-connected system has I_a as 20 A lagging V_{aN} by 30°, I_b as 50 A leading V_{bN} by 10° and I_c as 30 A in phase with V_{cN}. Determine the current along the neutral line.

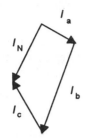

Figure 21.8   *Example*

We have $I_N = I_a + I_b + I_c$ and so can obtain I_N by a phasor diagram drawn to scale (Figure 21.8). V_a is at 0°, V_b at –120° and V_c at –240°; thus I_a is at 30°, I_b at –110° and I_c at –240°. Figure 21.8 gives I_N as about 37 A lagging by about 143°. An alternative to a scale drawing of the phasors is to resolve the current phasors into two right-angled directions and then use the Pythagoras theorem to calculate the result.

21.3.2 Delta connection

With delta connection, the alternator coils are connected so that the end of one coil is connected to the start of the next coil, as in Figure 21.9. With this system, the line-to-line between the a and b lines V_{ab} is the same as the phase voltage V_b. Likewise, the line-to-line voltage between the b and c lines is the same as the phase voltages V_c and the line-to-line voltage between the a and c lines is the same as the phase voltage V_a. Thus, for a balanced delta-connected system:

line voltage V_L = phase voltage V_p

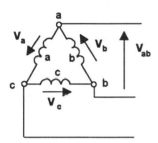

Figure 21.9   *Delta connection*

For the current at node a we must have the current I_a in the a line as:

$$I_a = I_{ba} - I_{ac}$$

where I_{ba} is the current through the b coil flowing from b to a and I_{ac} is the current through the a coil flowing from a to c. In general, the line current is thus the phasor difference between the two phase currents connected to that line. For a balanced load, the phase currents are the same for each phase. Figure 21.10 shows the resulting phasor diagram. Summing the components of I_{ba} and I_{ac} along the direction of I_a gives:

$$I_a = I_{ba} \cos 30° + I_{ac} \cos 30° = I_{ba} \times \sqrt{3}/2 + I_{ac} \times \sqrt{3}/2 = \sqrt{3} \times I_p$$

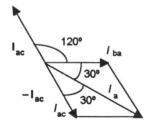

Figure 21.10   *Currents*

Thus, in general:

line current $I_L = \sqrt{3} \times$ phase current I_p

Thus, for a balanced delta-connected alternator with balanced loads:

Line current	Phase current	Line voltage	Phase voltage
$I_L = \sqrt{3}\, I_p$	$I_p = I_L/\sqrt{3}$	$V_L = V_p$	$V_p = V_L$

Example

What is the current in each phase of a delta-connected system when the line current drawn by a balanced load is 20 A?

Line current $I_L = \sqrt{3} \times I_p$ and so $I_p = I_L/\sqrt{3} = 20/\sqrt{3} = 11.5$ A.

Example

For the balanced three-phase delta-connected system shown in Figure 21.11, if the line voltage is 440 V and the three loads each have a resistance of 100 Ω, what are the sizes of (a) the phase voltage, (b) the phase current and (c) the line current?

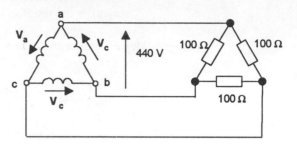

Figure 21.11   *Example*

(a) Phase voltage $V_p = V_L = 440$ V.
(b) Because the system is balanced, the phase current I_p will be the same as the current through a 100 Ω resistor. Each resistor has the line voltage across it and so $I_p = 440/100 = 4.4$ A.
(c) Line current $I_L = \sqrt{3} \times I_p = \sqrt{3} \times 4.4 = 7.6$ A.

Example

A balanced star-connected three-phase alternator has a phase voltage of 80 V and is connected to a balanced delta-connected load consisting of three 120 Ω resistors (Figure 21.12). Determine (a) the line voltage, (b) the voltage across a load resistor and (c) the current through a load resistor.

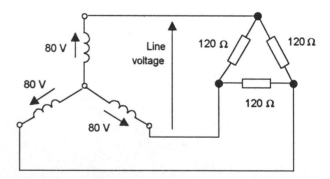

Figure 21.12   *Example*

Note that a star-connected alternator can be connected to a star-connected load or a delta-connected load and that a delta-connected alternator can be connected to a delta-connected load or a star-connected load.

(a) The line voltage given by a star-connected alternator is given by $V_L = \sqrt{3}\ V_p = \sqrt{3} \times 80 = 138.6$ V.

(b) The voltage V across a delta-connected load resistor is the same as the line voltage and so is 138.6 V.

(c) The current through a load resistor = V/R = 138.6/120 = 1.16 A.

21.4 Power in a balanced system

Consider a system having a balanced alternator and a balanced load. For a star-connected load, each of the load elements will carry the line current I_L and each will have the phase voltage V_p across it. The power developed in a load element is $P = I_L V_p \cos \phi$, where ϕ is the phase angle between the voltage and the current for the load impedance ($\cos \phi$ is the power factor). But $V_p = V_L/\sqrt{3}$ and so $P = I_L V_L/\sqrt{3} \cos \phi$. The total power for the three load elements is thus:

$$\text{total power} = 3I_L V_L/\sqrt{3} \cos \phi = \sqrt{3}\ I_L V_L \cos \phi$$

For a delta-connected load, each of the load elements carries the phase current I_p and each will have the line voltage V_L across it. The power developed in a load element is $P = I_p V_L \cos \phi$. But $I_p = I_p/\sqrt{3}$ and so $P = I_L V_L/\sqrt{3} \cos \phi$. The total power for the three load elements is thus:

$$\text{total power} = 3I_L V_L/\sqrt{3} \cos \phi = \sqrt{3}\ I_L V_L \cos \phi$$

which is the same as the expression obtained for a star connected load. The total apparent power for any balanced three-phase system is thus $\sqrt{3}\ I_L V_L$ and the total reactive power is $\sqrt{3}\ I_L V_L \sin \phi$.

Example

Determine the power consumed when a balanced three-phase supply with a line voltage of 440 V is connected to a star-connected load composed of three identical 100 Ω resistors (as in Figure 21.7).

Since we have resistors, $\phi = 0°$. The voltage across a resistor is the phase voltage $V_p = V_L/\sqrt{3} = 440/\sqrt{3} = 254$ V and the current through a resistor is the line current $I_L = V_p/R = 254/100 = 2.54$ A. Thus, total power = $\sqrt{3}\ I_L V_L \cos \phi = \sqrt{3} \times 440 \times 2.54 = 1936$ W.

Example

Determine the power consumed when a balanced three-phase supply with a line voltage of 440 V is connected to a delta-connected load composed of three identical 100 Ω resistors (as in Figure 21.11).

Since we have resistors, $\phi = 0°$. The voltage across a resistor is the line voltage and thus 440 V. The phase current I_p will be the same as the current through a 100 Ω resistor and so $I_p = 440/100 = 4.4$ A. The line current is thus $I_L = \sqrt{3} \times I_p = \sqrt{3} \times 4.4 = 7.6$ A. Thus, total power = $\sqrt{3}\ I_L V_L \cos \phi = \sqrt{3} \times 7.6 \times 440 = 5792$ W.

Example

What is the power consumed when a balanced three-phase alternator with a line voltage of 440 V supplies a balanced delta-connected load if each element has an impedance of 50 Ω and a power factor of 0.8?

The voltage across a resistor is the line voltage and thus 440 V. The phase current I_p will be the same as the current through a load element and so $I_p = 440/50 = 8.8$ A. The line current is thus $I_L = \sqrt{3} \times I_p = \sqrt{3} \times 8.8 = 15.24$ A. Thus, total power $= \sqrt{3}\, I_L V_L \cos\phi = \sqrt{3} \times 15.24 \times 440 \times 0.8 = 9292$ W.

Example

A 10 kW, 440 V three-phase a.c. motor has an efficiency of 80% and a power factor of 0.6. What will be the power input and the line current required?

The electrical input power $= 10/0.80 = 12.5$ kW. Since, for both star and delta connections, power $= \sqrt{3}\, I_L V_L \cos\phi$ then, with $V_L = 440$ V, we have:

$$I_L = \frac{P}{\sqrt{3}\ V_L \cos\phi} = \frac{12.5 \times 10^3}{\sqrt{3} \times 440 \times 0.6} = 27.3 \text{ A}$$

21.4.1 Power measurement

A *wattmeter* can be used to measure true power. Such an instrument is like a moving coil galvanometer but, instead of there being a permanent magnet, the magnetic field is provided by a fixed coil (Figure 21.13(a)). The current in the moving coil then gives rise to a force on that coil which results in a deflection of the coil proportional to the product of the magnetic field and the current and so to the product of the currents through the fixed and moving coil. Since the moving coil is connected (Figure 21.13(b)) so that the current through it is proportional to the voltage, the product is a measure of the power.

When, with a three-phase system, there is a balanced load, the total power consumed can be measured by the use of just one wattmeter. Figure 21.14 shows how it can be connected for a star-connected load; it can be used also with a delta-connected load. Since the total power consumed is three times the power consumed by a single element, connection of a wattmeter to measure the power through a single element means that the result has just to be multiplied by three to give the total power.

The power in any three-phase system, whether balanced or unbalanced can be measured using three wattmeters. A wattmeter is connected to each element and the total power is then just the sum of the three wattmeter readings. Figure 21.15 shows the arrangement for a star-connected system; a similar arrangement can also be used for a delta-connected system.

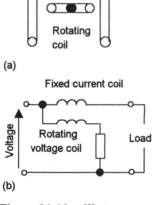

(a)

(b)

Figure 21.13 *Wattmeter*

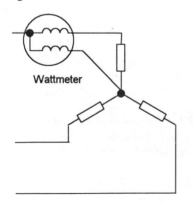

Figure 21.14 *Single wattmeter with balanced load*

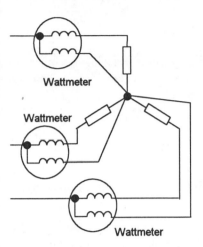

Figure 21.15   *Three wattmeters*

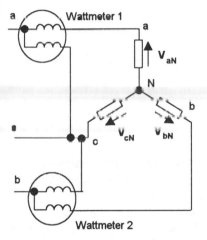

Figure 21.16   *Two wattmeters*

It is, however, possible to measure the power in a balanced or unbalanced system by using just two wattmeters. Figure 21.16 shows the arrangement with a star-connected system. The current-carrying coils are placed in series with any two of the three lines and the voltage coils between these lines and the other line.

For wattmeter 1, the instantaneous current through its current coil is i_a and the voltage across it is $v_{aN} - v_{cN}$. Thus, the instantaneous power is $P_1 = i_a(v_{aN} - v_{cN})$ and similarly for wattmeter 2, the instantaneous power is $P_2 = i_b(v_{bN} - v_{cN})$. Hence:

$$P_1 + P_2 = i_a(v_{aN} - v_{cN}) + i_b(v_{bN} - v_{cN}) = i_a v_{aN} + i_b v_{bN} - (i_a + i_b)v_{cN}$$

But at N we have $i_a + i_b + i_c = 0$ and thus $i_a + i_b = -i_c$ and:

$$P_1 + P_2 = i_a v_{aN} + i_b v_{bN} + i_c v_{cN}$$

Thus the sum of the two wattmeter readings is the sum of the instantaneous powers developed in each of the three elements.

For a balanced star-connected system with each of the elements having current lagging the corresponding phase voltage by ϕ, the phase angle between $\mathbf{I_a}$ and $\mathbf{V_{aNc}}$ is $(30° - \phi)$. Thus:

$$P_1 = V_L I_L \cos (30° - \phi)$$

The phase angle between $\mathbf{I_b}$ and $\mathbf{V_{bNa}}$ is $(30° + \phi)$. Thus:

$$P_2 = V_L I_L \cos (30° + \phi)$$

and so:

$$P_1 + P_2 = V_L I_L[\cos (30° - \phi) + \cos (30° + \phi)]$$

$$P_1 - P_2 = V_L I_L[\cos (30° - \phi) - \cos (30° + \phi)]$$

We can expand the angle terms using $\cos (A - B) = \cos A \cos B + \sin A \sin B$ and $\cos (A + B) = \cos A \cos B - \sin A \sin B$:

$$P_1 + P_2 = V_L I_L \times 2 \cos 30° \cos \phi$$

$$P_1 - P_2 = V_L I_L \times 2 \sin 30° \cos \phi$$

Since $\cos 30° = \tfrac{1}{2}\sqrt{3}$ and $\sin 30° = \tfrac{1}{2}$:

$$P_1 + P_2 = \sqrt{3} \, V_L I_L \cos \phi$$

$$P_1 - P_2 = V_L I_L \sin \phi$$

The sum of the wattmeter readings thus gives the total power consumed. Dividing the two equations:

$$\frac{P_1 - P_2}{P_1 + P_2} = \frac{\sin\phi}{\sqrt{3}\cos\phi} = \frac{\tan\phi}{\sqrt{3}}$$

If we divide the trigonometric relationship $\sin^2 A + \cos^2 A = 1$ by $\cos^2 A$ we obtain $\tan^2 A + 1 = 1/\cos^2 A$ and so $\cos^2 A = 1/(\tan^2 A + 1)$. Thus the above equation gives:

$$\cos^2\phi = \frac{1}{\left[1 + 3\left(\dfrac{P_1 - P_2}{P_1 + P_2}\right)^2\right]}$$

Hence the power factor can be determined from the wattmeter readings.

Example

Two wattmeters are used to measure the power consumed by a three-wire, balanced star-connected system (as in Figure 21.16). If the meters gave readings of 6 kW and –2 kW. What is the total power consumed and the power factor?

The total power is $P_1 + P_2 = 6 - 2 = 4$ kW and:

$$\cos^2\phi = \frac{1}{\left[1 + 3\left(\dfrac{P_1 - P_2}{P_1 + P_2}\right)^2\right]} = \frac{1}{\left[1 + 3\left(\dfrac{6 + 2}{6 - 2}\right)^2\right]} = 0.020$$

Hence the power factor is 0.14.

Problems

1 Determine the phase voltage for (a) a three-phase balanced star-connected system, (b) a three-phase balanced delta-connected system if they have a line voltage of 440 V.

2 A balanced 415 V, 50 Hz three-phase supply is connected to a star-connected load. Determine (a) the phase voltage, (b) the phase current, (c) the line current if (i) each element has a resistance of 30 Ω, (ii) each element has a resistance of 24 Ω, (iii) each element is a coil of resistance of 7 Ω and inductance 30 mH.

3 A balanced 415 V three-phase supply is connected to a balanced delta-connected load. Determine (a) the phase voltage, (b) the phase current, (c) the line current if (i) each element has a resistance of 18 Ω, (ii) each element has a resistance of 50 Ω.

4 A balanced star-connected alternator has a phase voltage of 180 V and is connected to a balanced star-connected load, each load element being a resistance of 150 Ω. Determine (a) the line voltage, (b) the phase current, (c) the line current.

5 A four-wire star-connected three-phase system has the line currents of I_a 10 A in phase with V_a, I_b 15 A in phase with V_b and I_c 12 A in phase with V_c. Determine the neutral current.

6 A balanced 415 V, 50 Hz three-phase supply is connected to a balanced star-connected load. Determine the total power dissipated

by the load if (i) each element having a resistance of 12 Ω, (ii) each element is a coil with a resistance of 7 Ω and inductance 30 mH.

7 A balanced 400 V, 50 Hz three-phase supply is connected to a balanced star-connected load. Determine the total power dissipated by the load if (i) each element having a resistance of 50 Ω, (ii) each element is a coil with a resistance of 10 Ω and inductance 20 mH, (iii) each element is a coil with a resistance of 30 Ω and a reactance of 40 Ω.

8 A balanced 415 V three-phase supply is connected to a three-phase motor with balanced coils. If the motor supplies 1100 W with a power factor of 0.8 lagging when the line current is 3.4 A, what is the power input to the motor and its efficiency?

9 A balanced 440 V, 50 Hz three-phase supply is connected to a balanced star-connected load, each element having a resistance of 40 Ω and a reactance of 30 Ω. Determine (a) the phase voltage, (b) the phase current, (c) the line current, (d) the power supplied, (e) the power factor.

10 A balanced 440 V, 50 Hz three-phase supply is connected to a balanced delta-connected load, each element consisting of a resistance of 50 Ω in series with a capacitance of 50 μF. Determine (a) the phase voltage, (b) the phase current, (c) the line current, (d) the power supplied, (e) the power factor.

11 A balanced 400 V three-phase supply is connected to a three-phase motor with balanced coils. If the motor has an efficiency of 90% and supplies 30 kW with a power factor of 0.9 lagging determine the line current?

12 A 415 V, three-phase motor requires a line current of 15 A at a power factor of 0.8 lagging. Determine the total power supplied.

13 A 4 kW three-phase motor is operating with an efficiency of 0.7 and is supplied by a 415 V, three-phase supply at a power factor of 0.8 lagging. Determine the line current.

14 Two wattmeters are used to measure the power consumed by a three-wire, balanced star-connected system (as in Figure 21.16). If the meters gave readings of 6.5 kW and –2.1 kW, what is the total power consumed and the power factor?

15 Two wattmeters are used to measure the power consumed by a three-phase balanced motor (as in Figure 21.16). If the meters gave readings of 300 kW and 100 kW, what is the total power consumed and the power factor?

16 A balanced 400 V, 50 Hz three-phase supply is connected to a balanced star-connected load, each element having a resistance of 10 Ω and a reactance of 30 Ω. Two wattmeters are used to measure the power consumed (as in Figure 21.16). What will be the readings on the meters?

22 Transformers

A transformer is a device for taking an alternating current at one voltage and transforming it to alternating current at another voltage; it basically consists of two coils, called the primary and secondary coils, linked by a common magnetic circuit (Figure 22.1). Section 11.4.1 introduced the basic principles; the following section reviews those principles and the chapter then goes on to discuss the basic constructional form of single-phase, three-phase and autotransformers and their use for power transmission and distribution, isolation and impedance matching.

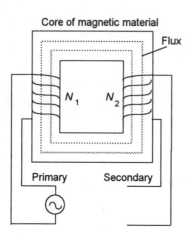

Figure 22.1 *Basic transformer*

22.1.1 Basic principles

For an ideal transformer, i.e. one where all the magnetic flux is confined to the core and links both the primary and secondary coils and the exciting current needed to establish the magnetic flux in the core is negligible, an a.c. input to the primary coil produces an alternating magnetic flux and induces the same e.m.f. in each turn of both coils. Thus, if N_1 is the number of turns in the primary coil and N_2 the number of turns in the secondary coil, then (induced e.m.f. in primary coil)/(induced e.m.f. in secondary coil) = N_1/N_2. When there is no load connected to the secondary coil, the voltage V_2 between the terminals of the coil is the same as the induced e.m.f. If there is no load then there is no current and so no energy taken from the secondary coil and so no energy is taken from the primary coil. This can only be the case if the induced e.m.f. is equal to and opposing the input voltage V_1. Thus:

$$\frac{V_1}{V_2} = \frac{N_1}{N_2} = \text{turns ratio}$$

If there is a resistive load connected to the secondary (Figure 22.2), a secondary current flows and hence power is dissipated. This current in the secondary coil produces its own alternating flux in the core and results in an alternating e.m.f. being induced in the primary coil and so a current in the primary coil. If the power losses in a transformer are negligible, then when there is a current in the primary coil the power supplied to the primary coil must equal the power taken from the secondary coil:

$$I_1 V_1 = I_2 V_2$$

where I_1 is the current in the primary coil and I_2 that in the secondary coil. We can state the above equation as:

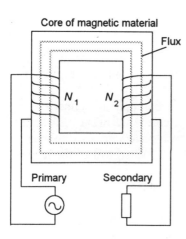

Figure 22.2 *Basic transformer*

input volt-amperes = output volt-amperes

The equation can be rearranged as $V_1/V_2 = I_2/I_1$ and so:

$$\frac{V_1}{V_2} = \frac{I_2}{I_1} = \frac{N_1}{N_2} = \text{turns ratio}$$

and:

$$I_1 N_1 = I_2 N_2$$

The product of the current through a coil and its number of turns is called its *ampere-turns*. Thus the input ampere-turns = the output ampere-turns.

Figure 22.3 shows the symbols that are used for transformers in circuit diagrams.

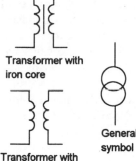

Transformer with iron core

Transformer with air core

General symbol

Figure 22.3 *Symbols*

Example

An ideal transformer is used to light a 12 V, 15 W lamp from the 240 V mains supply. Determine the turns ratio required and the current taken from the mains supply.

Turns ratio $V_1/V_2 = 240/12 = 20$. The current taken by the lamp $I_2 = 15/12 = 1.25$ A. Thus, since $I_2/I_1 =$ turns ratio, we have $I_1 = 1.25/20 = 0.0625$ A.

Example

An ideal transformer has 20 primary turns and 80 secondary turns and supplies a secondary current of 1.5 A. What is (a) the value of the ampere-turns of the transformer, (b) the primary current?

(a) Value of ampere-turns = $I_2 N_2 = 80 \times 1.5 = 120$ ampere-turns.
(b) $I_1 N_1 = I_2 N_2$ and so $I_1 = (N_2/N_1)I_2 = (80/20) \times 1.5 = 6.0$ A.

22.2 Transformer construction

For power transformers, two basic types of core construction are used. In the *core type* the windings are wound around two legs of a rectangular-shaped magnetic core, one winding on top of the other (Figure 22.4(a)), so that both the primary and the secondary have one half of each winding on each limb. The low-voltage winding is the innermost winding. In the *shell-type* the windings are wound around the centre leg of a three-legged core (Figure 22.4(b)). The low-voltage winding is the innermost winding.

22.2.1 Core material

When an alternating current is applied to the primary coil of a transformer, the magnetic state of the transformer core is taken through a complete magnetisation cycle, i.e. a hysteresis loop (see Section 15.5).

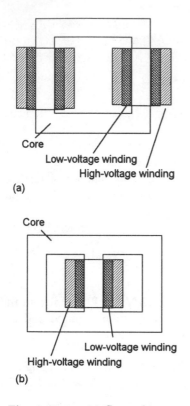

(a)

Core
Low-voltage winding
High-voltage winding

(b)

Core
Low-voltage winding
High-voltage winding

Figure 22.4 *(a) Core, (b) shell*

The energy consumed in this cycle is proportional to the area enclosed by the hysteresis loop. Hence, to minimise such losses, a material is required which has a hysteresis loop with a small enclosed area. Another source of loss is eddy currents. The changing magnetic field produced by the alternating current in the primary coil will, as well as inducing e.m.f.s in the coils, induce e.m.f.s and hence give rise to currents in the core. Such currents are termed *eddy currents*. A solid core can have a very low electrical resistance and so these eddy currents can dissipate large amounts of power. To minimise such losses, transformer cores are made of a stack of thin laminations, each electrically insulated from its neighbours. This increases the electrical resistance of the core and so reduces the size of eddy currents and hence eddy current losses.

The material used for power transformer cores is generally silicon steel, this material having a high value of permeability at low magnetic field strengths. For small transformers for use at high frequencies with low power levels, a compressed powdered ferromagnetic alloy, known as permalloy, is used.

22.2.2 Power rating

The power rating for a transformer is defined as:

$$\text{rating in VA} = V_2 I_{2FL}$$

where V_2 is the secondary voltage and I_{2FL} is the secondary current on full-load. Full-load is the load which allows the maximum power output. This maximum power output is determined by the rate at which the heat generated by internal losses can be dissipated. The major part of this is generated by the current through the primary and secondary windings, i.e. the power loss $= I^2R$; this loss is termed the *copper loss*. Thus there is a maximum current which can be obtained from the secondary without the transformer overheating. When a transformer is said to be operating at half load, it means that it is delivering half its rated power output and hence its secondary current is half the full load value.

Example

A 2.75 kVA single-phase transformer has a primary with 600 turns and a secondary with 100 turns. If the alternating primary voltage is 3300 V, what will be the minimum resistance which can be connected as the load?

The secondary voltage $= V_1 N_2/N_1 = 3300 \times (100/600) = 550$ V. Since $V_2 I_{2FL} = 3000$, then $I_{2FL} = 2750/550 = 5$ A. The value of the resistance to give this current is $R = V_2/I_{2FL} = 550/5 = 110$ Ω. This is the minimum value of resistance since any smaller value would give rise to a larger current.

22.2.3 Efficiency

The power losses that occur with a transformer are:

1 *Copper loss*
This is generated by the current through the primary and secondary windings and is thus power loss = $I_1^2 R_1 + I_2^2 R_2$, where I_1 is the current through a primary of resistance R_1 and I_2 is the current through a primary of resistance R_2.

2 *Iron loss*
This is the loss in the core resulting from hysteresis and eddy currents.

The efficiency of a transformer is defined as:

$$\text{efficiency} = \frac{\text{output power}}{\text{input power}} \times 100\%$$

The output power is $V_2 I_2 \cos \phi$ and equals the input power minus losses. We can thus express the above equation as:

$$\text{efficiency} = \frac{\text{input power} - \text{losses}}{\text{input power}} \times 100\%$$

where losses = copper loss + iron loss.

Example

A 40 kVA single-phase transformer has an iron loss of 400 W and a full load copper loss of 600 W. What is the efficiency of the transformer for a power factor of 0.8 at (a) full load, (b) half load?

(a) The total loss at full load = 400 + 600 = 1000 W. At a power factor of 0.8, output power = $V_2 I_2 \cos \phi$ = 40 000 × 0.8 = 32 000 W. The input power is the output power plus losses = 32 000 + 1000 = 33 000 W. Hence the efficiency = (32 000/33 000) × 100% = 97.0%.
(b) At half load, the current through the secondary will be half the full load value. Since the copper loss is proportional to I_2^2 then, at half load, the copper loss is reduced to $(\frac{1}{2})^2$ of 600 W, i.e. 150 W. The total loss is now 400 + 150 = 550 W. The output power = $V_2 I_2 \cos \phi$ = $\frac{1}{2}$ × 40 000 × 0.8 = 16 000 W. The input power is output power plus losses = 16 000 + 550 = 16 550 W. Hence the efficiency = (16 000/16 550) × 100% = 96.7%.

22.3 Auto-transformers

The *auto-transformer* has just one coil on a magnetic core (Figure 22.5). The full coil is the primary and a section of it is tapped off as the secondary. The primary and secondary are thus not only magnetically linked but electrically linked, the electrical circuit being as shown in Figure 22.6. Since all the turns link the same flux:

$$\frac{V_1}{V_2} = \frac{N_1}{N_2} = \text{turns ratio}$$

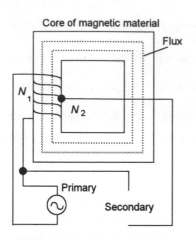

Figure 22.5 *Auto-transformer*

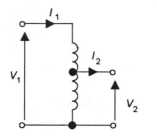

Figure 22.6 *Auto-transformer*

The secondary tapping can be a slider so that N_2 can be varied and so the output voltage can be varied. The current through the common part of the coil is $I_2 - I_1$ and thus that part of the winding has to carry less current than the upper part. As a consequence, smaller cross-section copper wire can be used for this part and hence give a cost saving.

The advantages of auto-transformers over double-wound transformers are:

1   Less copper is required since some of the windings are shared and as a result carry less current. The volume of copper in a winding is proportional to the number of turns N and the cross-sectional area of the wire. Since the cross-sectional area that can be used depends on the current I carried, the volume of copper is proportional to NI. Thus, for a double-wound transformer the volume of copper will be proportional to $N_1 I_1 + N_2 I_2$ and so, since $N_1 I_1 = N_2 I_2$, to $2N_1 I_1$. For an auto-transformer, the volume of copper is proportional to $(N_1 - N_2)I_1 + N_2(I_2 - I_1) = N_1 I_1 + N_2 I_2 - 2N_2 I_1$ and, since $N_1 I_1 = N_2 I_2$, to $2N_1 I_1 - 2N_2 I_2$. Thus (volume of copper in auto-transfer)/(volume in double-wound transformer) $= (2N_1 I_1 - 2N_2 I_2)/2N_1 I_1 = 1 - N_2/N_1$.

2   A higher efficiency since there is better magnetic linkage between the primary and secondary turns than occurs with a double-wound transformer.

3   The weight and volume of the transformer is less.

4   A continuously variable output voltage can be obtained.

The disadvantages are:

1   There is no electrical isolation of the secondary from the primary.

2   A break in the secondary part of the coil will result in the full primary voltage being applied to the secondary circuit.

Example

An ideal auto-transformer has primary and secondary voltages of 500 V and 400 V respectively. Determine (a) the primary and secondary currents and the current in the secondary section of the winding when the load is 30 kVA, (b) the factor by which such a transformer has less copper than the comparable double-wound transformer.

(a) $30\ 000 = I_1 V_1 = I_2 V_2$ and so $I_1 = 30\ 000/500 = 60$ A and $I_2 = 30\ 000/400 = 75$ A. The current in the secondary section of the winding $= I_2 - I_1 = 75 - 60 = 15$ A.
(b) (Volume of copper in auto-transfer)/(volume in double-wound transformer) $= 1 - N_2/N_1 = 1 - 400/500 = 0.2$.

22.4 Three-phase transformers

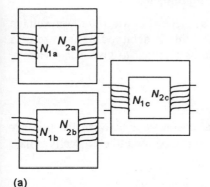

(a)

(b)

Figure 22.7 *Three-phase transformers. (a) single-phase (b) single three-legged core*

Power generation and distribution around a country is generally by means of three-phase a.c,. and thus, three-phase transformers are needed to step-down the voltage of many kilo-volts produced at the generators to, for example, the value that can be used in the home of 240 V. A three-phase transformer consists of three single-phase transformers, one for each phase a, b and c, and these can be either separate (Figure 22.7(a)) or combined on one core (Figure 22.7(b)).

The primaries and secondaries of the three single-phase units can be connected in either star or delta configurations. There are thus four possible connections for a three-phase transformer:

1. *Primary–star and secondary–delta* (Figure 22.8(a))
   The input line voltage V_{L1} is $\sqrt{3}\ V_1$ and the output line voltage V_{L2} is the secondary voltage V_2. Hence $V_{L1}/V_{L2} = \sqrt{3} \times$ turns ratio.

2. *Primary–delta and secondary–star* (Figure 22.8(b))
   The input line voltage V_{L1} is the primary voltage V_1 and the output line voltage V_{L2} is $\sqrt{3}\ V_2$. Hence $V_{L1}/V_{L2} = \sqrt{3}/$turns ratio.

3. *Primary–delta and secondary–delta*
   The input line voltage V_{L1} is the primary voltage V_1 and the output line voltage V_{L2} is the secondary voltage V_2. Hence $V_{L1}/V_{L2} =$ turns ratio.

4. *Primary–star and secondary–star*
   This is generally not used.

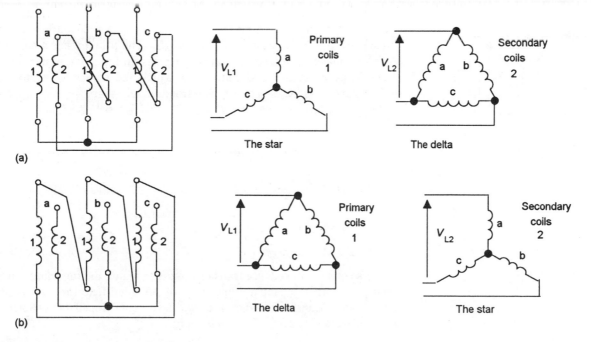

(a)

(b)

Figure 22.8 *(a) Star–delta connections, (b) delta–star connections*

22.5 Uses of transformers

The following are some of the uses of transformers.

22.5.1 Power supplies for electronic equipment

The a.c. supply voltage has often to be stepped down to a lower voltage for use with electronic equipment and this is achieved by the use of a step-down transformer. It might then be rectified to give a low voltage d.c. supply.

22.5.2 Isolation

Because there are no direct electrical connections between the primary and secondary coils of a transformer, it can be used to electrically isolate one circuit from one another.

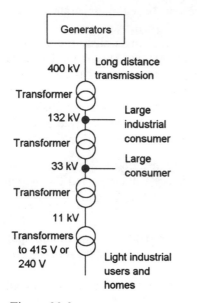

22.5.3 Power transmission

Figure 22.9 *Power transmission*

The power in a single-phase circuit is given by $VI \cos \phi$. Thus for the transmission of a given power and power factor, VI is constant and so doubling V means halving I. The power loss when a r.m.s. current I passes through a conductor of resistance R is I^2R, thus if I is halved then the power loss is quartered. Thus the higher the voltage of transmission the lower the power losses. For this reason, power in Britain is transmitted over long distances at a voltage of 400 kV. Step-down three-phase transformers are used to bring the voltage down to levels at which it is used in industry and in homes (Figure 22.9).

22.5.4 Impedance transforming

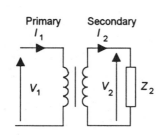

Figure 22.10 *Impedance transforming*

One use of a transformer is to alter the apparent value of an impedance. Consider the situation where a transformer has a secondary load of impedance Z_2 (Figure 22.10), then $Z_2 = V_2/I_2$. The primary has a current I_1 for a voltage V_1 and so the transformer has an input impedance $Z_1 = V_1/I_1$. Thus Z_1 is the impedance into which the source generator is working but Z_2 is the impedance seen from the secondary.

For a transformer we have $V_1/V_2 = N_1/N_2$ and $I_2/I_1 = N_1/N_2$. Thus:

$$V_1 = \frac{N_1}{N_2}V_2 \quad \text{and} \quad I_1 = \frac{N_2}{N_1}I_2$$

$$\frac{V_1}{I_1} = \left(\frac{N_1}{N_2}\right)^2 \frac{V_2}{I_2}$$

$$Z_1 = \left(\frac{N_1}{N_2}\right)^2 Z_2$$

A use of such impedance transforming is to match the impedance of a loudspeaker to the output impedance of an amplifier, the maximum power being transferred when the two have equal impedances. Such a process is known as *impedance matching*.

Example

A step-down transformer with a turns ratio of 3:1 is connected to a load of 100 Ω. What will be the input resistance of the transformer?

$$R_1 = (N_1/N_2)^2 R_2 = 3^2 \times 100 = 900 \ \Omega.$$

Example

The output stage of an amplifier has an impedance of 135 Ω and is to be used to drive a loudspeaker. Determine the turns ratio of the transformer which will match this impedance to that of a loudspeaker of impedance 15 Ω to give maximum power transfer.

$$Z_1 = (N_1/N_2)^2 Z_2 = 135 = (N_1/N_2)^2 \times 15, \text{ so } (N_1/N_2) = \sqrt{(135/15)} = 3.$$

Activities

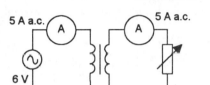

Figure 22.11   *Activity 1*

1 Use two C-cores and coils, e.g. two 120 + 120 turn coils, to form a transformer and hence investigate the behaviour of the transformer using the circuit shown in Figure 22.11. Examine, for different turn ratios, e.g. 1:1, 2:1, 1:2, how the primary current depends on the current in the secondary coil by varying the secondary load resistance.

2 Examine a range of transformers and explain their constructional forms.

Problems

1 An ideal transformer has 1000 primary turns and 3500 secondary turns. If the a.c. mains voltage of 240 V is connected to the primary, what voltage will be produced at the secondary?

2 An ideal transformer has a primary a.c. input of 100 V and gives a secondary output of 75 V across a 25 Ω resistor. What is (a) the turns ratio, (b) the primary current?

3 An ideal transformer has a turns ratio of 10:3. If there is an a.c. input of 250 V to the primary and a resistance of 15 Ω across the secondary, determine (a) the secondary e.m.f., (b) the secondary current, (c) the primary current, (d) the power input required.

4 A 5 kVA ideal transformer has a primary with 500 turns and a secondary with 40 turns. When there is a primary input of an a.c. voltage of 3 kV what will be (a) the primary and secondary currents when the transformer is on full load, (b) the secondary e.m.f.?

5 A step-down ideal transformer has a secondary output of 240 V. When the primary current is 3.0 A, the secondary load is 10 Ω. What will be the primary voltage?

6 An ideal transformer has a primary coil with 1200 turns and the current through it is 0.5 A when the input a.c. voltage is 100 V. The secondary has 300 turns. What will be (a) the secondary voltage, (b) the ampere-turns?

7 A 100 VA ideal transformer transforms a primary voltage of 240 V to a secondary voltage of 12 V. Determine the primary and secondary currents when the transformer is on full load.

8   A 500 VA single-phase transformer has an iron loss of 20 W and a full load copper loss of 30 W. Determine the efficiency when it has a resistive load at (a) full-load, (b) half load.

9   A 100 kVA single-phase transformer has an iron loss of 800 W and a full load copper loss of 1000 W. Determine the efficiency at a power factor of 0.8 when it is operating at (a) full load, (b) half load.

10 A 50 kVA single-phase transformer operating at a power factor of 0.8 has an efficiency of 98.0% on full load and 96.9% on quarter load. Determine the iron loss and the full load copper loss.

11 A 5 kVA single-phase transformer has an iron loss of 50 W and a full load copper loss of 120 W. What is the efficiency when it is operating with a resistive load at half load?

12 The primary and the secondary voltages of an auto-transformer are 500 V and 400 V respectively. Determine the current in the secondary when the current in the primary is 80 A.

13 The primary and the secondary voltages of an auto-transformer are 500 V and 450 V respectively. When the current in the primary is 45 A, determine (a) the current in the secondary and (b) the current in the common part of the coil?

14 A three-phase, step-down transformer, turns ratio per phase 12, is connected in delta-star configuration. If the primary voltage is 6.6 kV and primary current 10 A, what are the secondary line current and voltage?

15 The output stage of an amplifier has an impedance of 50 Ω and is to be connected to a circuit of impedance 1000 Ω. Determine the turns ratio of the transformer which will match the amplifier impedance to that of the circuit to give maximum power transfer.

16 The output stage of an amplifier has an impedance of 50 Ω and is to be connected to a circuit of impedance 800 Ω. Determine the turns ratio of the transformer which will match the amplifier impedance to that of the circuit to give maximum power transfer.

17 An electronic circuit with an output impedance of 1.2 kΩ is to be connected to a circuit of impedance 10 Ω. Determine the turns ratio of the transformer which will enable maximum power transfer between the circuits.

18 A load of impedance 10 Ω is supplied with power via a step down transformer having a turns ratio of 5. What will be the impedance perceived in the primary?

19 What will be the power delivered to a load of resistance 150 Ω when (a) it is directly connected to an alternator of 10 V and internal resistance 50 Ω, (b) a transformer is used to match the impedances? What turns ratio will be required for the transformer?

20 An electronic circuit has an output impedance of 10 kΩ and delivers an a.c. voltage of 10 V. If it is matched to a load by a step-down single-phase transformer of turns ratio 20, what will be the impedance of the load and the power dissipated in it?

23 Motors

23.1 Introduction

This chapter is an introduction to d.c. motors and a.c. induction motors; their basic form of construction, principles of operation, operating characteristics and general applications are discussed.

The basic principles involved in explaining the action of motors are:

1   A force is exerted on a conductor in a magnetic field, which has a component at right angles to it, when a current passes through it (see Section 11.5). For a conductor of length L carrying a current I in a magnetic field of flux density B at right angles to the conductor, the force $F = BIL$.

2   When a conductor moves in a magnetic field then an e.m.f. is induced across it (see Section 11.2). The induced e.m.f. e is equal to the rate at which the magnetic flux Φ swept through by the conductor changes (Faraday's law) and is in such a direction as to oppose the change producing it (Lenz's law), i.e. $e = -d\Phi/dt$.

23.2 D.c. motors

The basic principle of a d.c. motor is a loop of wire which is free to rotate in the field of a magnet (Figure 23.1). When a current is passed through the loop, the resulting forces acting on its sides at right angles to the field cause forces to act on those sides to give rotation.

Coils of wire are mounted in slots on a cylinder of magnetic material called the *armature* which is mounted on bearings and free to rotate. It is mounted in the magnetic field produced by *field poles*. This magnetic field might be produced by, for small motors, permanent magnets or a current in, so-termed, *field coils*. Whether permanent magnet or field coils, these generally are part of the outer casing of the motor and are termed the *stator*. Figure 23.2 shows the basic elements of a d.c. motor with the magnetic field produced by field coils. In practice there will be more than one armature coil and more than one set of stator poles. The ends of the armature coil are connected to adjacent segments of a segmented ring called the *commutator* with electrical contacts made to the segments through fixed carbon contacts called *brushes*. They carry direct current to the armature coil. As the armature rotates, the commutator reverses the current in each coil as it moves between the field poles. This is necessary if the forces acting on the coil are to remain acting in the same direction and so the rotation continue.

Figure 23.3 shows the principle of a four-pole d.c. motor with the magnetic field produced by current carrying coils and having an array of armature conductors.

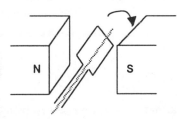

Figure 23.1 *D.c. motor principle*

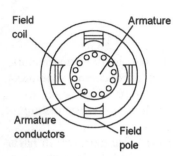

Field coil

Armature

Armature conductors

Field pole

Figure 23.3 D.c. motor

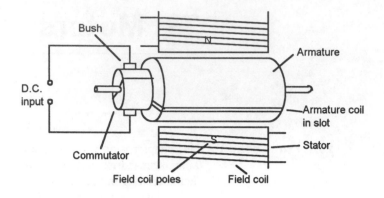

Bush

D.C. input

Commutator

Field coil poles

N

Armature

Armature coil in slot

Stator

Field coil

Figure 23.2 D.c. motor

23.2.1 Basic theory

Consider a d.c. motor with a flux density B at right angles to an armature loop of length L and carrying a current i_a (Figure 23.4). The force acting on the conductor is Bi_aL. The forces result in a torque about the coil axis of Fb, with b being the breadth of the coil. Thus:

$$\text{torque per turn} = BLi_ab = \Phi i_a$$

where Φ is the flux linked per armature turn. In practice there will be more than one armature loop and more than one set of poles, the torque will, however, be proportional to Φi_a and so we can write:

$$\text{torque acting on armature} = k_t\,\Phi i_a$$

where k_t is the torque constant.

Since an armature coil is rotating in a magnetic field, electromagnetic induction will occur and a back e.m.f. v_b will be induced which is equal to the rate at which the flux linked by the coil changes and hence, for an armature rotating with an angular velocity ω, is proportional to $\Phi\omega$. Thus we can write:

$$\text{back e.m.f. } v_b = k_v\Phi\omega$$

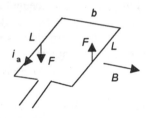

Figure 23.4 Force on a single armature turn

where k_v is the back e.m.f. constant.

We can consider a d.c. motor to be the armature coil, represented by a resistor R in series with an inductance L, in series with a source of back e.m.f. (Figure 23.5). If we are just concerned with steady-state conditions we can neglect the inductance of the armature coil. The voltage providing the current i through the resistance is the applied armature voltage V_a minus the back e.m.f., i.e. $V - v_b$. Hence:

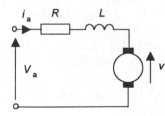

Figure 23.5 Armature circuit

$$V_a = v_b + Ri_a$$

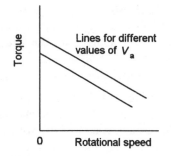

Figure 23.6 *Torque–speed characteristic*

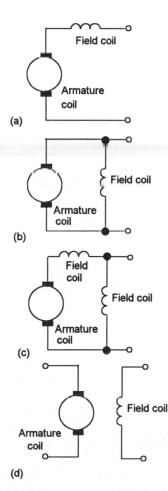

Figure 23.7 *D.c. motors: (a) series, (b) shunt, (c) compound, (d) separately wound*

$$i_a = \frac{V_a - v_b}{R} = \frac{V_a - k_v \Phi \omega}{R}$$

and so the torque T is:

$$T = k_t \Phi i_a = \frac{k_t \Phi}{R}(V_a - k_v \Phi \omega)$$

For a constant value of flux, e.g. a permanent magnet motor, graphs of the torque against the rotational speed ω are given by $T = k_1 V_a - k_2 \omega$ and are a series of straight lines for different voltage values (Figure 23.6). The starting torque, i.e. the torque when $\omega = 0$, is proportional to the applied voltage and the torque then decreases with increasing speed. The power developed in rotating the armature is ωT (see Section 8.4.1).

If the load driven by a d.c. motor is increased, the speed will drop and hence the current through the armature. The relationship between the speed and the armature current determines how the motor will react to load changes.

Example

A permanent d.c. motor develops a torque of 4 N m when the armature current is 2 A. What will be the torque when the armature current is 0.5 A?

Torque = $k_t \Phi i_a = k i_a$ and thus $k = 4/2 = 2$ N m/A. When the current is 0.5 A, then torque = $2 \times 0.5 = 1$ N m.

23.2.2 D.c. motors with field coils

D.c. motors with field coils are classified as series, shunt, compound and separately excited according to how the field windings and armature windings are connected (Figure 23.7).

1   *Series wound motor*
    With the series wound motor the armature and fields coils are in series and thus carry the same current. Thus the flux Φ depends on the armature current and so the torque acting on the armature = $k_t \Phi i_a = k i_a^2$. When $\omega = 0$ then $i_a = V/R$ and so the starting torque = $k(V/R)^2$. Thus, such a motor exerts the highest starting torque and has the greatest no-load speed. Such motors are used where high starting torques are required, e.g. hoists and car engine starters. As the speed is increased, so the torque decreases. Since Ri is small, $V_a = v_b + Ri \simeq v_b$ and so, since $v_b = k_v \Phi \omega$ and Φ is proportional to i, we have V_a proportional to $i\omega$. To a reasonable approximation V_a is constant and so the speed is inversely proportional to the current. The speed thus drops quite markedly when the load is increased. Reversing the polarity of the supply to the coils has no effect on the direction of rotation of the motor; it will continue rotating in the same direction since both the field and armature currents have been reversed. Such d.c. motors are used where large starting torques are

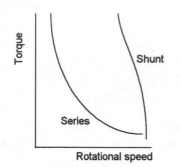

Figure 23.8 *Torque–speed characteristics*

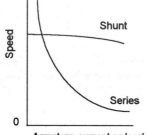

Figure 23.9 *Speed–load characteristics*

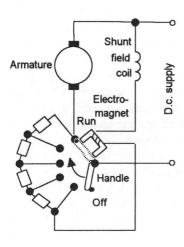

Figure 23.10 *Starter*

required, e.g. car starters and hoists. If the core is laminated, to reduce eddy currents, the series motor can be used with a single-phase a.c. supply and is then known as a *universal motor*. Such motors are used in portable power tools and domestic appliances such as food mixers and vacuum cleaners.

2 *Shunt wound motor*
With the shunt wound motor the armature and field coils are in parallel. The field coil is wound with many turns of fine wire and so has a more larger resistance than the armature coil; thus, with a constant supply voltage, the field current is virtually constant. Thus, to a reasonable approximation, we have $T = k_1V - k_2\omega$. It provides the lowest starting torque and a much lower no-load speed. Since Ri is small, $V_a = v_b + Ri \simeq v_b$ and so, since $v_b = k_v\Phi\omega$ and Φ is virtually constant, we have V_a proportional to ω. With V_a virtually constant, the motor gives almost constant speed regardless of load and such motors are very widely used because of this characteristic. To reverse the direction of rotation, either the armature or field supplied must be reversed.

3 *Compound motor*
The compound motor has two field windings, one in series with the armature and one in parallel. Compound wound motors aim to get the best features of the series and shunt wound motors, namely a high starting torque and constant speed regardless of load.

4 *Separately excited motor*
The separately excited motor has separate control of the armature and field currents. We have a torque $T = (k_t\Phi/R)(V_a - k_v\Phi\omega)$. If the field current is kept constant then Φ is constant, and if V_a is constant, then $T = k_1 - k_2\omega$. As the speed increases the torque drops; generally the drop in torque is relatively small.

Figure 23.8 indicates the torque–speed characteristics of the series and shunt-wound motors and Figure 23.9 the way the speed depends on the armature current and hence the load.

23.2.3 Motor starting

D.c. motors develop a torque at standstill and so are self-starting. They do, however, require a starting resistance to limit the starting current. The starting current $i = (V_a - v_b)/R$ and since there is initially no back e.m.f. to limit the current the starting current can be very large. Figure 23.10 shows the form of a d.c. motor starter in which an external resistance is inserted when the motor is started up and then gradually taken out as the speed increases. The handle is then held in the run position by the electromagnet, this being energised by the current through the field coil. If the field coil becomes disconnected, the electromagnet becomes de-energised and a spring returns the handle to the off position.

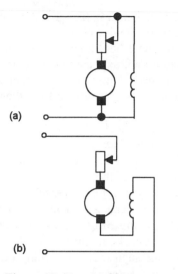

(a)

(b)

Figure 23.11 *(a) Shunt motor (b) series motor*

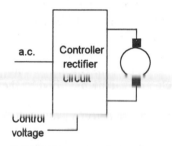

a.c.

Controller rectifier circuit

Control voltage

Figure 23.12 *Speed control using SCRs*

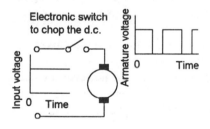

Electronic switch to chop the d.c.

Input voltage

Armature voltage

0 Time

Figure 23.13 *PWM*

23.3 Three-phase induction motor

Example

A 10 kW d.c. motor has an armature resistance of 0.1 Ω and is connected to a 100 V d.c. supply. Determine (a) the starting current if no starting resistance is in the armature circuit, (b) the series starting resistance to limit the starting current to twice the rated value.

(a) The starting current = V_s/R_a = 100/0.1 = 1000 A.
(b) The rated current = P/V = 10 000/100 = 100 A. To limit the current to 200 A, $R = V_s/I_a$ = 100/200 = 0.5 Ω and so a starting resistance of 0.5 – 0.1 = 0.4 Ω is required.

23.2.4 Speed control of d.c. motors

The speed of a permanent magnet motor depends on the current through the armature coil and thus can be controlled by changing the armature current. With a field coil motor the speed can be changed by either varying the armature current or the field current; generally it is the armature current that is varied. Thus speed control can be obtained by controlling the voltage applied to the armature. Figure 23.11 shows circuits that can be used when a variable resistor is used. This method is, however, very inefficient since the controller resistor consumes large amounts of power. Another alternative is to control the voltage by using an electronic circuit.

With an alternating current supply, silicon-controlled rectifiers (SCRs) can be used. A control voltage (Figure 23.12) is used to switch the rectifiers on and off and so control the average value of the voltage supplied to the motor. With a d.c. supply, a constant d.c. supply voltage is chopped by means of an electronic circuit so that the average value is varied (Figure 23.13). This is called *pulse width modulation* (PWM).

Example

A 240 V d.c. shunt motor running at 12 rev/s has an armature resistance of 0.5 Ω and takes an armature current of 20 A. What resistance should be placed in series with the armature to reduce the speed to 10 rev/s?

Since $v_b = k_v\Phi\omega = k\omega$ then v_{12}/v_{10} = 12/10. $V_a = v_b + Ri$ and so v_{12} = 240 – 20 × 0.5 = 230 V and hence v_{10} = 192 V. For the armature circuit 240 – 192 = $I_a(R_a + R)$ = 20(0.5 + R) and so R = 1.9 Ω.

Alternating current motors can be classified into two groups: induction and synchronous motors. Induction motors tend to be cheaper than synchronous motors and are very widely used.

The action of an *induction motor* depends on the principle that when a magnetic field moves past a conductor, the conductor is set in motion and endeavours to follow the field. We can explain this as being due to the relative motion between magnetic field and conductor inducing an

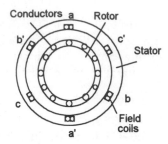

Figure 23.14 *Three-phase induction motor*

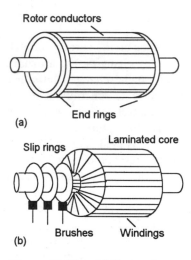

Figure 23.15 *(a) Squirrel-cage rotor, (b) wound rotor*

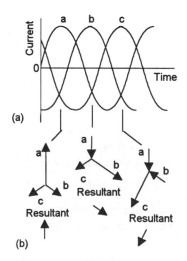

Figure 23.16 *The rotating field, (a) the currents, (b) the fluxes due to separate phases and the resultants*

e.m.f. in the conductor and hence an eddy current in it. The direction of the current is such as to produce a magnetic field which opposes the motion producing it (Lenz's law). This magnetic field interacts with the magnetic field responsible for its production and, as a consequence, the conductor moves to reduce the relative velocity between the magnetic field and conductor. Thus if we produce a rotating magnetic field then we can produce rotation.

One method of producing a rotating magnetic field involves the use of a balanced three-phase supply. The *three-phase induction motor* has a stator, i.e. the stationary bit, with three windings aa', bb', cc' located 120° apart, each winding being connected to one of the three lines a, b, c of the supply (Figure 23.14). Because the three phases reach their maximum currents at different times, the magnetic field can be considered to rotate round the stator poles, completing one rotation in one full cycle of the current. The rotating part of the motor is termed the *rotor*. Figure 23.15(a) shows the form of one such rotor, this being the simplest form and termed the *squirrel-cage rotor*. It has copper or aluminium bars that fit into slots in end rings to form complete electrical circuits. Note that there are no external electrical connections to this rotor. Another form is the *wound-rotor*; this consists of rotor windings that are mirror images of the stator windings; Figure 23.15(b) shows a simplified version of the rotor. These windings are usually star-connected and the ends of the three rotor wires connected to slip rings on the rotor shaft and shorted by brushes riding on the slip rings. The rotor currents are thus, in this situation, accessible and can be modified by the inclusion of external resistors in order to give speed control. The squirrel-cage induction motor is simpler, cheaper and more rugged than the wound-rotor motor.

For the cage motor shown in Figure 23.14 there are one set of stator windings per phase, i.e. two poles per phase. The magnetic flux produced per winding is proportional to the current in that winding and thus the magnetic fluxes for the three sets of windings vary with time in the manner shown in Figure 23.16(a).

Figure 23.16(b) shows the sizes and directions of these fluxes at a number of times. The resultant magnetic fluxes rotate and complete a cycle in the same time as the current in a winding. If, however, there are p poles per phase then the frequency of rotation is f/p, where f is the frequency of the current in a phase. This frequency is termed the *synchronous speed* n_s, i.e. $n_s = f/p$.

When the rotor is rotating at low speeds compared with the synchronous speed, there is a high relative velocity of the magnetic field past a rotor and so a large e.m.f. is induced and hence large eddy current and consequently a large torque. When the rotor speeds up, the relative velocity becomes less and so the torque less. If the rotor ever reached the synchronous speed then there would be no relative velocity and hence no torque. The speed of the rotor n_r relative to that of the magnetic field, i.e. the synchronous speed n_s, is termed the *slip*:

$$\text{slip} = n_s - n_r$$

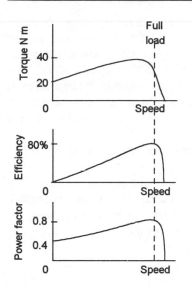

Figure 23.17 *Characteristics*

The term *per unit slip* is used for $(n_s - n_r)/n_s$ and the term *percentage slip* for $[(n_s - n_r)/n_s] \times 100$. Since $n = \omega/2\pi$ we can also express per unit slip as $(\omega_s - \omega_r)/\omega_s$. The torque acting on the rotor is proportional to the rotor current. This is proportional to the induced e.m.f. and hence the rate of change of flux. The rate of change of flux depends on the relative velocity. Thus the torque is proportional to the per unit slip.

With an unloaded motor there is little torque needed to overcome friction and thus little torque is needed to get the rotor speed close to the synchronous speed. The operating speed is thus close to the synchronous speed. When a load is applied, the rotor speed falls by a very small amount and so such a motor is suitable for reasonably constant speed applications. Figure 23.17 shows the characteristics of a typical squirrel-cage induction motor. The efficiency is the ratio of the mechanical output power to the electrical input power.

Example

A two-pole, 50 Hz, three-phase induction motor has a rotor speed of 40 rev/s, what is the unit slip?

The number of pairs of poles per phase is 1. Hence $n_s = f/p = 50/1 = 50$ rev/s. The rotor speed $n_r = 40$ rev/s and the per unit slip is $(50 - 40)/50 = 0.2$.

Example

A four-pole, 50 Hz, three-phase induction motor has a no-load rotor speed of 24 rev/s when a torque of 2.0 N m is produced. What is the rotor speed with an external load and a torque of 3.0 N m?

The number of pairs of poles per phase is 2 and so $n_s = f/p = 50/2 = 25$ rev/s. The rotor speed $n_r = 24$ rev/s. Hence, at no-load the per unit slip $= (25 - 24)/25 = 0.04$. The torque T is proportional to the per unit slip and thus $T = 2.0 = k \times 0.04$, where k is a constant $= 50$. When there is a load $T = 3.0 = ks$ and the per unit slip $s = 3.0/50 = 0.06$. Hence $0.06 = (n_s - n_r)/n_s = (25 - n_r)/25$ and so $n_r = 23.5$ rev/s.

23.3.1 Starting methods

Three-phase squirrel-cage motors can be started by directly connecting them to the voltage supply. However, initially the current can be much larger than the value that occurs when the motor is running at full speed. To overcome this problem *star–delta starting* is often used. The motor is started with the stator windings connected in star configuration and then, when the rotor is close to full speed, they are reconnected in delta configuration. Figure 23.18 shows the circuit used with small motors. Initially the main contactor and the star connector are closed to give the star configuration of the coils. Then, when running speed has been attained, the star contactor is opened and the delta contactor closed to give the delta configuration. The star and delta contactors are interlocked so that it is impossible for both to be simultaneously closed.

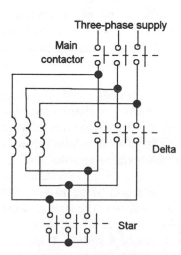

Figure 23.18 *Star–delta starter*

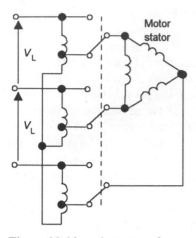

Figure 23.19 *Auto-transformer starter*

For a motor with an impedance of Z per phase when not running, when started in a star-connected configuration the phase current I_p = the line current $I_L = V_p/Z$. In a delta-connected configuration, the phase current = V_L/Z and, since $I_L = \sqrt{3}\ I_p$ and $V_L = \sqrt{3}\ V_p$, thus $I_L = \sqrt{3}\ (V_L/Z) = \sqrt{3}\ (\sqrt{3}\ V_p/Z) = 3V_p/Z$. Thus the line current in star configuration is one-third that in delta configuration. Starting the motor in star-configuration thus gives a smaller current. This smaller current gives a smaller torque, but switching to delta enables higher torque.

An alternative starting method which allows the motor to remain connected in delta configuration the entire time is to use an *auto-transformer* (Figure 23.19) to give a lower voltage for starting. The auto-transformer is star-connected and the figure shows the switch positions when the motor is started. When the motor is up to speed, the switches are moved to the top positions and so the entire input line voltage is then applied.

23.4 Single-phase induction motor

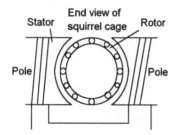

Figure 23.20 *Single-phase induction motor*

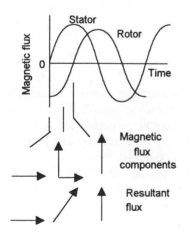

Figure 23.21 *Rotating magnetic field*

Figure 23.20 shows the basic form of a *single-phase squirrel-cage induction motor*. When an alternating current passes through the stator windings an alternating magnetic field is produced. This magnetic field does not rotate but just builds up in one direction as the current increases in the positive half cycle, then decreases to zero before building up in the opposite direction in the negative half cycle before decreasing again to zero. Such a magnetic field induces e.m.f.s in the conductors of the rotor and currents flow in the rotor. If the rotor is stationary, the forces on the current carrying conductors of the rotor in the magnetic field of the stator result in no net torque. However, if the rotor is already rotating a net torque is produced. This is because the current in the rotating rotor produces a magnetic field which, when added to that of the stator results in a rotating magnetic field (Figure 23.21).

The rotating rotor conductors have the maximum voltage induced in them when they cut the magnetic flux of the stator at right angles. However, because of the inductance of the rotor, the current in a rotor conductor will be almost 90° out-of-phase with the voltage. The result is that the magnetic field of the rotor is virtually 90° out-of-phase with the stator magnetic field. It is the addition of the stator magnetic field and this 90° shifted rotor magnetic field which gives rise to a rotating magnetic field (Figure 23.21).

The motor is not self-starting. A number of methods are used to make the motor self-starting and give the initial impetus to start it; one is to use an auxiliary starting winding (Figure 23.22) to give the rotor an initial push. The start winding is in a plane 90° displaced from the main stator winding and has a higher resistance-to-reactance ratio than the main winding, as a result of using finer wire than the main winding, so that the currents in the two coils are out of phase by about 90°. The motor then behaves like a two-phase motor at start-up. The start winding is switched out when the motor is running; a centrifugal switch is mounted on the rotor and opens the switch when a particular speed is reached. Such a motor is termed a *split-phase single-phase motor*.

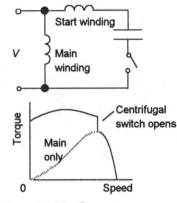

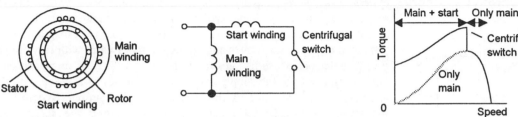

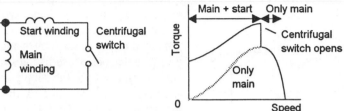

Figure 23.22 *Split-phase motor*

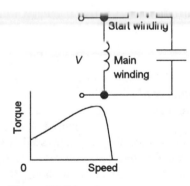

Figure 23.23 *Capacitor-start*

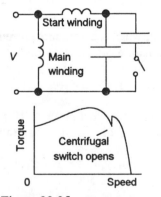

Figure 23.24 *Capacitor-run*

Figure 23.25 *Capacitor-start capacitor-run*

Higher starting torque can be obtained if a capacitor is connected in series with the start winding in order to introduce the required phase difference between the currents. Such a motor is termed a *capacitor-start motor* when the capacitor is switched out when the motor is running (Figure 23.23), *capacitor-run* when it remains in circuit for running (Figure 23.24), and *capacitor-start, capacitor-run* when one capacitor is switched out of circuit when running and another capacitor remains in circuit for running (Figure 23.25).

Split-phase motors are used for such applications as fans, blowers, washing machines and, in general, applications requiring low or medium starting torque. Capacitor-start motors are used for 'hard-to-start' applications such as pumps and refrigerators. Capacitor-run motors are used for fans and pumps. Capacitor-start, capacitor-run motors are used for high starting torque applications such as compressors and pumps.

In the induction motor, the rotor rotates at a speed determined by the frequency of the alternating current applied to the stator. For a constant frequency supply to a two-pole single-phase motor the magnetic field will alternate at this frequency. This speed of rotation of the magnetic field is termed the *synchronous speed*. The rotor will never quite match this frequency of rotation, typically differing from it by about 1 to 3%. This difference is termed *slip*. Thus for a 50 Hz supply the speed of rotation of the rotor will be almost 50 revolutions per second.

The rotation of the field with a three-phase motor is much smoother than with the single-phase motor and the three-phase motor having the great advantage over the single-phase motor of being self-starting.

23.4.1 The single-phase shaded-pole motor

Figure 23.26 shows the basic form of a *shaded-pole motor*. It consists of a laminated core with the pole pieces divided into two parts, one part having a shading ring round it. When an alternating current is applied to the coil, an alternating magnetic flux is produced in the core. The alternating flux in the core induces an e.m.f. in the shading ring and hence a current which gives rise to magnetic flux in a direction opposing the change producing it (Lenz's law). The flux in the shaded part thus lags behind that in the unshaded part. The result is to give a rotating magnetic field and hence rotation of the rotor.

Shaded-pole motors have relatively low power ratings, typically up to a few tens of watts, develop only low torques and are not very efficient. They are used for loads requiring low starting torques, e.g. driving small cooling fans with electronic equipment, hair dryers and toys.

23.5 Synchronous motors

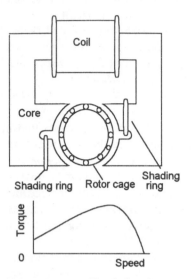

Figure 23.26   *Shaded-pole motor*

Synchronous motors have stators similar to those used with induction motors but have a rotor which is a permanent magnet or an electromagnet energised by d.c. (Figure 23.27). The magnetic field produced by the stator rotates and so the magnet rotates with it. With one pair of poles per phase of the supply, the magnetic field rotates through 360° in one cycle of the supply and so the frequency of rotation with this arrangement is the same as the frequency of the supply. Synchronous motors are used when a precise speed is required the speed being controlled by controlling the frequency of the voltage supply to the stator.

Applying the voltage at the required frequency to the stator is likely to result in no rotation. This is because the inertia of the rotor is such that by the time it reacts to the field and moves in one direction, the field is likely to have rotated so far that it is in the opposite direction and the rotor experiences a torque in the opposite direction; the stator field is rotating so fast that the rotor poles cannot catch up with it and lock into its rotation. The motor is thus not inherently self-starting and some system has to be employed to start it and get the rotor up to the speed at which the rotation of the rotor synchronises with that of the stator field. One method is to disconnect the rotor field winding from its d.c. supply and short circuit it. The arrangement then becomes an induction motor and can be accelerated up to speed before the d.c. supply is reconnected. Another method is to use a variable frequency supply for the stator; start with a low frequency and increase it up to the required frequency.

23.6 Speed control with a.c. motors

A.c. motors have the great advantage over d.c. motors of being cheaper, more rugged, reliable and maintenance free. However speed control is generally more complex. The speed of an a.c. motor depends on the speed of the rotating magnetic field and hence to the frequency of the a.c. supply. Thus, speed control of a.c. motors can be based on the provision of a variable frequency supply. With one method, the a.c. is first rectified to d.c. by a *converter* and then *inverted* back to a.c. again but at a frequency that can be selected. Another method that is often used for operating slow-speed motors is the *cycloconverter*. This converts a.c. at one frequency directly to a.c. at another frequency without the intermediate d.c. conversion.

Problems

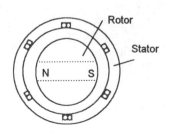

Figure 23.27   *Three-phase synchonous motor*

1   A d.c. motor develops a torque of 6 N m when the armature current is 10 A. What is the torque when the armature current drops to 2 A?
2   The armature of a d.c. motor has a resistance of 0.1 Ω and takes a current of 60 A from a 100 V supply. What is the induced e.m.f. in the armature?
3   A 240 V d.c. shunt motor takes an armature current of 15 A. If the resistance of the armature circuit is 0.8 Ω, what is the induced e.m.f. in the armature windings?
4   Explain why a d.c. shunt motor cannot normally be switched on straight from the supply but requires a starting circuit.

5 A d.c. motor has a load torque proportional to the square of its speed. Show that the power input is proportional to the cube of the speed if the field coil magnetic flux is proportional to the current.

6 A 5 kW d.c. shunt motor has an armature resistance of 0.2 Ω and is connected to a 100 V d.c. supply. Determine (a) the starting current if no starting resistance is in the armature circuit, (b) the series starting resistance to limit the current to 1.5 the rated value.

7 A 5 kW, 100 V d.c. shunt motor requires 5 V to send the rated current through the armature when it is stationary. Determine (a) the armature current that will occur when the motor is started up by the rated voltage and (b) the series resistor that will limit the starting current to twice the rated running value.

8 A 500 V series d.c. motor has a speed of 15 rev/s and takes a current of 20 A. The resistance of the field coil is 0.2 Ω and that of the armature 0.4 Ω. Determine the resistance to be connected in series to reduce the speed to 10 rev/s.

9 A 230 V series d.c. motor takes a current of 30 A. The resistance of the field coil is 0.4 Ω and that of the armature 0.2 Ω. Determine the resistance required in series to reduce the speed by 40% if the current at the new speed is 20 A. Assume that the magnetic flux produced by the field coil is proportional to the current.

10 A series d.c. motor has a field coil resistance of 1.0 Ω and an armature coil resistance of 0.5 Ω. It takes a current of 20 A from a 250 V supply when running at a speed of 1000 rev/min. At what speed will it run when a 3.5 Ω resistance is placed in series with the motor if the same current is taken as before?

11 A d.c. shunt motor has an armature of resistance 0.5 Ω and takes an armature current of 20 A when connected to a 220 V supply and running at 750 rev/min. What resistance should be connected in series with the armature to reduce the speed to 500 rev/min?

12 A two-pole, 50 Hz, three-phase induction motor has a rotor speed of 2900 rev/min, what is the unit slip?

13 A two-pole, 50 Hz, three-phase induction motor has a rotor speed of 23 rev/s when it produces a torque of 4.0 N m. What will be the speed when the torque is 1.5 N m?

14 A four-pole, 50 Hz, three-phase induction motor has a no-load rotor speed of 1470 rev/min with a torque of 1.5 N m. What will be the speed when an external load results in a torque of 2.5 N m?

15 A four-pole, 50 Hz, three-phase induction motor has a full-load slip of 5%. What is (a) the synchronous speed, (b) the rotor speed?

16 An eight-pole, 50 Hz, three-phase induction motor has a full-load slip of 2.5%. What is (a) the synchronous speed, (b) the rotor speed?

17 Explain how, with a split-phase, single-phase, induction motor, sufficient phase displacement is obtained between the currents in the two windings to enable starting.

18 Explain the function of the capacitor in the capacitor-start, single-phase, induction motor.

19 What is the function of the shading ring in the shaded-pole, single-phase motor?

24 Direct stress

24.1 Introduction

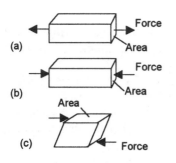

Figure 24.1 *(a) Tension, (b) compression, (c) shear*

The term *direct stress* is used when the area of material being stressed is at right angles to the line of action of the external forces, as when the material is in tension or compression (Figure 24.1). When the forces being applied are in the same plane as the area being stressed then the stresses are termed shear stresses. This chapter is a consideration of the action of tensile and compressive forces on materials, extending the consideration beyond that given in Chapter 6. Chapter 25 looks at shear stress.

Often members of structures might be made up of a length of one material joined to a length of another or involve a material for which there are changes in cross-section or, as in reinforced concrete, might involve axial rods inside another material. This chapter thus considers such composite structural members and, in addition, the effects of temperature changes on structural members which are restrained from expanding or contracting as a result of the changes in temperature.

24.2 Composite members

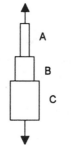

Figure 24.2   *Elements in series*

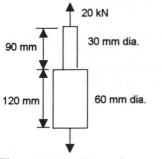

Figure 24.3   *Example*

Consider a composite member with two, or more, elements in series, such as in Figure 24.2 where we have three rods connected end-to-end with the rods being of different cross-sections and perhaps different materials. At equilibrium the same forces act on each of the series members and so the forces stretching member A are the same as those stretching member B and the same as those stretching member C. The total extension of the composite bar will be the sum of the extensions arising for each series element.

Example

A rod is formed with one part of it having a diameter of 60 mm and length 120 mm and the other part a diameter of 30 mm and length 90 mm (Figure 24.3) and is subject to an axial stretching force of 20 kN. What will be the stresses in the two parts of the rod and the total extension if both parts are of the same material and it has a modulus of elasticity of 200 GPa?

Each part will experience the same force and thus the stress on the larger diameter part is $20 \times 10^3/(\tfrac{1}{4}\pi \times 0.060^2) = 7.1$ MPa and the stress on the smaller diameter part is $20 \times 10^3/(\tfrac{1}{4}\pi \times 0.030^2) = 28.3$ MPa.

Since E = stress/strain = $(F/A)/(e/L)$ then extension $e = FL/AE$. The extension of the 60 mm diameter part is $(20 \times 10^3 \times 0.120)/(\tfrac{1}{4}\pi$

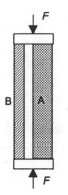

Figure 24.4 *A compound member*

$\times\ 0.060^2 \times 200 \times 10^9) = 4.2 \times 10^{-6}$ m. The extension of the 30 mm diameter part is $(20 \times 10^3 \times 0.090)/(\frac{1}{4}\pi \times 0.030^2 \times 200 \times 10^9) = 12.7 \times 10^{-6}$ m. The total extension is thus 16.9×10^{-6} m.

24.2.1 Members in parallel

Figure 24.4 shows a member made up of two parallel rods, A of one material and B of another material, the load being applied to rigid plates fixed across the ends of the two. With such a compound bar, the load F applied is shared by the members. Thus if F_A is the force acting on member A and F_B is the force acting on member B, we have:

$$F_A + F_B = F$$

If σ_A is the resulting stress in element A and A_A is its cross-sectional area then $\sigma_A = F_A/A_A$ and if σ_B is the stress in element B and A_B is its cross-sectional area then $\sigma_B = F_B/A_B$. Thus:

$$\sigma_A A_A + \sigma_B A_B = F$$

Since the elements A and B are the same initial length and must remain together when loaded, the strain in A of ε_A must be the same as that in B of ε_B. Thus, assuming Hooke's law is obeyed:

$$\frac{\sigma_A}{E_A} = \frac{\sigma_B}{E_B}$$

where E_A is the modulus of elasticity of the material of element A and E_B that of the material of element B.

Example

A square section reinforced concrete column has a cross-section of 450 mm $\times$ 450 mm and contains four steel reinforcing bars, each of which has a diameter of 25 mm (Figure 24.5). Determine the stresses in the steel bars and in the concrete when the total load on the column is 2 MN. The steel has a modulus of elasticity of 210 GPa and the concrete a modulus of elasticity of 14 GPa.

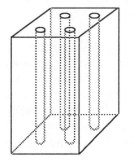

Figure 24.5 *Example*

The area of the column that is steel is $4 \times \frac{1}{4}\pi \times 25^2 = 1963$ mm². The area of the column that is concrete is $450 \times 450 - 1963 = 200\ 537$ mm². The ratio of the stress on the steel σ_s to that on the concrete σ_c is given by $\sigma_s/E_s = \sigma_c/E_c$ and so:

$$\sigma_S = \frac{210}{14}\sigma_c = 15\sigma_c$$

The force F on the column is related to the stresses and areas of the components by:

$$F = \sigma_S A_S + \sigma_c A_c$$

$$2 \times 10^6 = 15\sigma_c \times 1963 \times 10^{-6} + \sigma_c \times 200\ 537 \times 10^{-6}$$

Hence the stress on the concrete is 8.7 MPa and that on the steel is 130.4 MPa.

Example

A steel rod of diameter 25 mm and length 300 mm is in a copper tube of the same length, internal diameter 30 mm and external diameter 45 mm with the ends of both rod and tube fixed to rigid end plates (Figure 24.6). If the arrangement is subject to a compressive axial load of 80 kN what will be the stress in both materials? The modulus of elasticity of the steel is 210 GPa and that of the copper 105 GPa.

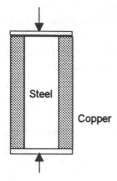

Figure 24.6   *Example*

The cross-sectional area of steel A_S is $\tfrac{1}{4}\pi \times 0.025^2 = 490.8 \times 10^{-6}$ m² and the cross-sectional area of the copper A_c is $\tfrac{1}{4}\pi(0.045^2 - 0.030^2)$ $= 883.6 \times 10^{-6}$ m². The ratio of the stress on the steel σ_s to that on the copper σ_c is given by $\sigma_s/E_s = \sigma_c/E_c$ and so:

$$\sigma_S = \frac{210}{105}\sigma_c = 2\sigma_c$$

The force F on the composite is related to the stresses and areas of the components by:

$$F = \sigma_S A_S + \sigma_c A_c$$

$$80 \times 10^3 = 2\sigma_c \times 883.6 \times 10^{-6} + \sigma_c \times 490.8 \times 10^{-6}$$

Hence, the stress in the copper is 35.4 MPa and that in the steel is 70.8 MPa.

24.3 Thermal strain

For most materials, an increase in temperature results in expansion. If, however, the material is fixed in such a way that it cannot expand then compressive stresses occur as a consequence of an increase in temperature. If a material is constrained from contracting when the temperature falls, tensile stresses are produced. Consider a bar of initial length L. If the temperature is raised by θ and the bar is free to expand, the length increases to:

$$L_\theta = L(1 + a\theta)$$

where a is the coefficient of linear expansion of the bar material. The change in length of the bar is thus:

$$L_\theta - L = L(1 + a\theta) - L = La\theta$$

If this expansion is prevented, it is as if a bar of length $L(1 + a\theta)$ has been compressed to a length L and so the resulting compressive strain ε is:

$$\varepsilon = \frac{La\theta}{L(1 + a\theta)}$$

Since $a\theta$ is small compared with 1, we can make the approximation:

$$\varepsilon = a\theta$$

If the material has a modulus of elasticity E and Hooke's law is obeyed, the stress σ produced is:

$$\sigma = a\theta E$$

Example

A bar of mild steel is constrained between two rigid supports. Calculate the stress developed in the bar as a result of a rise in temperature of 8°C. The coefficient of linear expansion is 11×10^{-6} /K and the modulus of elasticity is 210 GPa.

$$\text{Stress} = Ea\theta = 210 \times 10^9 \times 11 \times 10^{-6} \times 8$$

Hence the stress is 18.5×10^6 Pa = 18.5 MPa and compressive.

24.3.1 Composite bars

Consider the effect of a temperature change on a compound bar made of two materials having different coefficients of expansion, as in Figure 24.7(a) which consists of two members A and B, say a circular bar inside a circular tube. The two materials have coefficients of expansion a_A and a_B and modulus of elasticity values E_A and E_B. The two members are of the same initial length L and attached rigidly together at their ends so that at any temperature they will still be the same length. If the two members had not been fixed to each other, when the temperature changes A would have increased its length by $La_A\theta_A$ and B its length by $La_B\theta_B$ to give the situation illustrated by Figure 24.7(b). There will now be a difference in length between the two members at temperature θ of $(a_A - a_B)\theta L$. However, because the materials are so constrained that they must expand by the same amount, the expansion of each will differ from that occurring if they were free. Thus when the two members are rigidly fixed together (Figure 24.7(c)), this difference in length is eliminated by compressing member B with a force F and extending A with a force F. Because the composite bar has no net force acting on it, then the forces acting on material A must be opposite and equal to those acting on material B. The extension e_A of A due to this force is:

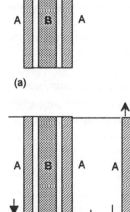

(a)

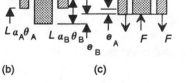

(b) (c)

Figure 24.7 *Composite bar*

$$e_A = \frac{\sigma_A L}{E_A}$$

where A_A is its cross-sectional area. The contraction e_B of B due to this force is:

$$e_B = \frac{\sigma_B L}{E_B}$$

where A_B is its cross-sectional area. But $e_A + e_B = (a_A - a_B)\theta L$ and so:

$$(a_A - a_B)\theta = \frac{\sigma_A}{E_A} + \frac{\sigma_B}{E_B}$$

The above analysis gives the stresses acting on the materials due solely to a temperature change. If such a compound bar is also subject to loading then the total stress on a material is obtained by using the *principle of superposition*. This states that the resultant stress or displacement at a point in a bar subject to a number of loads can be determined by finding the stress or displacement caused by each load considered acting separately on the bar and then adding the contributions caused by each load to obtain the resultant stress. Thus with a composite bar the stress in a member is the sum of the stresses obtained by considering the thermal stress and the loading separately, with due regard being paid as to whether they are compressive or tensile. A tensile stress is normally regarded as being positive and a compressive stress as negative. Thus a tensile loading stress of 80 MPa combined with a compressive thermal stress of 20 MPa would mean a total stress of $80 - 20 = 60$ MPa, a tensile stress.

Example

A steel tube with an external diameter of 35 mm and an internal diameter of 30 mm has a brass rod of diameter 20 mm inside it and rigidly joined to it at each end. At 15°C, when the materials were joined, there were no stresses in the materials. What will be the longitudinal stresses produced when the temperature is raised to 100°C? The brass has a modulus of elasticity of 120 GPa and a coefficient of linear expansion of 18×10^{-6} per °C, the steel a modulus of elasticity of 210 GPa and a coefficient of linear expansion of 11×10^{-6} per °C.

$$(a_A - a_B)\theta = \frac{\sigma_s}{E_s} + \frac{\sigma_b}{E_b}$$

$$(18 - 11)10^{-6} \times (100 - 15) = \frac{\sigma_s}{120 \times 10^9} + \frac{\sigma_b}{210 \times 10}$$

$$595 \times 10^3 = \frac{\sigma_s}{120} + \frac{\sigma_b}{210}$$

Since the force on A will be opposite and equal to that on B:

$$\sigma_s A_s = \sigma_b A_b$$

$$\sigma_x \times \frac{1}{4}\pi(0.035^2 - 0.030^2) = \sigma_b \times \frac{1}{4}\pi \times 0.020^2$$

and so $\sigma_s = 1.23\sigma_b$. Substituting for σ_s in the earlier equation gives:

$$595 \times 10^3 = \frac{1.23\sigma_b}{120} + \frac{\sigma_b}{210}$$

Hence $\sigma_b = 42$ MPa and $\sigma_s = 52$ MPa. The steel is in tension and the brass in compression.

Example

If the heated compound bar in the above example is then subject to a compressive axial load of 50 kN, what will be the stresses in the copper and steel elements.

Considering just the effects of the 50 kN force:

$$\frac{\sigma_b}{120 \times 10^9} = \frac{\sigma_s}{210 \times 10^9}$$

Hence $1.75\sigma_b = \sigma_s$. Since $\sigma_s A_s + \sigma_b A_b = F$:

$$\sigma_s \times \tfrac{1}{4}\pi(0.035^2 - 0.030^2) + \sigma_b \times \tfrac{1}{4}\pi 0.020^2 = 50 \times 10^3$$

$$1.75\sigma_b \times \tfrac{1}{4}\pi(0.035^2 - 0.030^2) + \sigma_s \times \tfrac{1}{4}\pi 0.020^2 = 50 \times 10^3$$

The compressive stress in the brass due to the load is 66 MPa and the compressive stress in the steel due to the load is 116 MPa. Thus the resultant stress, taking into account the thermal stresses, is for the steel a compressive stress of $-116 + 52 = -64$ MPa and for the brass a compressive stress of $-116 - 42 = -158$ MPa.

Problems

1 A steel bolt (Figure 24.8) has a diameter of 25 mm and carries an axial tensile load of 50 kN. Determine the average tensile stress at the shaft section *aa* and the screwed section *bb* if the diameter at the root of the thread is 21 mm.

2 A brass rod of length 160 mm and diameter 30 mm has an axial hole of diameter 20 mm drilled in it to a depth of 70 mm. What will be the change in length of the rod if it is subject to an axial compressive load which produces a maximum stress of 125 MPa? The modulus of elasticity of the brass is 100 GPa.

3 A timber beam with a rectangular cross-section 150 mm × 150 mm is reinforced by a steel plate of thickness 6 mm and the same width of 150 mm as the beam being bolted to one face (Figure 24.9). When the composite is subject to an axial load, what will be the stress in the steel when the stress in the timber is 6 MPa and the value of the axial load? The steel has a modulus of elasticity of 200 GPa and the timber a modulus of elasticity of 8 GPa.

4 A square section, 500 mm × 500 mm, reinforced concrete column has a reinforcement of four steel rods, each of diameter 25 mm, embedded axially in the concrete. Determine the compressive

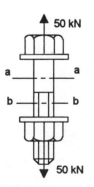

Figure 24.8 *Problem 1*

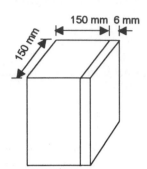

150 mm 6 mm

150 mm

Figure 24.9   *Problem 3*

stresses in the concrete and the steel when the column is subject to a compressive load of 1000 kN. The modulus of elasticity of the steel is 200 GPa and that of the concrete 14 GPa.

5   A cast iron pipe of external diameter 300 mm and internal diameter 250 mm is filled with concrete and the composite is subject to an axial load of 900 kN. What will be the stresses in the cast iron and concrete? The cast iron has a modulus of elasticity of 140 GPa and the concrete a modulus of elasticity of 14 GPa.

6   A steel pipe of external diameter 500 mm and wall thickness 12 mm is filled with concrete and the composite is subject to an axial load. What is the maximum allowable load if the stress in the steel is not to exceed 120 MPa and that in the concrete not to exceed 8 MPa? The steel has a modulus of elasticity of 200 GPa and the concrete a modulus of elasticity of 14 GPa.

7   A steel rod is rigidly clamped at both ends when the temperature is 0°C. What will be the stress produced in the rod by a rise in temperature to 60°C? The steel has a modulus of elasticity of 200 GPa and a coefficient of expansion of 12×10^{-6} /K.

8   Determine the change in stress per degree in a rigidly clamped steel rod when the temperature changes. The steel has a modulus of elasticity of 200 GPa and a coefficient of expansion of 12×10^{-6} /K.

9   A compound tube has a length of 750 mm and is fixed between two rigid supports. It consists of a copper tube of external diameter 100 mm and internal diameter 87 mm encasing a steel tube of external diameter 87 mm and internal diameter 75 mm. Determine the stresses set up in the tubes as a result of the temperature being increased by 40°C. The steel has a modulus of elasticity of 210 GPa and a coefficient of linear expansion of 12×10^{-6} per °C and the copper a modulus of elasticity of 130 GPa and a coefficient of linear expansion of 17×10^{-6} per °C.

10   A brass rod of diameter 25 mm is enclosed centrally in a steel tube with internal diameter 25 mm and external diameter 40 mm, both having a length of 1.0 m and rigidly fastened at the ends. Determine the stresses in the rod and tube resulting from a temperature increase of 100°C. The steel has a modulus of elasticity of 200 GPa and a coefficient of linear expansion of 12×10^{-6} per °C and the brass a modulus of elasticity of 100 GPa and a coefficient of linear expansion of 19×10^{-6} per °C.

25 Shear stress

25.1 Introduction

This chapter is about the stresses produced as a result of forces being applied to materials which result in shear. The forces are applied in such a way that one layer of the material tends to slide over another. Shear stresses are involved with fastenings, e.g. lapped plates held together by rivets or bolts when the forces pull on the sheets. Another situation in which shear forces are involved is when rods are twisted.

25.2 Shear stress and strain

Figure 25.1 shows how a material can be subject to shear. The forces are applied in such a way as to tend to slide one layer of the material over an adjacent layer. With shear, the area over which forces act is in the same plane as the line of action of the forces. The force per unit area is called the *shear stress*:

$$\text{shear stress} = \frac{\text{force}}{\text{area}}$$

The unit of shear stress is the pascal (Pa).

With tensile and compressive stresses, changes in length are produced; with shear stress there is an angular change ϕ. Shear strain is defined as being the angular deformation:

$$\text{shear strain} = \phi$$

The unit used is the radian and, since the radian is a ratio, shear strain can be either expressed in units of radians or without units.

For the shear shown in Figure 25.1 tan $\phi = x/L$ and, since for small angles, tan ϕ is virtually the same as ϕ expressed in radians:

$$\text{shear strain} = \frac{x}{L}$$

Area over which
force applied *A*

L

F

Figure 25.1   *Shear*

Example

Figure 25.2 shows a component that is attached to a vertical surface by means of an adhesive. The area of the adhesive in contact with the component is 100 mm². The weight of the component results in a force of 30 N being applied to the adhesive–component interface. What is the shear stress?

Shear stress = force/area = $30/(100 \times 10^{-6})$ = 0.3×10^6 Pa = 0.3 MPa.

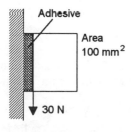

Adhesive

Area
100 mm²

30 N

Figure 25.2   *Example*

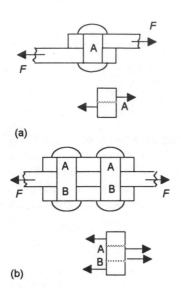

(a)

(b)

Figure 25.3 *(a) Lap joint, (b) double cover butt joint*

25.2.1 Shear with joints

Figure 25.3 shows examples of shear occurring in fastenings, in this case riveted joints. In Figure 25.3(a), a simple lap joint, the rivet is in shear as a result of the forces applied to the plates joined by the rivet. The rivet is said to be in *single shear*, since the bonding surface between the members is subject to just a single pair of shear forces; there is just one shear surface A subject to the shear forces. In Figure 25.3(b) the rivets are used to produce a double cover butt joint; the rivets are then said to be in *double shear* since there are two shear surfaces A and B subject to shear forces.

Example

What forces are required to shear a lap joint made using a 25 mm diameter rivet if the maximum shear stress it can withstand is 250 MPa?

The joint is of the form shown in Figure 25.3(a). The rivet is in single shear and thus, since shear stress = force/area:

$$\text{force} = \text{shear stress} \times \text{area} = 250 \times 10^6 \times \tfrac{1}{4}\pi \times 0.025^2$$

$$= 1.2 \times 10^5 \text{ N} = 120 \text{ kN}$$

Example

What forces are required to shear a double cover lap joint made with two 25 mm diameter rivets if the maximum shear stress for a rivet is 400 MPa?

The joint is of the form shown in Figure 25.3(b). The rivets are in double shear. We can think of the applied force being halved with half being applied to each shear surface (alternatively we can think of the area that has to be sheared for a rivet to fail being double the cross-sectional area of a rivet). Since shear stress = force/area:

$$\text{force} = \text{shear stress} \times \text{area}$$

$$= 2 \times 300 \times 10^6 \times \tfrac{1}{4}\pi \times 0.025^2$$

$$= 3.9 \times 10^5 \text{ N} = 390 \text{ kN}$$

Example

Calculate the maximum load that can be applied to the coupling shown in Figure 25.4 if the pin has a diameter of 8 mm and the maximum shear stress it can withstand is 200 MPa.

The pin is in double shear, at A and B, and thus the forces applied to each shear surface are ½F. Since shear stress = force/area:

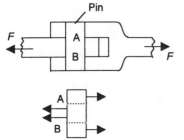

Figure 25.4 *Example*

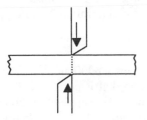

Figure 25.5   *Cropping a plate*

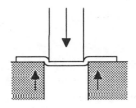

Figure 25.6   *Punching a hole*

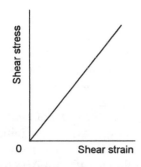

Figure 25.7   *Shear stress proportional to shear strain*

force = shear stress × area = $2 \times 200 \times 10^6 \times \frac{1}{4}\pi \times 0.008^2$

$$= 20 \times 10^3 \text{ N} = 20 \text{ kN}.$$

25.2.2 Shear strength

The *shear strength* of a material is the maximum shear stress that the material can withstand before failure occurs. Every time a guillotine is used to crop a material, shear stresses are being applied at a value equal to the maximum shear stress for that material (Figure 25.5). The area over which the shear forces are being applied is the cross-sectional area of the plate being cropped.

Similarly, when a punch is used to punch holes in a material (Figure 25.6), shear stresses are being applied at a value equal to the shear strength of the material. In this case the area over which the punch force is applied is the plate thickness multiplied by the perimeter of the hole being punched.

Example

A plate of mild steel 1.0 m wide and 0.8 mm thick is to be cropped by a guillotine. What is the force which the guillotine has to apply if the shear strength of the steel is 200 MPa?

Since shear strength = force/area then:

force = shear strength × area = $200 \times 10^6 \times 1.0 \times 0.008$

$$= 160 \times 10^3 \text{ N} = 160 \text{ kN}$$

Example

What is the maximum diameter hole that can be punched in an aluminium plate of thickness 14 mm if the punching force is limited to 50 kN? The shear strength of the aluminium is 90 MPa.

Shear strength = (punch force)/(area being sheared) and so:

area = force/shear strength = $(50 \times 10^3)/(90 \times 10^6)$

But the area is $\pi d \times 14 \times 10^{-3}$, where d is the diameter of the hole, and so $d = 13$ mm.

25.2.3 Shear modulus

Within some limiting stress, metals usually have the shear stress proportional to the shear strain and we have the 'Hooke's law for shear' (Figure 25.7). For such a situation, the gradient of the shear stress–shear strain graph is the *shear modulus* (or *modulus of rigidity*):

shear modulus $G = \dfrac{\text{shear stress}}{\text{shear strain}}$

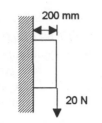

Figure 25.8   *Example*

The SI unit for the shear modulus is the pascal (Pa). For mild steel a typical value of the shear modulus is 75 GPa, for aluminium 28 GPa.

Example

A rubber mounting pad of vertical face area 8000 mm² is firmly fastened to a vertical wall (Figure 25.8) so that it projects 200 mm from the wall. When a downward force of 20 N is applied to the free vertical edge of the pad a vertical deflection of 1 mm is obtained. What is the shear modulus of the material?

The shear stress is shear force/area = $20/(8000 \times 10^{-6})$ = 2500 Pa and the shear strain is $1/200 = 0.005$. Hence the shear modulus is:

$$G = \frac{\text{shear stress}}{\text{shear strain}} = \frac{2500}{0.005} = 5 \times 10^5 \text{ Pa} = 0.5 \text{ MPa}$$

25.3 Torsion

Twisting a material involves shear forces. Thus if an area on the surface of a rod is considered, as in Figure 25.9, then twisting the rod is accomplished by shear forces acting on the area. *Torsion* is the term used for the twisting of a structural member when it is acted on by torques so that rotation is produced about the longitudinal axis of one end of the member with respect to the other. Thus, while the modulus of elasticity (see Section 6.3.1) is a measure of the stiffness of a material, the greater the modulus, the more difficult it is to bend the material; the shear modulus is a measure of the rigidity or ease of twisting of a material, the greater the modulus, the more difficult it is to twist a material. The shear modulus is thus of importance in the design of shafts in such situations as the drive shaft of a motor car.

25.3.1 Torsion stress

For simplification in deriving equations for torsion we will make the following assumptions:

1   The shaft has a uniform circular cross-section, is straight and initially unstressed, and the shaft material is uniform throughout.

2   The shear stress is proportional to shear strain.

3   The axis of the twisting moment is the axis of the shaft.

4   Each circular section is rotated different amounts and results in shear forces.

(a)

(b)

Figure 25.9   *Twisting a rod*

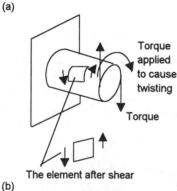

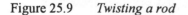

Figure 25.10   *Twisting of a cylindrical shaft*

Consider such a shaft of radius r and length L. If the angle of twist is θ then the situation is as shown in Figure 25.10; AC is the initial position of a line along the surface of the bar and BC is its new position as a result of the end of the bar being rotated through the arc AB. The arc AB subtends an angle θ and so arc AB = $r\theta$. But AB also equals $L\phi$, where ϕ is the resulting shear strain; if you think of a strip on the surface of the shaft it becomes sheared as shown in Figure 25.11.

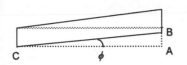

Figure 25.11 *Shear of strip on the surface*

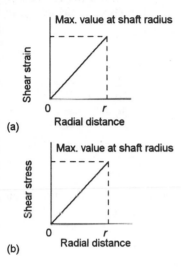

(a)

(b)

Figure 25.12 *Shear strain and shear stress*

Thus, $\phi L = r\theta$ and so the shear strain is:

$$\phi = \frac{r\theta}{L}$$

This states that for a given angle of twist per unit length, i.e. θ/L, the shear strain is proportional to the distance r from the central axis (Figure 25.12(a)). The shear stress τ at this radius is thus:

$$\tau = \frac{Gr\theta}{L}$$

where G is the shear modulus. The shear stress is, for a particular angle of twist per unit length and material, proportional to the distance r from the central axis (Figure 25.12(b)).

Example

A steel tube of length 4 m has an external diameter of 10 mm and an internal diameter of 6 mm. Determine the maximum and minimum shear stresses in the tube when, with one end fixed, the other is rotated through 30°. The steel has a shear modulus of 80 GPa.

With $\theta = 30° = \pi/6$ radians:

$$\tau = \frac{G\theta r}{L} = \frac{80 \times 10^9 \times (\pi/6)r}{4} = 1.05 \times 10^{10} r \text{ Pa}$$

The maximum shear stress occurs at the maximum radius of 5 mm and thus is 52.5 MPa and the minimum shear stress occurs at the minimum radius of 3 mm and thus is 31.5 MPa.

25.3.2 The general equation for torsion of a circular shaft

A shaft may be considered to be made up of a number of thin concentric layers (Figure 25.13). Consider the torsion of such an element of the shaft with a radius r and thickness δr (Figure 25.14). The element is considered to be thin enough for the assumption to be made that the shear stress is uniform throughout its thickness. Thus for an element of area δA the shear force is:

$$\text{shear force} = \text{area} \times \text{shear stress} = \delta A \times \tau$$

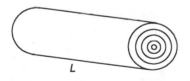

Figure 25.13 *Solid tube as a large number of concentric elements*

where τ is the shear stress acting on the element. The shear stress acts circumferentially to the thin element across the area of the cross-section. Thus the shear force on the total area of the element wall is:

$$\text{shear force} = 2\pi r \delta r \times \tau$$

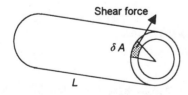

Figure 25.14 *Shear force on an element*

This shear force is acting at a radius r in a tangential direction. It gives a torque T about the axis of the element of:

$$T = \text{shear force} \times \text{radius} = 2\pi r \delta r \times \tau \times r = 2\pi r^2 \delta r \times \tau$$

Using $\tau = Gr\theta/L$, we can write the above equation as:

$$T = 2\pi r^2 \delta r \times \frac{Gr\theta}{L} = \frac{G\theta}{L} \times 2\pi r^3 \delta r$$

This gives the relationship for a thin element. For a solid rod we need to sum of all the contributions from each of the thin elements which go to make up the solid rod:

$$T = \text{sum of all the } (G\theta/L) \times 2\pi r^3 \, \delta r \text{ terms}$$

The sum of all the $2\pi r^3 \, \delta r$ terms is a geometric term and is given the name of the *polar second moment of area J* about the shaft axis. The units of J are m^4. Thus:

$$T = \frac{G\theta J}{L}$$

The above equation and that for shear stress $\tau = Gr\theta/L$ are often combined and referred to as the *general equation for the torsion of circular cross-section shafts*:

$$\frac{T}{J} = \frac{\tau}{r} = \frac{G\theta}{L}$$

For a solid shaft the polar second moment of area about its axis is $J = \pi D^4/32$, where D is its diameter.

Example

What is the minimum diameter of a solid shaft if it is to transmit a torque of 30 kN and the shear stress in the shaft is not to exceed 80 MPa?

For a solid shaft the polar second moment of area about its axis is given by $J = \pi D^4/32$ and thus:

$$T = \frac{J\tau}{r} = \frac{2J\tau}{D} = \frac{\pi D^3 \tau}{16}$$

$$D^3 = \frac{16T}{\pi\tau} = \frac{16 \times 30 \times 10^3}{\pi \times 80 \times 10^6}$$

Thus the minimum diameter shaft is 0.124 m.

Example

Sketch a graph of the shear stress across a section of a 50 mm diameter solid shaft when subject to a torque of 200 N m.

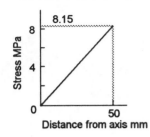

Figure 25.15 *Example*

For a solid shaft the polar second moment of area about its axis is $J = \pi D^4/32 = \pi \times 0.050^4/32 = 6.14 \times 10^{-7}$ m^4. Thus:

$$\tau = \frac{Tr}{J} = \frac{200r}{6.14 \times 10^{-7}} = 3.26 \times 10^8 r \text{ Pa}$$

When $r = 0$ then the shear stress is 0; when $r = 25$ mm then the shear stress is 8.15 MPa. Since the shear stress is proportional to the distance r from the axis, the graph is as shown in Figure 25.15.

Example

Calculate the maximum shear stress produced for a 6 mm diameter bolt when it is tightened by a spanner which applies a torque of 7 N m.

For a solid shaft the polar second moment of area about its axis is $J = \pi D^4/32$. Thus:

$$T = \frac{T\tau}{R} = \frac{2J\tau}{D} = \frac{\pi D^3 \tau}{16}$$

Hence:

$$\tau = \frac{16T}{\pi D^3} = \frac{16 \times 7}{\pi \times 0.006^3} = 165 \times 10^6 \text{ Pa} = 165 \text{ MP}$$

25.3.3 Polar second moment of area

If we make the elements which constitute a rod infinitesimally thin, then the equation for the polar second moment of area can be written as:

$$J = \int_0^r 2\pi r^3 \, dr$$

Thus, with a solid shaft of diameter D:

$$J = 2\pi \int_0^{D/2} r^3 \, dr = \frac{\pi D^4}{32}$$

For a hollow shaft with external diameter D and internal diameter d:

$$J = 2\pi \int_{d/2}^{D/2} r^3 \, dr = \frac{\pi}{32}(D^4 - d^4)$$

Example

Calculate the external diameter of a hollow shaft needed to transmit a torque of 30 kN m if the shear stress is not to exceed 80 MPa and the shaft has an external diameter twice the internal diameter.

For a hollow shaft with $D = 2d$:

$$J = \frac{\pi(D^4 - d^4)}{32} = \frac{\pi(D^4 - 0.5^4 D^4)}{32} = \frac{0.9375\pi D^4}{32}$$

But:

$$J = \frac{TR}{\tau} = \frac{TD}{2\tau} = \frac{30 \times 10^3 D}{2 \times 80 \times 10^6}$$

Thus:

$$\frac{0.9375\pi D^4}{32} = \frac{30 \times 10^3 D}{2 \times 80 \times 10^6}$$

Hence $D = 0.127$ m.

25.3.4 Polar section modulus

For a shaft of radius R, the relationship between the torque T and the maximum shear stress τ_{max} is:

$$\frac{T}{J} = \frac{\tau_{max}}{R}$$

and so:

$$T = \frac{J}{R}\tau_{max} = Z_p\tau_{max}$$

The *polar section modulus of section* Z_p is a purely geometric term and is defined as:

$$Z_p = \frac{J}{R}$$

For a solid shaft where $J = \pi D^4/32$, with $D = 2R$, then $Z_p = \pi D^3/16$.

Example

Determine the polar section modulus for a uniform solid shaft of diameter 40 mm.

For a solid shaft where $J = \pi D^4/32$, with $D = 2R$, then:

$$Z_p = J/R = \pi D^3/16 = \pi \times 0.040^3/16 = 1.26 \times 10^{-5} \text{ m}^3.$$

25.3.5 Transmission of power

For a rotating shaft of radius r, the distance travelled by a point on its surface in one revolution is $2\pi r$ and when it rotates through n revolutions the distance travelled is $2\pi rn$. If is rotating at n revolutions per second then the distance travelled per second is $2\pi rn$. If the shaft is being rotated by a torque T, the force acting at a radius r is $F = T/r$. The work done per second, i.e. the power, will be the product of the force and the distance moved per second by the point to which the force is applied. Thus:

power $= F \times 2\pi rn = 2\pi nT$

The angular velocity ω is $2\pi n$. Thus:

power transmitted $= \omega T$

With the angular velocity in rad/s and the torque in N m, the power is in watts.

Example

Determine the power that can be transmitted by a solid steel shaft of diameter 100 mm which is rotating at 5 rev/s if the shear stress in the shaft is not to exceed 70 MPa.

For a solid shaft, $J = \pi D^4/32 = \pi \times 0.100^4/32 = 9.82 \times 10^{-6}$ m^4. The maximum torque that can act is given by $J\tau/R = 2J\tau/D = 2 \times 9.82 \times 10^{-6} \times 70 \times 10^6/0.100 = 13.7$ kN m. Thus:

max. power $= 2\pi nT = 2\pi \times 5 \times 13.7 \times 10^3 = 430$ kW

Example

The drive shaft of a car is a tube with an external diameter of 50 mm and an internal diameter of 47 mm. Determine the maximum shear stress in the shaft when it is transmitting a power of 70 kW and rotating at 80 rev/s.

The maximum torque T is given by power $P = \omega T$ as:

$$T = \frac{P}{2\pi n} = \frac{70 \times 10^3}{2\pi \times 80} = 139 \text{ N m}$$

For the tube, $J = \pi(D^4 - d^4)/32 = \pi(0.050^4 - 0.047^4)/32 = 1.35 \times 10^{-7}$ m^4. The maximum shear stress σ is thus:

$$\sigma = \frac{Tr}{J} = \frac{139 \times 0.025}{1.35 \times 10^{-7}} = 25.7 \text{ MPa}$$

Problems

1 A horizontal cantilever projects 80 mm from its clamped point. The cantilever has a uniform cross-section of 40 mm by 20 mm. If a force of 50 kN is applied vertically at the free end of the cantilever, in the plane of the end face, what is (a) the shear stress, (b) the shear strain? The shear modulus is 76 GPa.

2 Calculate the maximum thickness of plate that can have a hole punched in it by a punch of diameter 20 mm if the shear strength of

the plate is 200 MPa and the maximum force that can be exerted by the punch is 100 kN.

3 What force is needed to shear a mild steel plate of thickness 0.7 mm and width 1.0 m if the shear strength of the steel is 200 MPa?

4 Calculate the forces required to shear the rivets in (a) a lap joint and (b) a double cover butt joint if the rivets have a diameter of 25 mm. The shear strength of the rivet material is 330 MPa.

5 Determine the shear stress in each rivet in a lap joint made with six rivets, each of 8 mm diameter, when the joint is subject to a shear force of 1.5 kN.

6 A double-cover butt joint has eight rivets, each of 10 mm diameter. What will be the maximum shear force that can be applied to the joint if the shear stress in a rivet must not exceed 20 MPa?

7 A solid circular steel shaft is subject to a torque of 1 kN m. What should the diameter of the shaft be if the maximum shear stress is to be 70 MPa?

8 Calculate the torque that can be transmitted by a solid circular shaft of diameter 130 mm if the maximum shear stress is to be 80 MPa.

9 A solid steel shaft has a diameter of 60 mm. Determine the maximum torque that can be applied to the shaft if the maximum permissible shear stress is 40 MPa.

10 A solid steel shaft has a diameter of 60 mm. Determine the maximum torque that can be applied to the shaft if the maximum permissible twist per unit length is 1° per metre. The steel has a shear modulus of 80 GPa.

11 Calculate the angle of twist produced in a solid circular shaft of diameter 40 mm and length 1.0 m when the shear stress is 40 MPa and the shear modulus is 80 GPa.

12 Calculate the maximum shear stress produced in a bolt of diameter 20 mm when it is tightened by a spanner which exerts a force of 60 N with a radius of action of 150 mm.

13 Determine the maximum shear stress produced in a 6 mm diameter bolt when it is tightened by a spanner which applies a torque of 7 N m.

14 Determine the external diameter of a tube needed to transmit torque of 40 kN m if it has an external diameter which is twice its internal diameter and the shear stress is not to exceed 80 MPa.

15 A tubular shaft has an internal diameter which is half its external diameter and is subject to a torque of 50 kN m. What external diameter will be required if the shear stress in the shaft is not to exceed 80 MPa?

16 A tubular shaft has an internal diameter of 120 mm and an external diameter of 160 mm. Calculate the shear stress produced at the outer surface of the shaft when the applied torque is 40 kN m.

17 A tubular shaft has an external diameter of 50 mm and an internal diameter of 25 mm. When subject to a torque of 1.5 kN m it is twisted through 1° per metre. Determine (a) the maximum shear stress in the shaft, (b) the shear modulus of the material, (c) the maximum shear strain.

18 A tubular drive shaft has an external diameter of 90 mm and an internal diameter of 64 mm. Determine the maximum and minimum shear stresses in the shaft when it is subject to a torque of 5 kN m.

19 A shaft has a polar section modulus of 2.0×10^{-4} m^3. What will be the maximum torque that can be used with the shaft if the shear stress in the shaft is not to exceed 50 MPa?

20 Calculate the maximum shear stress in a solid circular shaft of diameter 120 mm when it is transmitting a power of 10 kW at 60 rev/min.

21 A solid shaft of diameter 80 mm rotates at 5 rev/s and transmits a power of 50 kW. Determine the maximum shear stress in the shaft.

22 A tubular shaft has an external diameter of 150 mm and an internal diameter of 100 mm. Determine the maximum power that can be transmitted by the shaft when rotating at 3 rev/s if the shear stress in the shaft is not to exceed 50 MPa.

23 A tubular shaft has an inner diameter of 30 mm and an external diameter of 42 mm and has to transmit a power of 60 kW. What will be the limiting frequency of rotation of the shaft if the shear stress in the shaft is not to exceed 50 MPa?

24 The motor of an electric fan delivers 150 W to rotate the fan blades at 18 rev/s. Determine the smallest diameter of solid shaft that can be used if the shear stress in the shaft is not to exceed 80 MPa.

25 The specification for a shaft states that it must be a solid circular steel shaft capable of transmitting 500 kW when rotating at 240 rev/min. The shear stress must not exceed 50 MPa. Calculate the diameter of the shaft that will meet the specification and the angle of twist that will occur per metre length of shaft. The shear modulus of the steel is 80 GPa.

26 Structures

26.1 Introduction

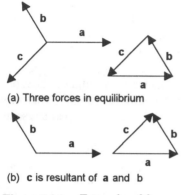

(a) Three forces in equilibrium

(b) **c** is resultant of **a** and **b**

Figure 26.1 *Triangle of forces*

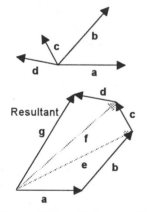

Resultant

Figure 26.2 *Polygon of forces*

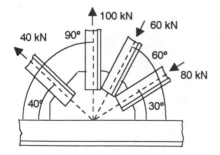

100 kN

60 kN

40 kN   90°

60°

80 kN

40°   30°

Figure 26.3 *Example*

This chapter extends the principles of equilibrium considered in chapter 5 to the analysis of the forces involves in structures.

26.1.1 Resultant of more than two forces

The triangle law (Section 5.2.2) can be used to determine when three forces are in equilibrium. If three forces **a**, **b** and **c** are in equilibrium then **a** and **b** must have a resultant which is in the opposite direction to **c** since only then will there be equilibrium (Figure 26.1). Consider now the problem of determining the sum of more than two forces. Suppose we have forces **a**, **b**, **c** and **d**. We can use the triangle law to find the sum **e** of forces **a** and **b**. We can then use the triangle law to find the sum **f** of **e** and **c**; then use it to find the sum **g** of **f** and **d**. Figure 26.2 illustrates the above procedure. In fact what we have done is take the forces in sequence and draw the arrows tail to head. The resultant of all the forces is the force required to complete the polygon shape and link the head of the last force to the tail of the start force. Thus if we have the forces **a**, **b**, **c**, **d** and −**g** then they will be in equilibrium and form the closed polygon.

An alternative to drawing the polygon is to resolve all the forces into their vertical and horizontal components. We can then easily determine the sum of the vertical components and the sum of the horizontal components. We then have replaced all the forces by just two components and from these can determine the resultant.

Example

Determine, by drawing the polygon of forces, the resultant force acting on the gusset plate as a result of the forces shown in Figure 26.3.

Figure 26.4 shows the resulting polygon when we take the forces in succession, starting from the extreme right and working anti-clockwise. The resultant is represented by the line needed to complete the polygon and is thus about 134 kN in the direction indicated.

Example

Determine, by considering the resolved components, the resultant force acting on the bracket shown in Figure 26.5 due to the three forces indicated.

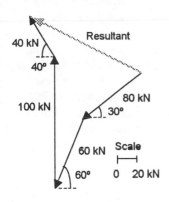

40 kN

40°

Resultant

100 kN

80 kN

30°

60 kN Scale

60° 0 20 kN

Figure 26.4 *Example*

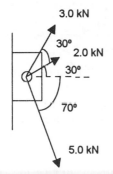

3.0 kN

30°

2.0 kN

30°

70°

5.0 kN

Figure 26.5 *Example*

For the 3.0 kN force: the horizontal component = 3.0 cos 60° = 1.5 kN and the vertical component = 3.0 sin 60° = 2.6 kN. For the 2.0 kN force: the horizontal component = 2.0 cos 30° = 1.7 kN and the vertical component = 2.0 sin 30° = 1.0 kN. For the 5.0 kN force: the horizontal component = 5.0 cos 70° = 1.7 kN and the vertical component = −5.0 sin 70° = −4.7 kN. The minus sign for a force is because it is acting downwards and in the opposite direction to the other vertical components which we have taken as being positive. All the horizontal components are in the same direction. Thus:

sum of horizontal components = 1.5 + 1.7 + 1.7 = 4.9 kN

sum of vertical components = 2.6 + 1.0 − 4.7 = −1.1 kN

Figure 26.5 shows how we can use the parallelogram rule (see Section 5.3), or the triangle rule, to find the resultant with these two components. Since the two components are at right angles to each other, the resultant can be calculated using the Pythagoras theorem. Thus:

$$(\text{resultant})^2 = 4.9^2 + 1.1^2$$

Hence the resultant has a magnitude of 5.0 kN. The resultant is at an angle θ downwards from the horizontal given by tan θ = 1.1/4.9 and so θ = 12.7°.

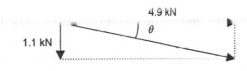

4.9 kN

θ

1.1 kN

Figure 26.6 *Example*

26.2 Pin-jointed frameworks

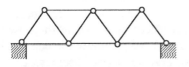

(a) A Warren bridge truss

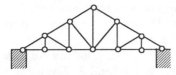

(b) A Howe roof truss

Figure 26.7 *Trusses*

The term *framework* is used for an assembly of members which have sectional dimensions which are small compared with their length. A framework composed of members joined at their ends to give a rigid structure is called a *truss* and when the members all lie in the same plane a *plane truss*. Bridges and roof supports are examples of trusses (Figure 26.7), the structural members being typically I-section beams, bars or channels which are fastened together at their ends by welding, riveting or bolts.

Several assumptions are made in analysing simple trusses:

1 Each member can be represented as a straight line representing its longitudinal axes with external forces only applied at the ends of members. The joints between members are treated as points located at the intersection of the members. The weight of a member is assumed to be small compared with the forces acting on it.

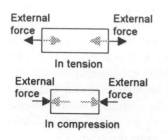

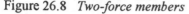

Figure 26.8 *Two-force members*

2 All members are assumed to be *two-force members*. For such a member, equilibrium occurs under the action of just two forces with the forces being of equal size and having the same line of action but in opposite directions so that a member is subject to either just tension or just compression (Figure 26.8). A member which is in tension is called a *tie*; a member that is in compression is called a *strut*. The convention is adopted of labelling tensile forces by positive signs and compressive forces by negative signs, this being because tensile forces tend to increase length whereas compressive forces decrease length.

3 All the joints are assumed to behave as *pin-jointed* and permit each end of a member to rotate freely about the joint. Thus the joint is capable of supporting a force in any direction. Welded and riveted joints can usually be assumed to behave in this way. Pin-jointed members can only be in tension or compression.

26.3 Bow's notation

Bow's notation is a useful method of labelling the forces in a truss. The spaces between the members and their external forces and reactions are labelled using letters or numbers when working in a consistent direction, e.g. clockwise. The spaces inside the truss are then labelled when working in the same direction. The internal forces are labelled by the two letters or number on each side of them. Thus, in Figure 26.9, letters are used to label the spaces and so the force in the member linking junctions 1 and 2 is F_{AF} and the force in the member linking junctions 3 and 6 is F_{GH}. In the illustration of Bow's notation in Figure 26.9, the joints were labelled independently of the spaces between forces. However, the space labelling can be used to identify the joints without the need for independent labelling for them. The joints are labelled by the space letters or numbers surrounding them when read in a clockwise direction. Thus, in Figure 26.9, junction 1 could be identified as junction AFE and junction 3 as junction BCHG.

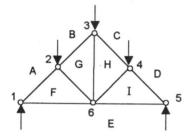

Figure 26.9 *Bow's notation*

26.3.1 Force polygons

Bow's notation aids the construction of force polygons. Consider the set of concurrent forces shown in Figure 26.10. The diagram has been labelled with Bow's notation with letters used to identify the spaces between the forces when working in a consistent direction, in this case clockwise, round the point at which the forces act. To draw the polygon of forces we start with, say, force F_{BC} and draw an arrow to represent it. The tail of this vector is labelled as B and its head as C. The point C now becomes the starting point for drawing the arrow to represent force F_{CD}, it having a tail C and head D. The point D now becomes the starting point for drawing the arrow to represent force F_{DE}, it having a tail D and head E. The point E now becomes the starting point for drawing the arrow to represent force F_{EA}, it having a tail E and head A. The point A now becomes the starting point for drawing the arrow to represent force F_{AB}, it having a tail A and a head B to complete the polygon (Figure 26.11).

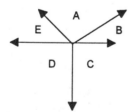

Figure 26.10 *Bow's notation and concurrent forces*

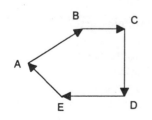

Figure 26.11 *Polygon*

The above method can be extended to enable a force diagram to be completed for a complete plane truss. Each joint in a structure will be in equilibrium if the structure is in equilibrium and thus we can analyse a structure by considering the equilibrium of each joint. Force diagrams are drawn for the forces at each joint, the joints being taken in sequence, recognising that some elements of the polygon for one joint will also figure in the polygons for other joints so that we can combine all the polygons for each joint in one force diagram. The following example illustrates this.

Example

Determine the forces in the truss members shown in Figure 26.12.

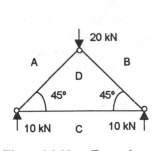

Figure 26.12   *Example*

Figure 26.12 has been labelled using Bow's notation. Starting with joint ABD we draw a line to represent the 20 kN force, labelled its tail as A and head as B (Figure 26.13). All we know about force F_{DB} is its direction so we draw a line at 45°, starting at B and ending at D. All we know about the third force acting at this joint, i.e. F_{DA}, is its direction so we draw a line at 45°, starting at D and ending back at A. In order to give a closed triangle then a scale diagram gives F_{BC} = 14 kN and F_{DA} = 14 kN. For joint BCD we can start with the force F_{DB} that we have already drawn. Then force F_{BC} is 10 kN starting at B and ending at C. Force F_{CD} is at right angles and so enables the triangle DBC to be completed. From the scale diagram, F_{CD} is 10 kN. For joint CAD the forces give the triangle CAD. In drawing this composite force diagram we do not put arrows on the lines to represent the forces. This is because we will use the same lines to represent the oppositely-directed forces at each end of a particular member.

An easy way to determine whether a member is in tension or compression is to imagine what would happen if it was removed. If the joints at each end would move closer together then the member is in compression and termed a *strut*, if the joints at each end would move further away then the member is in tension and termed a *tie*. Thus, for Figure 26.12, we have members AD and BD in compression with force 14 kN and CD in tension with 10 kN.

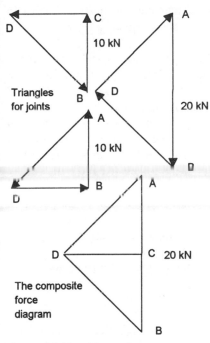

Figure 26.13   *Example*

26.4 Method of joints

Each joint in a structure will be in equilibrium if the structure is in equilibrium, thus the analysis of trusses by the *method of joints* involves considering the equilibrium conditions at each joint, in isolation from the rest of the truss, by analysis of the horizontal and vertical components of the forces. The procedure is:

1   Draw a labelled line diagram of the framework.

2   Determine any unknown external forces or reactions at supports by considering the truss at a single entity, ignoring all internal forces in truss members.

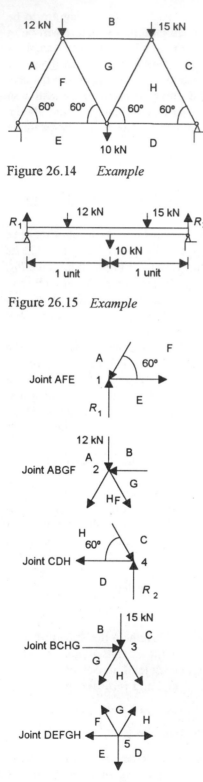

Figure 26.14   *Example*

Figure 26.15   *Example*

Figure 26.16   *Example*

3   Consider a junction in isolation from the rest of the truss and the forces, both external and internal, acting on that junction. The sum of the components of these forces in the vertical direction must be zero, as must be the sum of the components in the horizontal direction. Solve the two equations to obtain the unknown forces. Because we only have two equations at a junction, the junctions to be first selected for this treatment should be where there are no more than two unknown forces.

4   Then consider each junction in turn, selecting them in the order which leaves no more than two unknown forces to be determined at a junction.

Example

Determine the forces acting on the members of the truss shown in Figure 26.14. The ends of the truss rest on smooth surfaces.

The reactions at the supported ends will be vertical, because the surfaces are stated as being smooth. Note that if the span is not given, we assume an arbitrary length of 1 unit for a member and then the other distances related to this length.

Considering the truss as an entity we have the situation shown in Figure 26.15. Taking moments about the end at which reaction R_1 acts gives $12 \times 5 + 10 \times 10 + 15 \times 15 = 20R_2$ and so $R_2 = 19.25$ kN. Equating the vertical components of the forces gives $R_1 + R_2 = 12 + 10 + 15$ and so $R_1 = 17.75$ kN.

Figure 26.16 shows free-body diagrams for each of the joints in the framework. The directions of the forces in the members have been guessed; if the forces are in the opposite directions then, when calculated, they will have the a negative sign.

For joint 1, the sum of the vertical components must be zero:

$$17.75 - F_{AF} \sin 60° = 0$$

Hence $F_{AF} = 20.5$ kN. The sum of the horizontal components must be zero and so:

$$F_{AF} \cos 60° - F_{FE} = 0$$

Hence $F_{FE} = 10.25$ kN.

For joint 2, the sum of the vertical components must be zero:

$$12 + F_{FG} \sin 60° - F_{AF} \sin 60° = 0$$

With $F_{AF} = 20.5$ kN, then $F_{FG} = 6.6$ kN. The sum of the horizontal components must be zero and so:

$$F_{BG} - F_{FG} \cos 60° - F_{AF} \cos 60° = 0$$

Hence $F_{BG} = 13.6$ kN.

For joint 4, the sum of the vertical components must be zero:

$$F_{CH} \sin 60° - R_2 = 0$$

Hence $F_{CH} = 22.2$ kN. The sum of the horizontal components must be zero:

$$F_{DH} - F_{CH} \cos 60° = 0$$

Hence $F_{DH} = 11.1$ kN.

For joint 3, the sum of the vertical components must be zero:

$$15 + F_{GH} \sin 60° - F_{CH} \sin 60° = 0$$

Hence $F_{GH} = 4.9$ kN. The sum of the horizontal components must be zero and so:

$$F_{GH} \cos 60° + F_{CH} \cos 60° - F_{BG} = 0$$

This is correct to the accuracy with which these forces have already been calculated.

For joint 5, the sum of the vertical components must be zero:

$$10 - F_{FG} \sin 60° - F_{GH} \sin 60° = 0$$

This is correct to the accuracy with which these forces have been calculated. The sum of the horizontal components must be zero:

$$F_{FE} + F_{FG} \cos 60° - F_{DH} - F_{GH} \cos 60° = 0$$

This is correct to the accuracy with which these forces have already been calculated.

The directions of the resulting internal forces are such that member AF is in compression, BG is in compression, CH is in compression, DH is in tension, FE is in tension, FG is in tension and GH is in tension. Thus the internal forces in the members of the truss are $F_{AF} = -20.5$ kN, $F_{BG} = -13.6$ kN, $F_{CH} = -22.2$ kN, $F_{DH} = +11.1$ kN, $F_{FE} = +10.25$ kN, $F_{FG} = +6.6$ kN, $F_{GH} = +4.9$ kN.

26.5 Method of sections

This method is simpler to use than the method of joints when all that is required are the forces in just a few members of a truss. We imagine the structure to be cut at some particular place. The place chosen should be one which cuts the member in which the force is to be determined. Just one side of the cut is considered and the conditions for equilibrium applied to the external forces and the internal forces acting on the cut members. This generally involves taking moments about joints and considering the force components in the horizontal and vertical directions. Since this can lead to only three equations, no more than three members of the truss should be cut by the section. The procedure for using this method is thus:

1   Draw a line diagram of the structure.

2   Label the diagram using Bow's notation.

3   Put a straight line through the diagram to section it. No more than three members should be cut by the line and they should include those members for which the internal forces are to be determined.

4   Consider one of the parts isolated by the section and then write equations for equilibrium of that part: take moments about some joint and also sum the vertical and the horizontal components. To eliminate the forces in members whose lines of action intersect, take moments about their point of intersection. If necessary any unknown external forces or reactions at supports can be determined by considering the truss as an entity, ignoring all internal forces in members.

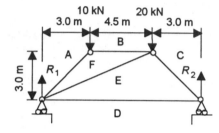

Figure 26.17   *Example*

Example

Determine, using the method of sections, the internal force F_{FE} for the plane pin-jointed truss shown in Figure 26.17.

The reactions at the supports will be vertical because the truss rests on rollers. Considering the truss as an entity, then taking moments about the left-hand end:

$$10.5R_2 = 10 \times 30 + 20 \times 7.5$$

Hence, $R_2 = 17.1$ kN. Since the vertical components must also balance:

$$R_1 + R_2 = 10 + 20$$

and so $R_1 = 12.9$ kN.

To determine F_{FE} we consider sectioning the truss along the line XX (Figure 26.18) and will consider the equilibrium of the left-hand section of the truss (Figure 26.19). Taking moments about Q:

$$7.5 \times 12.9 = 10 \times 4.5 + 3.0F_{DE}$$

Hence, $F_{DE} = 17.25$ kN. Taking moments about P:

$$3.0F_{DE} = 10 \times 3.0 + F_{FE} \times 3.0 \cos \theta$$

The triangle including angle θ has an angle at the reaction point of $(90° - \theta)$. This is also the angle in the triangle with the reaction point and points P and Q. Hence tan $(90° - \theta) = 7.5/3.0$ and $\theta = 21.8°$. Hence, $F_{FE} = 7.8$ kN.

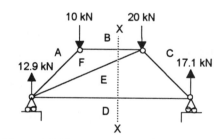

Figure 26.18   *Example*

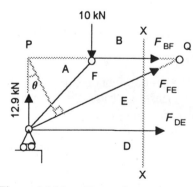

Figure 26.19   *Example*

Example

Figure 26.20 shows a truss bridge carrying a load of 20 kN. What will be the forces acting in the member X due to this load?

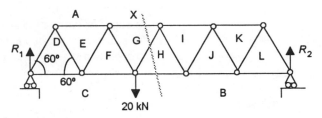

Figure 26.20 *Example*

Since the force is required for only one member, the method of sections is the simplest method to use. Taking the length of each member to be 1 unit and considering the truss as an entire entity, then taking moments about the left-hand end:

$$5R_2 = 20 \times 2$$

Hence, $R_2 = 8$ kN. Since the sum of the vertical components must balance, i.e. $R_1 + R_2 - 20$, then $R_1 = 12$ kN.

With the section as shown in Figure 26.20 and considering just the right-hand section (Figure 26.21), then taking moments about P:

$$F_{AG} \times 1 \sin 60° = 8 \times 3 \times 1$$

Hence, $F_{AG} = 27.6$ kN. The force in the member is compressive and thus written as -27.6 kN.

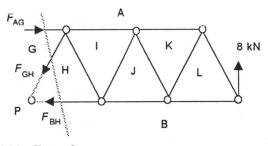

Figure 26.21 *Example*

26.6 Redundancy

The arrangement of members in a truss to form triangles gives a stable structure (Figure 26.22(a)). However, the arrangement of four members to form a rectangular framework (Figure 26.22(b)) gives an unstable structure since a small sideways force can cause the structure to collapse. To make the rectangle stable, a diagonal member is required and this converts it into two triangular frameworks (Figure 26.22(c)). The addition of two diagonal members (Figure 26.22(d)) gives a rigid structure but one of the diagonal members is 'redundant' in not being

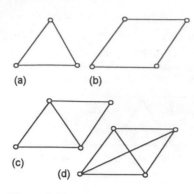

(a) (b)

(c)

(d)

Figure 26.22 *Frameworks*

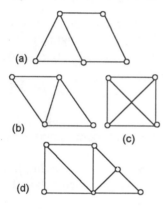

(a)

(b)

(c)

(d)

Figure 26.23 *Example*

required for stability. Note that the analysis of members in trusses might show that a member is unloaded. An unloaded member must not be confused with a redundant member; it may be that the member becomes loaded under different conditions.

There is a relationship between the number of members and the number of joints needed for internal stability with a pin-jointed truss. If m is the number of members and j the number of joints, then a stable pin-jointed structure can be produced if:

$$m + 3 = 2j$$

If $m + 3$ is greater than $2j$ then 'redundant' members are present, if $m + 3$ is less than $2j$ then the structure is unstable.

Many structures are designed to include 'redundant' members so that the structure will be 'fail safe' and not collapse if one or more members fail or so that they can more easily cope with different loading conditions.

Example

Of the structures shown in Figure 26.23, which can be stable, which unstable and which contain redundant members?

(a) $m = 6$, $j = 5$, hence $m + 3 < 2j$ and it is unstable.
(b) $m = 5$, $j = 4$, hence $m + 3 = 2j$ and it can be stable.
(c) $m = 6$, $j = 4$, hence $m + 3 > 2j$ and there is redundancy.
(d) $m = 9$, $j = 6$, hence $m + 3 = 2j$ and it can be stable.

Problems

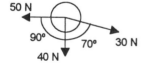

Figure 26.24 *Problem 3*

1 Three forces act at a point on an object. One of the forces is 6 N horizontally to the left, another 3 N at 70° anticlockwise to the 6 N force, and the third 4 N at 150° anticlockwise to the 6 N force. Determine the resultant force.

2 Determine the resultant force acting on an object if it is acted on by four forces acting in the same plane of 1 N in a westerly direction, 3 N in a south-westerly direction, 6 N in a north-easterly direction and 5 N in a northerly direction.

3 Determine the size and direction of the force needed to produce equilibrium for the force system shown in Figure 26.24.

4 For each of the following systems of forces as a result of considering the components of each force, determine the resultant force: (a) 2 N in an easterly direction, 3 N at 60° west of north and 2 N due south, (b) 4 N in a north-easterly direction, 3 N due west and 5 N at 30° south of east, (c) 2.8 N in a north-easterly direction, 4 N at 60° south of west and 6 N at 30° south of east.

5 Forces of 10 N, 12 N and 20 N act in the same plane on an object in the directions west, 30° west of north, and north respectively. Determine, from the components of the forces, the resultant force.

6 Three forces act in the same plane on the same point on an object. If the forces are 4 N in a direction due north, 7 N in a south-easterly

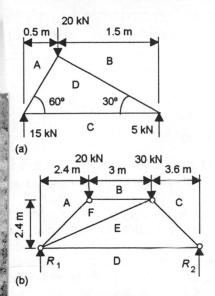

(a)

(b)

Figure 26.25   *Problem 8*

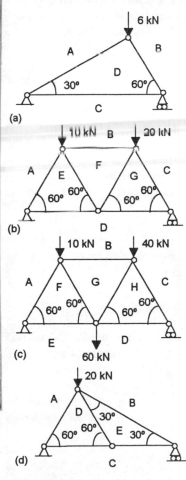

(a)

(b)

(c)

(d)

Figure 26.26   *Problem 9*

direction and 4 N in a direction 60° south of west, by considering the components of the forces determine the resultant force.

7 Forces of 1 N, 2 N, 3 N, 4 N and 5 N act in the same plane on an object in the directions north, north-east, east, 60° west of south, and due west respectively. Determine, by considering the components of the forces, the resultant force.

8 Determine by scale drawing, either vector diagrams for each joint or force diagrams, the forces acting in the members of the pin-jointed frameworks shown in Figure 26.25.

9 Using the method of joints, determine the magnitude and nature of the forces in each member of the plane pin-jointed frameworks shown in Figure 26.26.

10 Using the method of section, determine the forces in the members marked X in Figure 26.27.

11 Determine whether the trusses shown in Figure 26.28 can be stable, unstable or contain redundant members.

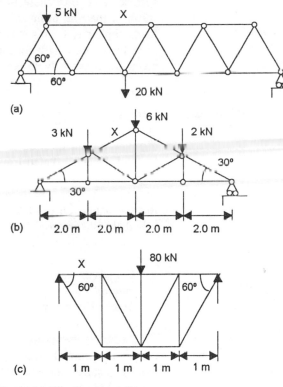

(a)

(b)

(c)

Figure 26.27   *Problem 10*

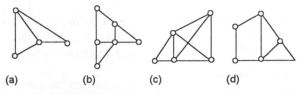

(a)   (b)   (c)   (d)

Figure 26.28   *Problem 11*

27 Beams

27.1 Introduction

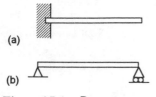

(a)

(b)

Figure 27.1 *Beams*

This chapter is about the bending of beams; a *beam* is defined as being any structural member which carries loads at right angles to its axis and so results in bending.

Two common forms of beam are cantilevers and the simply supported beam. With a *cantilever* (Figure 27.1(a)), the beam is rigidly fixed at one end and the other end is free to move. With the *simply supported beam* (Figure 27.1(b)), the beam is supported at its ends on rollers or smooth surfaces or one of these is combined with a pin at the other end. The result is that the beam is free to bend under the action of forces. The loads applied to beams may be point loads or loads distributed over part of or the entire length. In this chapter the bending of these two forms of beam are discussed with both point loads and distributed loads.

27.2 Bending

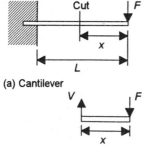

(a) Cantilever

(b) Forces for vertical equilibrium

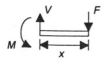

(c) Vertical and moment equilibrium

Figure 27.2 *Cantilever*

Consider a cantilever (Figure 27.2(a) which has a concentrated load F applied at the free end and an imaginary cut through the beam at a distance x from the free end. Now consider the section of beam to the right of the cut isolated from the rest of the beam. With the entire beam in equilibrium we must also have any section also in equilibrium. For the section of beam to be in vertical equilibrium, we must have a vertical force V acting on it such that $V = F$ (Figure 27.2(b)). This force V is called the *shear force* because the combined action of V and F on the section is to shear it. In general:

> *The shear force at a transverse section of a beam is the algebraic sum of the external forces acting at right angles to the axis of the beam on one side of the section concerned.*

In addition to vertical equilibrium we must also have the section of beam in rotational equilibrium. For the section of the beam to be in moment equilibrium and not rotate, we must have a moment M applied (Figure 27.2(c)) at the cut so that $M = Fx$. This moment is termed the *bending moment*:

> *The bending moment at a transverse section of a beam is the algebraic sum of the moments about the section of all the forces acting on one (either) side of the section concerned.*

The conventions most often used for the signs of shear forces and bending moments are:

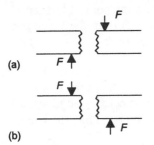

Figure 27.3   *Shear force:*
(a) positive, (b) negative

(a) Sagging

(b) Hogging

Figure 27.4   *Bending moment:*
(a) positive, (b) negative

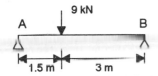

Figure 27.5   *Example*

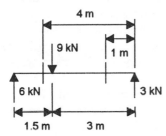

Figure 27.6   *Example*

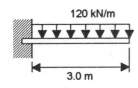

Figure 27.7   *Example*

1   *Shear force*
When the shear forces on either side of a section are clockwise (Figure 27.3(a)), i.e. the left-hand side of the beam is being pushed upwards and the right-hand side downwards, the shear force is taken as being positive. When the shear forces on either side of a section are anticlockwise (Figure 27.3(b)), i.e. the left-hand side of the beam is being pushed downwards and the right-hand side upwards, the shear force is taken as being negative.

2   *Bending moment*
Bending moments are positive if they give rise to sagging (Figure 27.4(a) and negative if they give rise to hogging (Figure 27.4(b)).

Example

Determine the shear force and bending moment at points 1 m and 4 m from the right-hand end of the beam shown in Figure 27.5. Neglect the weight of the beam.

The reactions at the ends A and B can be found by taking moments about A. Thus, $R_B \times 4.5 = 9 \times 1.5$ and so $R_B = 3$ kN. Considering the vertical equilibrium gives $R_A + R_B = 9$ and thus $R_A = 6$ kN. Figure 27.6 shows the forces acting on the beam.

If we make an imaginary cut in the beam at 1 m from the right-hand end, then the force on the beam to the right of the cut is 3 kN upwards and that to the left is $9 - 6 = 3$ kN downwards. The shear force is thus negative and -3 kN.

If we make an imaginary cut in the beam at 4 m from the right-hand end, then the force on the beam to the right of the cut is $9 - 3 = 6$ kN downwards and that to the left is 6 kN upwards. The shear force is thus positive and $+6$ kN.

The bending moment at a distance of 1 m from the right-hand end of the beam, when we consider that part of the beam to the right, is 3×1 kN m. Since the beam is sagging the bending moment is $+3$ kN m. At a distance of 4 m from the right-hand end of the beam, the bending moment is $3 \times 4 - 9 \times 0.5 = +7.5$ kN m.

Example

A uniform cantilever of length 3.0 m (Figure 27.7) has a weight per metre of 120 kN. Determine the shear force and bending moment at distances of 1.0 m and 3.0 m from the free end if no other loads are carried by the beam.

Note the symbol used for a distributed load, i.e. a load which is spread over a length of beam rather than being concentrated at just one point.

At 1.0 m from the free end, there is 1.0 m of beam to the right and it has a weight of 120 kN (Figure 27.8(a)). Thus the shear force is $+120$ kN; it is positive because the forces are clockwise. The weight of this section can be considered to act at its centre of gravity

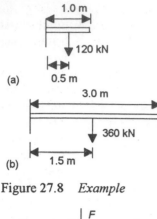

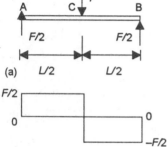

Figure 27.8 *Example*

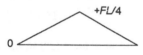

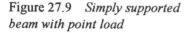

Figure 27.9 *Simply supported beam with point load*

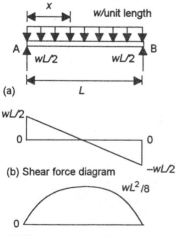

Figure 27.10 *Simply supported beam with distributed load*

which, because the beam is uniform, is at its midpoint. Thus the 120 kN weight force can be considered to be 0.5 m from the 1.0 m point and so the bending moment is $-120 \times 0.5 = -60$ kN m; it is negative because there is hogging.

At 3.0 m from the free end, there is 3.0 m of beam to the right and it has a weight of 360 kN (Figure 27.8(b)). Thus the shear force is $+360$ kN. The weight of this section can be considered to act at its midpoint, a distance of 1.5 m from the free end. Thus the bending moment is $-360 \times 1.5 = -540$ kN m.

27.2.1 Bending moment and shear force diagrams

Figures which graphically show how the variations of the shear forces and bending moments along the length of a beam are termed *shear force diagrams* and *bending moment diagrams*. The two quantities are plotted above the centre line of the beam if positive and below it if negative. The following show such diagrams for commonly occurring situations.

1   *Simply supported beam with point load at mid-span*
    Figure 27.9(a) shows the beam and the forces concerned, the weight of the beam being neglected. For a central load F, the reactions at each end will be $F/2$.

    Consider the shear forces. At point A, the forces to the right are $F - F/2$ and so the shear force at A is $+F/2$; it is positive because the forces are clockwise about A. This shear force value will not change as we move along the beam from A until point C is reached. To the right of C we have just a force of $F/2$ and this gives a shear force of $-F/2$; it is negative because the forces are anticlockwise about it. To the left of C we have just a force of $F/2$ and this gives a shear force of $+F/2$; it is positive because the forces are clockwise about it. Thus at point C, the shear force takes on two values. For points between C and B, the forces to the left are constant at $F/2$ and so the shear force is constant at $-F/2$. Figure 27.9(b) shows the shear force diagram.

    Consider the bending moments. At point A, the moments to the right are $F \times L/2 - F/2 \times L = 0$. The bending moment is thus 0. At point C the moment to the right is $F/2 \times L$ and so the bending moment is $+FL/2$; it is positive because sagging is occurring. At point B the moment to the right is zero, likewise that to the left $F \times L/2 - F/2 \times L = 0$. Between A and C the bending moment will vary, e.g. at one-quarter the way along the beam it is $FL/8$. In general, between A and C the bending moment a distance x from A is $Fx/2$ and between C and B is $Fx/2 - F(x - L/2) = F/2(L - x)$. Figure 27.9(c) shows the bending moment diagram. The maximum bending moment occurs under the load and is $FL/4$.

2   *Simple supported beam with uniformly distributed load*
    Figure 27.10 shows a simple supported uniform beam which carries a uniformly distributed load of w/unit length. The reactions at each end will be $wL/2$.

    Consider the shear force a distance x from the left-hand end of the beam. The load acting on the left-hand section of beam is wx. Thus the shear force is $V = wL/2 - wx = w(\frac{1}{2}L - x)$. When $x = \frac{1}{2}L$, the shear force is zero. When $x < \frac{1}{2}L$ the shear force is positive and when $x > \frac{1}{2}L$ it is negative. Figure 27.10(b) shows the shear force diagram.

Consider the bending moment. At A the moment due to the beam to the right is $-wL \times L/2 + wL/2 \times L = 0$. At the midpoint of the beam the moment is $-wL/2 \times L/4 + wL/2 \times L/2 = wL^2/8$; the bending moment is thus $+wL^2/8$. At the quarter-point along the beam, the moment due to the beam to the right is $-3L/4 \times 3L/8 + wL/2 \times 3L/4 = 3wL^2/32$. In general, the bending moment due to the beam at distance x is $M = -wx \times x/2 + wL/2 \times x = -wx^2/2 + wLx/2$. The bending moment is a maximum at $x = L/2$ and so $wL^2/8$ (you can show it is a maximum by differentiating to give $dM/dx = -wx + wL/2$ and thus $dM/dx = 0$ at $x = L/2$). Figure 27.10(c) shows the bending moment diagram.

3   *Cantilever with point load at free end*
Consider a cantilever which carries a point load F at its free end (Figure 27.11(a)), the weight of the beam being neglected. The shear force at any section will be $+F$, the shear force diagram thus being as shown in Figure 27.11(b). The bending moment at a distance x from the fixed end is $M = -F(L - x)$. The minus sign is because the beam shows hogging. The bending moment diagram is a line of constant slope F. At the fixed end, when $x = 0$, the bending moment is FL; at the free end it is 0.

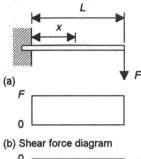

(a)

(b) Shear force diagram

(c) Bending moment diagram

Figure 27.11 *Cantilever with point load at free end*

4   *Cantilever with uniformly distributed load*
Consider a cantilever which has just a uniformly distributed load of w per unit length (Figure 27.12(a)). The shear force a distance x from the fixed end is $V = +w(L - x)$. Thus at the fixed end the shear force is $+wL$ and at the free end it is 0. Figure 27.12(b) shows the shear force diagram. The bending moment at a distance x from the fixed end is, for the beam to the right of the point, given by:

$$M = -w(L - x) \times (L - x)/2 = -\tfrac{1}{2}w(L - x)^2$$

This is a parabolic function. At the fixed end, where $x = 0$, the bending moment is $-\frac{1}{2}wL^2$. At the free end the bending moment is 0. Figure 27.12(c) shows the bending moment diagram.

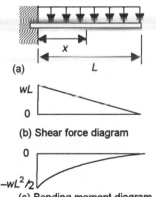

(a)

(b) Shear force diagram

(c) Bending moment diagram

Figure 27.12 *Cantilever with uniformly distributed load*

The following are some general points with regard to shear force and bending moment diagrams:

1   The point on a beam where the bending moment changes sign and is zero is called the *point of contraflexure* or the *inflexion point*.

2   Between point loads, the shear force is constant and the bending moment gives a straight line.

3   Throughout a length of beam with a uniformly distributed load, the shear force varies linearly and the bending moment is parabolic.

4   The bending moment is a maximum when the shear force is zero.

5   The shear force is a maximum when the slope of the bending moment diagram is a maximum and zero when the slope is zero.

6   For point loads, the shear force changes abruptly at the point of application of the load by an amount equal to the size of the load.

For a proof that the bending moment is a maximum when the shear is zero, consider a short length δx of a beam (Figure 27.13). At one side there is a bending moment M and shear force V and at the other side they have increased to $M + \delta M$ and $V + \delta V$. If there is a distributed load of w per unit length, then the segment has a weight of $w\delta x$ which can be considered as acting at the centre of the segment. Taking moments about A, for equilibrium we have:

$$M + \delta M + (w\delta x)\tfrac{1}{2}\delta x = M + V\delta x$$

Neglecting multiples of small quantities we have $\delta M = V\delta x$ and so in the limit as dx tends to zero:

$$V = \frac{dM}{dx}$$

The shear force is thus equal to the rate of change of the bending moment and so when the shear force is zero the slope of the bending moment is zero and hence a maximum or a minimum.

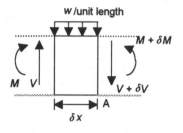

Figure 27.13   *Small segment*

27.3 Bending stress

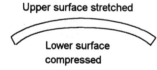

Figure 27.14   *Bending*

When a beam bends, one surface becomes extended and so in tension and the other surface becomes reduced in length and so in compression (Figure 27.14). This implies that between the upper and lower surface there is a plane which is unchanged in length when the beam is bent. This plane is called the *neutral plane* and the line where the plane cuts the cross-section of the beam is the *neutral axis*.

Consider a beam, or part of a beam, where it can be assumed that it is bent to form the arc of a circle (Figure 27.15). This occurs with a simply supported beam and is generally assumed to be reasonably realised in other forms of bending. Consider the section through the beam aa which is a distance y from the neutral axis. It has increased in length as a consequence of the beam being bent. The strain it experiences will be its change in length ΔL divided by its initial unstrained length L. But for circular arcs, the arc length is the radius of the arc multiplied by the angle it subtends. Thus, since aa is of radius $R + y$, we have:

$$L + \Delta L = (R + y)\theta$$

The neutral axis NA will, by definition, be unstrained and so its length $L = R\theta$. Hence, the strain on aa, which initially was length L and is increased in length by ΔL, is:

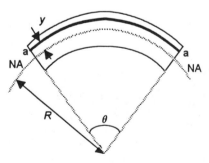

Figure 27.15   *Bending into an arc of a circle*

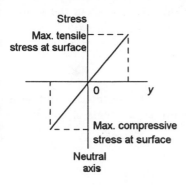

Figure 27.16 *Stress variation across beam section*

$$strain = \frac{\Delta L}{L} = \frac{(R+y)\theta - R\theta}{R\theta} = \frac{y}{R}$$

The strain thus varies linearly through the thickness of the beam being larger the greater the distance y from the neutral axis. For a uniform rectangular cross-section beam the neutral axis is located symmetrically between the two surfaces and thus the maximum strain occurs on the surfaces of the beam. Provided we can use Hooke's law the stress due to bending which is acting on aa is:

$$stress = E \times strain = \frac{Ey}{R}$$

With a uniform rectangular cross-section beam, the maximum bending stresses will be on the surfaces since these are the furthest distance from the neutral axis. Figure 27.16 shows how the stress will vary across the section of the beam.

Example

A uniform square cross-section steel strip of side 4 mm is bent into a circular arc by bending it round a drum of radius 4 m. Determine the maximum strain and stress produced in the strip. Take the modulus of elasticity of the steel to be 210 GPa.

The neutral axis of the strip will be central and so the surfaces will be 2 mm from it and the radius of the neutral axis will be 4.002 m. Thus:

$$max.\ strain = \frac{y}{R} = \frac{2 \times 10^{-3}}{4.002} = 0.5 \times 10^{-3}$$

This is the value of the compressive strain on the inner surface of the strip and that of the tensile strain on the outer surface. Hence:

$$max.\ stress = E \times max.\ strain$$

$$= 210 \times 10^9 \times 0.5 \times 10^{-3} = 105\ MPa$$

This will be tensile on the outer surface of the strip and compressive on the inner.

27.3.1 The general bending equations

Consider the element aa in Figure 27.15 to have an area δA in the cross-section of the beam and be at a distance y from the neutral axis. The element will be stretched as a result of the bending. The stress σ due to the bending acting on this element is Ey/R, where E is the modulus of elasticity of the material and so the forces stretching this element aa are:

$$force = stress \times area = \sigma \delta A = \frac{Ey}{R} \delta A$$

The moment δM of the force acting on this element about the neutral axis is the product of the force and its distance y from the axis:

$$\delta M = \frac{Ey}{R}\delta A \times y = \frac{E}{R}y^2 \delta A$$

The total bending moment M produced over the entire cross-section is the sum of all the moments produced by all the elements of area in the cross-section. Thus:

$$M = \frac{E}{R} \times \left(\text{sum of all } y^2\delta A \text{ terms} \right)$$

The sum of all the $y^2\mathrm{d}A$ terms is termed the *second moment of area I* of the section concerned and is purely determined by the geometry of the section. Thus the above equation can be written as:

$$M = \frac{EI}{R}$$

Since the stress σ on a layer a distance y from the neutral axis is yE/R then we can also write the equation as:

$$M = \frac{\sigma I}{y}$$

The above equations are generally combined and written as the *general bending formula*:

$$\frac{M}{I} = \frac{\sigma}{y} = \frac{E}{R}$$

Example

An I-section beam is 5.0 m long and supported at both ends. It has a second moment of area of 120×10^{-6} m⁴, a depth of 250 mm and a uniformly distributed weight of 30 kN/m. Calculate (a) the maximum bending moment, (b) the maximum bending stress and (c) the radius of curvature of the beam where the bending moment is a maximum. The modulus of elasticity for the beam is 210 GPa.

(a) The beam loading is like that shown in Figure 27.10 and so the maximum bending moment is at the beam centre and $wL^2/8$, where w is the weight per unit length and L the total length.

$$M = \frac{wL^2}{8} = \frac{30 \times 5^2}{8} = 93.8 \text{ kN m}$$

(b) $\sigma = \dfrac{My}{I} = \dfrac{93.8 \times 10^3 \times 125 \times 10^{-3}}{120 \times 10^{-6}} = 97.7 \text{ MPa}$

(c) $R = \dfrac{EI}{M} = \dfrac{210 \times 10^9 \times 120 \times 10^{-6}}{93.8 \times 10^3} = 268.7 \text{ m}$

27.3.2 Position of the neutral axis

It can be shown that the neutral axis passes through the centroid of a beam by considering the longitudinal forces acting on the beam. For a beam bent into the arc of a circle, as in Figure 27.15, the forces acting on a segment a distance y from the neutral axis are stress × area:

$$\text{force} = \sigma \delta A = \frac{Ey}{R} \delta A$$

This is the force acting on just one layer. The total longitudinal force will be the sum of all the forces acting on such segments, i.e. sum of all $(Ey/R)\delta A$ terms making up the area of the section. Thus, when we consider infinitesimally small areas, the total force is:

$$\text{total force} = \int \frac{Ey}{R} \, dA = \frac{E}{R} \int y \, dA$$

But the beam is only bent and so there is no longitudinal force stretching the beam. Thus, since E and R are not zero, we must have:

$$\int y \, dA = 0$$

The integral $\int y \, dA$ is called the *first moment of area* of the section. The only axis about which we can take such a moment and obtain 0 is an axis through the centre of the area of the cross-section, i.e. the centroid of the beam. Thus the neutral axis must pass through the centroid of the section when the beam is subject to just bending.

Example

Determine the position of the neutral axis for the T-section beam shown in Figure 27.17.

The neutral axis will pass through the centroid. We can consider the T-section to be composed of two rectangular sections. The centroid of each will be at its centre. Hence, taking moments about the base of the T-section:

$$\text{moment} = 250 \times 30 \times 115 + 100 \times 50 \times 50 = 1.11 \times 10^6 \text{ mm}^4$$

Hence the distance of the centroid from the base is (total moment)/(total area), and hence the neutral axis:

$$\text{distance from base} = \frac{1.11 \times 10^6}{250 \times 30 + 100 \times 50} = 89 \text{ mm}$$

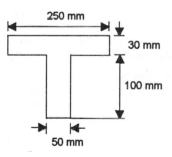

Figure 27.17 *Example*

27.3.3 Second moments of area

The *second moment of area I* of a section about an axis is:

$$I = \int y^2 \, dA$$

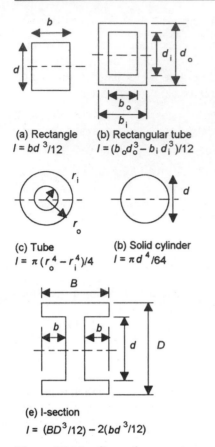

(a) Rectangle
$I = bd^3/12$

(b) Rectangular tube
$I = (b_o d_o^3 - b_i d_i^3)/12$

(c) Tube
$I = \pi(r_o^4 - r_i^4)/4$

(b) Solid cylinder
$I = \pi d^4/64$

(e) I-section
$I = (BD^3/12) - 2(bd^3/12)$

Figure 27.18 *Second moments of area*

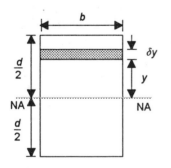

Figure 27.19 *Second moment of area*

The value of the second moment of area about an axis depends on the shape of the beam section concerned and the position of the axis. Figure 27.18 shows the second moments of area about the neutral axis (the axis passes through the centroid) for some common cross-sectional shapes of beams. The SI unit of the second moment of area is m⁴.

To illustrate the derivation of the second moment of area, consider a rectangular cross-section of breadth b and depth d (Figure 27.19). For a layer of thickness δy a distance y from the neutral axis, which passes through the centroid, the second moment of area for the layer is:

$$\text{second moment of area of strip} = y^2 \delta A = y^2 b \delta y$$

The total second moment of area for the section is thus:

$$\text{second moment of area} = \int_{-d/2}^{d/2} y^2 b \, dy = \frac{bd^3}{12}$$

If we had a second moment of area $I = \int y^2 \, dA$ of an area about an axis and then considered a situation where the area was moved by a distance h from the axis, the new second moment of area I_h would be:

$$I_h = \int (y + h)^2 \, dA = \int y^2 \, dA + 2h \int y \, dA + h^2 \int dA$$

But $\int y \, dA = 0$ and $\int dA = A$. Hence:

$$I_h = I + Ah^2$$

This is called the *theorem of parallel axes* and is used to determine the second moment of area about a parallel axis.

Example

Calculate the second moment of area about the neutral axis of a rectangular cross-section beam of depth 40 mm and breadth 100 mm.

For a rectangular section, the second moment of area about the neutral axis is $bd^3/12 = 100 \times 40^3/12 = 5.3 \times 10^5$ mm⁴.

Example

Determine the second moment of area about the neutral axis of the I-section shown in Figure 27.20.

One way of determining the second moment of area for such a section involves determining the second moment of area for the entire rectangle containing the section and then subtracting the second moments of area for the rectangular pieces 'missing' (Figure 27.21(a)). Thus for the rectangle containing the entire section, the second moment of area is $I = bd^3/12 = 50 \times 70^3/12 = 1.43 \times 10^6$ mm⁴. For the two 'missing' rectangles, each will have a second

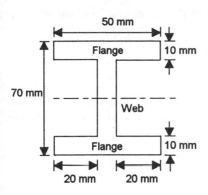

Figure 27.20 *Example*

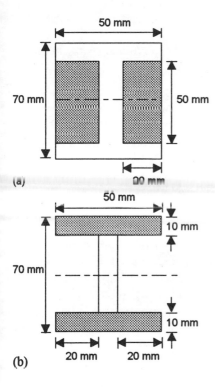

(a)

(b)

Figure 27.21 *Example*

moment of area of $20 \times 50^3/12 = 0.21 \times 10^6$ mm⁴. Thus the second moment of area of the I-section is:

$$1.43 \times 10^6 - 2 \times 0.21 \times 10^6 = 1.01 \times 10^6 \text{ mm}^4$$

Another way of determining the second moment of area of the I-section is to consider it as three rectangular sections, one being the central rectangular section, the web, and the others a pair of rectangular sections, the flanges, with their neutral axes displaced from the neutral axis of the I-section by 30 mm (Figure 27.21(b)). The central rectangular section has a second moment of area of $10 \times 50^3/12 = 0.104 \times 10^6$ mm⁴. Each of the outer rectangular areas will have a second moment of area given by the theorem of parallel axes as:

$$\frac{50 \times 10^3}{12} + 50 \times 10 \times 30^2 = 0.454 \times 10^6 \text{ mm}$$

Thus the second moment of area of the I-section is:

$$0.104 \times 10^6 + 2 \times 0.454 \times 10^6 = 1.01 \times 10^6 \text{ mm}^4$$

Example

A horizontal beam with a uniform rectangular cross-section of breadth 100 mm and depth 150 mm is 4 m long and rests on supports at its ends. It has negligible weight itself and supports a concentrated load of 10 kN at its midpoint. Determine the maximum tensile and compressive stresses in the beam.

The second moment of area is $I = bd^3/12 = 0.100 \times 0.150^3/12 = 2.8 \times 10^{-5}$ m⁴. The reactions at each support will be 5 kN and so the maximum bending moment, which will occur at the midpoint, is 10 kN m. The maximum bending stress will occur at the cross-section where the bending moment is a maximum and on the outer surfaces of the beam, i.e. $y = \pm75$ mm. Thus:

$$\sigma = \frac{My}{I} = \pm\frac{10 \times 10^3 \times 0.075}{2.8 \times 10^{-5}} = \pm26.8 \text{ MPa}$$

27.3.4 Section modulus

For a beam which has been bent, the maximum stress σ_{max} will occur at the maximum distance y_{max} from the neutral axis. Thus:

$$M = \frac{I}{y_{max}}\sigma_{max}$$

The quantity I/y_{max} depends only on the shape of the cross-section concerned and is termed the *section modulus Z*. Thus:

$$M = Z\sigma_{max}$$

For a rectangular cross-section beam, the second moment of area $I = bd^3/12$ and the maximum stress occurs at the surfaces which are $d/2$ from the neutral axis. Thus $Z = (bd^3/12)/(d/2) = bd^2/6$.

Example

A beam has a section modulus of 2×10^6 mm³, what will be the maximum bending moment that can be used if the stress must not exceed 6 MPa?

$M = Z\sigma_{max} = 2 \times 10^6 \times 10^{-9} \times 6 \times 10^6 = 12$ kN m.

Example

Determine the maximum stress set up in a rectangular cross-section beam of depth 150 mm and width 50 mm when a bending moment of 500 N m is applied.

For a rectangular cross-section beam $Z = bd^2/6$ and so $\sigma_{max} = M/Z = 500/(0.05 \times 0.15^2/6) = 2.7 \times 10^6$ Pa = 2.7 MPa.

Example

An I-section beam has a section modulus of 25×10^{-5} m³. What will be the maximum bending stress produced when the beam is subject to a bending moment of 30 kN m?

$\sigma_{max} = M/Z = 30 \times 10^3/(25 \times 10^{-5}) = 120$ MPa.

Example

A rectangular cross-section timber beam of length 4 m rests on supports at each end and carries a uniformly distributed load of 10 kN/m. If the stress must not exceed 8 MPa, what will be a suitable depth for the beam if its width is to be 100 mm?

For the simply supported beam with a uniform distributed load over its full length, the maximum bending moment $= wL^2/8 = 10 \times 4^2/8 = 20$ kN m. Hence:

$$Z = \frac{M}{\sigma_{max}} = \frac{20 \times 10^3}{8 \times 10^6} = 2.5 \times 10^{-3} \text{ m}^3$$

For a rectangular cross-section $Z = bd^2/6$ and thus:

$$d = \sqrt{\frac{6Z}{b}} = \sqrt{\frac{6 \times 2.5 \times 10^{-3}}{0.100}} = 0.387 \text{ m}$$

A suitable beam might thus be one with a depth of 400 mm.

Activities

1 Attach electrical resistance strain gauges to the surface of a beam and determine the surface strain when the beam is bent as either a cantilever or simply supported. Hence obtain a value for the surface stress and compare the result with that predicted by theory.

Problems

1 A beam of length 4.0 m and negligible weight rests on supports at each end and a concentrated load of 500 N is applied at its midpoint. Determine the shear force and bending moment at distances of (a) 1.0 m, (b) 2.5 m from the right-hand end of the beam.

2 A cantilever has a length of 2 m and a concentrated load of 8 kN is applied to its free end. Determine the shear force and bending moment at distances of (a) 0.5 m, (b) 1.0 m from the fixed end. Neglect the weight of the beam.

3 A uniform cantilever of length 4.0 m has a weight per metre of 10 kN. Determine the shear force and bending moment at 2.0 m from the free end if no other loads are carried by the beam.

4 A beam of length 3.0 m rests on supports at each end. A load of 3 kN is applied to the beam centre. What will be the resulting bending moment and shear force a distance of (a) 0.5 m, (b) 1.0 m, (c) 1.5 m from one end?

5 A cantilever of length 2.0 m carries a load of 12 kN at its free end. What, and where, will be the resulting (a) maximum bending moment and (b) maximum shear force?

6 Calculate (a) the maximum bending moment and (b) the maximum shear force for a cantilever having a weight of 20 kN/m and a length of 3.0 m.

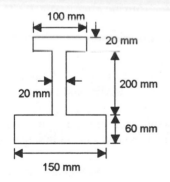

Figure 27.22 *Problem 7*

7 Determine the position of the neutral axis from the base for the non-symmetrical I-section shown in Figure 27.22.

8 Determine the position of the neutral axis of a T-section if the top of the T is a rectangle 100 mm by 10 mm and the stem of the T is a rectangle 120 mm by 10 mm.

9 Determine the second moment of area of an I-section, about its horizontal neutral axis when the web is vertical, if it has rectangular flanges each 120 mm by 10 mm, a web of thickness 12 mm and an overall depth of 150 mm.

10 Determine the second moment of area of a rectangular section of breadth 50 mm and depth 100 mm.

11 Calculate the second moments of area about an axis through the centroid for (a) a square section of side 250 mm, (b) a rectangular section 100 mm deep and 75 mm wide, (c) a circular section 50 mm diameter, (d) a tube section of internal diameter 50 mm and external diameter 75 mm, (e) a square hollow section of internal side 220 mm and external side 250 mm.

12 Steel strip is to be bent round a drum of radius 1 m. What is the maximum thickness of strip that can be bent in this way if the stress in the strip is not to exceed 100 MPa. The steel has a modulus of elasticity of 210 GPa.

13 Steel strip of thickness 5 mm is coiled on a drum of 1.5 m diameter. Calculate the maximum stress produced by the coiling. The modulus of elasticity is 210 GPa.

14 A strip of steel 2 mm thick passes over a pulley. Determine the diameter of the pulley needed if the stress in the steel strip must not exceed 100 MPa as a result of it bending to follow the circumference of the pulley. The steel has a modulus of elasticity of 210 GPa.

15 A steel strip 3 mm thick and 50 mm wide is to bend round a drum of diameter 4.0 m. What bending moment is required and what will be the maximum stresses produced? The modulus of elasticity of the steel is 210 GPa.

16 A beam has a rectangular cross section of width 60 mm and depth 100 mm. Determine the maximum bending moment that can be applied if the bending stresses are not to exceed ±150 MPa.

17 A beam has a rectangular cross section of width 40 mm and depth 160 mm. Determine the maximum bending stress if it is subject to a bending moment of 35 kN m.

18 A wooden joist of rectangular cross-section spans a gap of 5.0 m. The joist has a depth of 200 mm. Calculate the minimum width of joist required if the maximum tensile stress that the beam can be allowed to withstand is 14 MPa and the load to be carried is a centrally placed one of 20 kN.

19 A timber beam of thickness 100 mm and depth 300 mm protrudes 2.0 m from a wall in which it is fixed. What is the maximum load that can be applied to the free end if the maximum allowable stress is 7 MPa?

20 A beam has to support loading which results in a maximum bending moment of 25 kN m. If the maximum permissible bending stress is 7 MPa, what will be the required section modulus?

21 Calculate the maximum uniformly distributed load that a simply supported steel I-section can carry over a span of 6.0 m if the maximum permissible stress in the beam is 50 MPa and the I-section has a section modulus of 4×10^{-3} m^3.

22 An I-section girder has a section modulus of 100×10^{-5} m^3. What will be the maximum stress produced when such a beam is subject to a bending moment of 40 kN m?

23 A steel rod of 100 mm diameter is supported at both ends and has a span of 1.2 m. Determine the maximum bending stress in the rod when it supports a point load of 20 kN at mid-span. The weight of the rod can be ignored.

28 Circular motion

28.1 Introduction

This chapter is concerned with motion in a circle. Velocity is a vector quantity and has both size and direction. Thus, there is a constant velocity if equal distances are covered in the same straight line in equal intervals of time. Acceleration is the rate of change of velocity. Thus, there will be an acceleration if a velocity changes either as a result of equal distances in the same straight line not being covered in equal times or, if equal distances are being covered they are not in the same straight line, i.e. there is a change in direction. This is what happens with circular motion where we can have equal distances round the circumference of the circular path covered in equal times, i.e. constant speed, but the velocity is changing because the direction is continually changing and so there is an acceleration.

28.2 Centripetal force

Consider a point object of mass m rotating with a constant speed in a circular path of radius r (Figure 28.1(a)). At point A the velocity will be v in the direction indicated. At B, a time t later, the velocity will have the same size but be in a different direction. If the direction has changed by the angle θ in time t, then the amount by which the velocity has changed can be obtained resolving the velocity at B into two components, one in the same direction as the velocity at A of $v \cos \theta$ and the other at right angles to it of $v \sin \theta$, i.e. along the radial direction AC. The acceleration in the direction of the velocity at A is thus $(v \cos \theta - v)/t$. The acceleration at an instant, rather than the average over the time t, is obtained by considering the value the average value tends to as we make t small and hence θ small. So, since $(1 - \cos \theta)$ tends to 0, the instantaneous acceleration in the direction of the velocity at A $= v/t(\cos \theta - 1)$ tends to 0. As $v \sin \theta$ tends to $\delta\theta$, the acceleration in the direction AC $= (v \sin \theta - 0)/t$ tends to $\delta\theta/\delta t$. This is the rate at which angle is covered, i.e. the angular velocity ω, and so the instantaneous acceleration a in this direction at A is:

$$a = v\omega$$

Since $v = r\omega$ we can write the above equation in the two forms:

$$a = \frac{v^2}{r} = r\omega^2$$

The direction of this acceleration is towards the centre of the circle and hence is termed the *centripetal acceleration*.

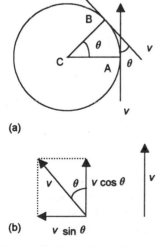

Figure 28.1 *Motion in a circle*

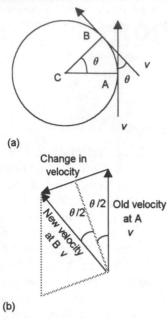

(a)

(b)

Figure 28.2   *Motion in a circle*

An alternative way of arriving at the same answer is to consider what has to be added, by vector means, to the velocity at A to give the velocity at B (Figure 28.2). For a small angle θ, to a reasonable approximation, the change in velocity $= 2v \sin \frac{1}{2}\theta = v\theta$. But the distance covered in time t is arc length AB; thus $v = $ (arc length AB)$/t = r\theta/t$ and so $\theta = vt/r$. The acceleration a is the change in velocity divided by the time; hence:

$$a = \frac{v\theta}{t} = \frac{v(vt/r)}{t} = \frac{v^2}{r} = \frac{(r\omega)^2}{r} = r\omega^2$$

This will tend to the instantaneous acceleration at A if we make t small and hence θ small. The direction of this instantaneous acceleration is the direction of the change in velocity vector, and this tends, as θ becomes small, to be towards the centre of the circle.

The equation $a = v^2/r$ gives the acceleration an object must experience, at right angles to its direction of motion, if it is to move in a circular path. The centripetal force necessary for this acceleration is thus:

$$F = ma = \frac{mv^2}{r} = m\omega^2 r$$

According to Newton's third law, to every action there is an opposite and equal reaction. In this case the reaction to the centripetal force is called the *centrifugal force* and acts in an outwards direction on the pivot C around which the motion is occurring.

Example

An object of mass 0.5 kg is whirled round in a horizontal circle of radius 0.8 m on the end of a rope. What is the tension in the rope when the object rotates at 4 rev/s?

$F = m\omega^2 r = 0.5 \times (2\pi \times 4)^2 \times 0.8 = 253$ N.

Example

Calculate the force acting on a bearing which is carrying a crank-shaft with an out-of-balance load of 0.10 kg at a radius of 100 mm and rotating at 50 rev/s.

Centripetal force $= m\omega^2 r = 0.10 \times (2\pi \times 50)^2 \times 0.100 = 987$ N. The force acting on the bearing is the reaction force, i.e. the centrifugal force, and is thus 987 N radially outwards.

Example

An object of mass 3 kg is attached to the end of a rope and whirled round in a vertical circle of radius 1 m. What are the maximum and minimum values of the tension in the rope when it rotates at 3 rev/s?

The centripetal force $F = m\omega^2 r = 3 (2\pi \times 3)^2 \times 1 = 1066$ N. At the top of the path, the centripetal force is provided by the tension in the rope plus the weight since they are acting in the same direction. Hence, $T + 3 \times 9.8 = 1066$ and so tension $T = 1037$ N. At the bottom of the path, the tension and the weight act in opposite directions and so $T - 3 \times 9.8 = 1066$ and the tension $T = 1095$ N. Thus the maximum tension is 1095 N and the minimum tension 1037 N.

28.2.1 Conical pendulum

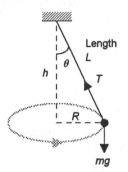

Figure 28.3 *Conical pendulum*

Consider a small body of mass m on the end of a length L of string which is whirled round in horizontal circle with the string being at some angle θ to the vertical line through the point of suspension (Figure 28.3); such an arrangement is called a conical pendulum. The only forces acting on the body are its weight mg and the tension T in the string.

The vertical component of the tension must be equal to mg and so $T \cos \theta = mg$. The horizontal component of the tension provides the centripetal force and so, when the object is rotating in a horizontal circle of radius R with an angular velocity ω, we have $T \sin \theta = m\omega^2 R$. Thus:

$$\frac{T \sin \theta}{T \cos \theta} = \tan \theta = \frac{\omega^2 R}{g}$$

But $\tan \theta = R/h$ and so:

$$h = \frac{g}{\omega^2}$$

Thus the height h of the pendulum depends only on the angular velocity. The conical pendulum forms the basis of the Watt engine governor (Figure 28.4). Two light rods are hinged at C to a vertical shaft which is rotated by the engine. At the ends of these rods are masses A and B. Two other rods DF and EF are hinged to AC and BC and also to a collar F which can slide up and down the shaft. A lever attached to F opens or closes a valve admitting steam to the engine. When the speed of rotation of the shaft increases, the rotation of the masses causes F to move up the shaft and so shut off some of the steam and slow the engine down. If the speed drops, F moves down the shaft and opens the valve to admit more steam and increase the speed of the engine.

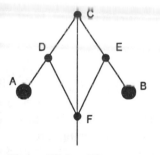

Figure 28.4 *Watt governor*

Example

What is the speed of rotation of a conical pendulum which rotates with a height of 200 mm?

$\omega^2 = g/h = 9.8/0.200$ and so $\omega = 7$ rad/s.

28.3 Cornering

Consider a vehicle of mass m rounding a horizontal corner of radius r (Figure 28.5). Since the reactive force R is mg at right angles to the plane, the maximum frictional force $F = \mu R = \mu mg$. Hence, when the centripetal force exceeds this frictional force, skidding will occur. Thus the maximum speed is given when $mv^2/r = \mu mg$ and so:

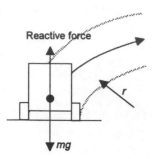

Figure 28.5 *Cornering on the flat*

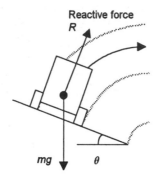

Figure 28.6 *A banked corner*

$$v = \sqrt{\mu r g}$$

Now consider a vehicle rounding a horizontal corner which is banked at an angle θ to the horizontal (Figure 28.6). If the vehicle is to traverse the corner without relying on frictional forces, the centripetal force has to be provided by a component $R \sin \theta$ of the reactive force. Thus, we have $R \sin \theta = mv^2/r$. But we must also have the vertical component of the reaction force equal to the weight mg and so $R \cos \theta = mg$. Thus:

$$\tan \theta = \frac{v^2}{rg}$$

Example

Calculate the maximum speed with which a vehicle can travel round a horizontal band of radius 40 m without skidding if the coefficient of friction between the tyres and the road is 0.6.

$$v = \sqrt{(\mu r g)} = \sqrt{(0.6 \times 40 \times 9.8)} = 15.3 \text{ m/s}.$$

Example

Calculate the angle of banking required on a corner of radius 60 m so that vehicles can travel round the corner at 20 m/s without any side thrust on the tyres.

Tan $\theta = v^2/rg = 20^2/(60 \times 9.8)$ and so $\theta = 34.2°$.

28.4 Centrifugal clutch

A *clutch* is a mechanism by which two shafts, one of which is rotating and the other stationary, can be smoothly engaged and torque transmitted from one shaft to another. A friction type of clutch relies upon sufficient friction being developed between the two surfaces brought into contact for the torque to be transmitted from one to the other without slipping occurring. For this to occur, a clamping force has to be provided to hold the friction surfaces together. For the clutch used in cars to connect the running engine drive shaft to the rest of the transmission system, springs are generally used. An alternative construction involves the clamping force being generated, wholly or partially, by a centrifugal operated mechanism.

Figure 28.7 shows the basic principles of operation of one form of such a centrifugal clutch. The weights are carried on the free ends of bell-crank levers. When the clutch rotates, the weights are thrown outwards as a result of the circular motion and the resulting pivoting about the pivot causes the pressure plate to become engaged with the driven plate and so torque is transferred. The net force between the plates is $F - P$, where F is the centrifugal force and P the force exerted by the spring. Thus the frictional force between the pads is $\mu(F - P)$, where μ is the coefficient of friction between the pad materials. The frictional torque is thus $\mu(F - P)R$, where R is the radius of rotation of the plates. If there are n such arrangements spaced round the shaft then the total

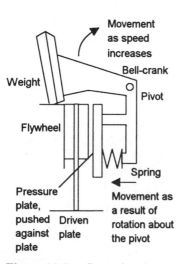

Figure 28.7 *Centrifugal clutch*

frictional torque T is $n\mu(F - P)R$. The power transmitted by a torque is $T\omega$ (see Section 8.4), where ω is the angular velocity, and so the power transmitted by the clutch is $n\mu(F - P)R\omega$.

Problems

1 An object of mass 0.3 kg is whirled round in a horizontal circle of radius 0.6 m on the end of a rope. What is the tension in the rope when the object rotates at 5 rev/s? How will the tension change if the angular speed is doubled?

2 An object of mass 5 kg on the end of rope is rotated in a horizontal circle of radius 4 m with a constant speed of 8 m/s. What is the tension in the rope?

3 An object of mass 500 g is whirled round on the end of a tethered string in a horizontal circle of radius 1 m. What is the maximum number of revolutions per second the object can be rotated at if the string will break when the tension in it reaches 40 N?

4 An object of mass 0.6 kg is whirled round in a vertical circle of radius 0.8 m at the end of a rope. What is the maximum and minimum tension in the rope when the mass rotates at 200 rev/min?

5 Calculate the greatest speed with which a car may pass over a humpback bridge without leaving the ground if the bridge forms the arc of a vertical circle of radius 8 m.

6 Calculate the maximum speed with which a vehicle can travel round a corner, unbanked and horizontal, of radius 50 m without skidding. The coefficient of friction between the tyres and road is 0.6.

7 Calculate the angle of banking required on a corner of radius 80 m so that vehicles can travel round the corner at 40 m/s without any side thrust on the tyres.

8 A train runs at 20 m/s round a curve of mean radius 360 m. If the distance between the rails is 1.4m, how much should the outer rail be raised above the inner rail so there is no side thrust on the rails?

9 At what speed can a car move round a corner of radius 75 m without side slip if the road is banked at an angle $\tan^{-1} 1/5$?

10 What is the centripetal force required if a body of mass 0.25 kg is to move in a horizontal circular path of radius 3.5 m with an angular speed of 4 rad/s?

11 A block is placed on a horizontal turntable at a distance of 50 mm from the central axle. When the turntable rotates, the block is just on the point of sliding when the speed of rotation is 1.4 rev/s. What is the coefficient of friction between the box and turntable surfaces?

12 A body rotates as a conical pendulum at the end of a string of length 400 mm. If the string is inclined at 60° to the vertical, at how many revolutions per second will the body rotate?

13 A body of mass 500 g rotates as a conical pendulum at the end of a string of length 1.5 m. If it rotates at 80 rev/min, determine the radius of the circle and the tension in the string.

14 A body of mass 1.8 kg rotates as a conical pendulum at the end of a string of length 0.50 m. If it rotates at 1 rev/s, determine the tension in the string and the vertical height of the point of suspension above the plane of rotation.

29 Angular dynamics

29.1 Introduction

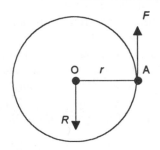

Figure 29.1   *Couple*

Dynamics is the study of objects in motion. This chapter thus follows on from Chapter 8, where the terms and equations used in describing angular motion were introduced. Here we now consider the torques, angular momentum and energy involved with angular motion.

The terms torque and couple were introduced in Section 8.4. If a force F is applied to the surface of a shaft of radius r (Figure 29.1), a reactive force R is set up which is equal in magnitude and opposite in direction to F, i.e. $R = -F$, and can be considered to act at O. This pair of oppositely directed, but equal in magnitude, forces which are not in the same straight line is called a *couple*. The turning moment of a couple is called the *torque T* and $T = Fr$.

29.2 Torque and angular acceleration

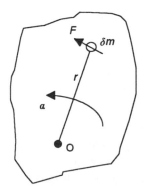

Figure 29.2   *Rotation of a rigid body*

Consider a force F acting on a small element of mass δm of a rigid body (note that the symbol δ in front of a quantity is used to indicate that it is a small bit of the quantity), the element being a distance r from the axis of rotation at O (Figure 29.2). The torque T acting on the element is Fr and since $F = \delta m \times a$ we can express the torque as:

$$\text{torque } T = \delta m \times ar$$

But $a = ra$, where a is the angular acceleration (Section 8.3). Thus $T = \delta m \times r^2a$. We can express this as:

$$\text{torque } T = Ia$$

where I is termed the *moment of inertia* and for the small element of mass m at radius r is given by:

$$\text{moment of inertia } I = \delta m \times r^2$$

The torque required to give a rotational acceleration for the entire rigid body will thus be the sum of the torques required to accelerate each element of mass in the body. Thus, if I is the moment of inertia of the entire body about O and is the sum of all the $\delta m \times r^2$ terms for the body:

$$\text{torque to give body angular acceleration } T = Ia$$

Example

What constant torque is required to cause a pulley wheel to rotate at 5 rev/s in 15 s starting from rest, if the wheel has a moment of inertia of 200 kg m² about its axle?

Using $\omega = \omega_0 + at$ then $2\pi \times 5 = 0 + 15a$ and so the angular acceleration a is 2.1 rad/s². Hence $T = Ia = 200 \times 2.1 = 420$ N m.

Example

A flywheel has a moment of inertia of 10 kg m² about its axle. What will be the angular acceleration produced by a torque of 5 N m if bearing friction is equivalent to a torque of 2 N m?

The accelerating torque is $5 - 2 = 3$ N m and so the angular acceleration $a = T/I = 3/10 = 0.3$ rad/s².

29.2.1 Moment of inertia

The *moment of inertia* of a body about some axis is a measure of the resistance of that body to rotation about the specified axis. For a small mass at the end of a light pivoted arm with radius of rotation r, we can consider the entire mass of the body to be located at the same distance r and so the moment of inertia is:

$I = mr^2$

For a solid body, we can consider the body to be made up of a large number of small masses, each of mass δm and at different distances r from the axis of rotation. The moment of inertia about this axis is then the sum of all the $r^2 \, \delta m$ terms. The following illustrates how we can carry out such summation.

For a uniform disc of radius r and total mass m, consider the disc to be composed of a number of rings, each of thickness δx and radius x, with values of x varying between 0 and r (Figure 29.3). For a ring of thickness δx and radius x, its area is effectively its circumference multiplied by the thickness and so is $2\pi x \, \delta x$. Since the mass per unit area of the disc is $m/\pi r^2$ then the mass δm of this ring element is:

$$\delta m = \frac{m}{\pi r^2} 2\pi x \delta x = \frac{2m}{r^2} x \delta x$$

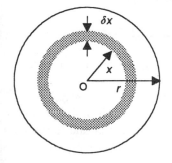

Figure 29.3   *Uniform disc*

Thus the moment of inertia of the disc about an axis at right angles to the disc plane through O is:

$$I = \int_0^r x^2 \, dm = \int_0^r \frac{2m}{r^2} x^3 \, dx = \tfrac{1}{2} mr^2$$

In a similar manner we can derive the moments of inertia of other bodies. Figure 29.4 gives the moments of inertia for some commonly encountered bodies.

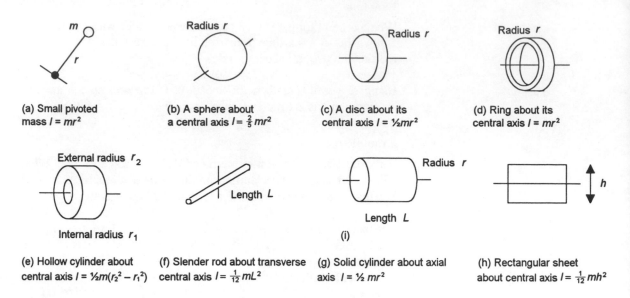

(a) Small pivoted mass $I = mr^2$

(b) A sphere about a central axis $I = \frac{2}{5} mr^2$

(c) A disc about its central axis $I = \frac{1}{2}mr^2$

(d) Ring about its central axis $I = mr^2$

(e) Hollow cylinder about central axis $I = \frac{1}{2}m(r_2^2 - r_1^2)$

(f) Slender rod about transverse central axis $I = \frac{1}{12}mL^2$

(g) Solid cylinder about axial axis $I = \frac{1}{2} mr^2$

(h) Rectangular sheet about central axis $I = \frac{1}{12}mh^2$

Figure 29.4 *Moments of inertia about the indicated axes*

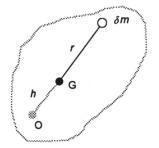

Figure 29.5   *Moment of inertia about a parallel axis*

The value of the moment of inertia for a particular body depends on the axis about which it is calculated. If I_G is the moment of inertia about an axis through the centre of mass G, consider what the moment of inertia would be about a parallel axis through O, a distance h away (Figure 29.5). The moment of inertia of an element δm of the mass about the axis through O is:

moment of inertia of element $= (r + h)^2 \, \delta m$

$$= (r^2 + h^2 + 2rh) \, \delta m$$

The total moment of inertia about the axis through O is thus:

$I_O = $ sum of $r^2 \, \delta m$ terms + sum of $h^2 \, \delta m$ terms + sum of $2rh \, \delta m$ terms

The sum of the $r^2 \, \delta m$ terms is the moment of inertia about an axis through the centre of mass G. The sum of $h^2 \, \delta m$ terms is mh^2. The sum of the $2rh \, \delta m$ terms is $2h \times$ sum of $r \, \delta m$ terms integral and is the total moment of the mass about the axis through the centre of mass and is thus zero. Hence:

$$I_O = I_G + mh^2$$

This is known as the *theorem of parallel axes*.

It is sometimes necessary to obtain the moment of inertia of a composite body about some axis. This can be done by summing the moments of inertia of each of the elemental parts about the axis, using the parallel axis theorem where necessary. This is illustrated in one of the following examples.

Example

Determine the moment of inertia about an axis through its centre of a disc of radius 100 mm and mass 200 g.

$I = \frac{1}{2}mr^2 = \frac{1}{2} \times 0.200 \times 0.100^2 = 1.0 \times 10^{-3}$ kg m^2.

Example

Determine the moment of inertia of a slender rod of mass m about an axis at right angles to its length of $2a$ and a distance b from its centre (Figure 29.6).

The moment of inertia of a slender rod of length L about an axis through its centre is $mL^2/12$. Thus, for $L = 2a$, we have $I = ma^2/3$. Using the parallel axis theorem, the moment of inertia about a parallel axis a distance b from the centre is thus $I = (ma^2/3) + mb^2$.

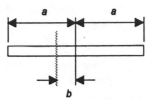

Figure 29.6   *Example*

Example

Determine the moment of inertia of the plate shown in Figure 29.7 about an axis at right angles to it and through its centre. The plate is 10 mm thick and has a radius of 250 mm and contains a central hole with a radius of 125 mm. The material has a density of 8000 kg/m^3.

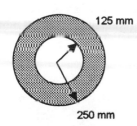

125 mm

250 mm

Figure 29.7   *Example*

This can be tackled by considering the plate as a composite body, i.e. a solid disc of radius 250 mm minus a disc of radius 125 mm. For the 250 mm diameter disc, the moment of inertia is:

$I = \frac{1}{2}MR^2 = \frac{1}{2} \times \pi \times 0.250^2 \times 0.010 \times 8000 \times 0.250^2$

$= 0.491$ kg m^2

For the 125 mm diameter disc:

$I = \frac{1}{2}MR^2 = \frac{1}{2} \times \pi \times 0.125^2 \times 0.010 \times 8000 \times 0.125^2$

$= 0.031$ kg m^2

Thus the moment of inertia of the disc with the hole is $0.491 - 0.031 = 0.460$ kg m^2.

Example

A record turntable is a uniform flat plate of radius 120 mm and mass 0.25 kg. What torque is required to uniformly accelerate the turntable to 33.3 rev/min in 2 s?

The angular acceleration required is given by $\omega = \omega_0 + at$ as $2\pi(33.3/60) = 0 + 2a$ and so $a = 1.74$ rad/s². The moment of inertia of the turntable is:

$$I = \tfrac{1}{2}mr^2 = \tfrac{1}{2} \times 0.25 \times 0.120^2 = 1.8 \times 10^{-3} \text{ kg m}^2$$

The torque required to accelerate the turntable is thus:

$$T = Ia = 1.8 \times 10^{-3} \times 1.74 = 3.13 \times 10^{-3} \text{ N m}.$$

29.2.2 Radius of gyration

Whatever the form of a body and however its mass is distributed, it is always possible to consider some radial distance k from the pivot axis at which all the mass m can be considered to act. The moment of inertia is then that of a point mass m at this distance k, i.e.

$$I = mk^2$$

k is called the *radius of gyration*. Thus, for a disc of radius r and mass m the moment of inertia about its central axis is $\tfrac{1}{2}mr^2$ and thus the radius of gyration for the disc is $k^2 = \tfrac{1}{2}r^2$ and so $k = r/\sqrt{2}$.

Example

Determine the moment of inertia about an axis of an object having a mass of 2.0 kg and a radius of gyration from that axis of 300 mm.

$$I = mk^2 = 2.0 \times 0.300^2 = 0.18 \text{ kg m}^2$$

Example

A drum with a mass of 60 kg, radius of 400 mm and radius of gyration of 250 mm has a rope of negligible mass wrapped round it and attached to a load of 20 kg (Figure 29.8). Determine the angular acceleration of the drum when the load is released.

For the forces on the load we have:

$$20g - T = 20a$$

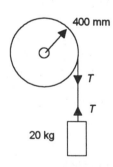

400 mm

T

T

20 kg

Figure 29.8   *Example*

The point of contact between the rope and the drum will have the same tangential acceleration and thus, using $a = ra$, we have $a = 0.400a$. Thus the above equation can be written as:

$$20g - T = 20 \times 0.400a$$

For the drum, the torque is $T \times 0.400$ and thus, using torque $= Ia$:

$$T \times 0.400 = Ia$$

Since $I = mk^2 = 60 \times 0.250^2$ then:

$$T \times 0.400 = 60 \times 0.250^2 \times a$$

and so, eliminating T between this and the earlier equation gives:

$$(20g - 20 \times 0.400a) \times 0.400 = 60 \times 0.250^2 \times a$$

The angular acceleration is thus 11.3 rad/s^2.

29.3 Angular momentum

The *angular momentum* of a body is the moment of the linear momentum about the axis of rotation. Thus, for a particle of mass m moving in a circular path of radius t with a linear velocity v, the liner momentum is mv and so the angular momentum is:

$$\text{angular momentum} = mvr = mr^2\omega$$

where ω is the angular velocity ($v = r\omega$). But mr^2 is the moment of inertia I and so:

$$\text{angular momentum} = I\omega$$

The angular momentum of a body is the product of the moment of inertia about the axis of rotation and the angular velocity. It is a vector quantity and has the unit of kg m^2/s or N m s.

Since torque $T = Ia$, where the angular acceleration a is the rate of change of angular velocity, then:

$$T = I \times \text{rate of change of } \omega = \text{rate of change of } I\omega$$

$$= \text{rate of change of angular momentum}$$

The principle of the conservation of momentum applies not only to linear momentum but also to angular momentum. If there is no externally applied torque to a system then angular momentum is conserved. An example of this is in the sudden meshing of two rotating gear wheels. If one wheel has a moment of inertia of I_1 and an angular velocity of ω_1 and the other a moment of inertia of I_2 and an angular velocity of ω_2, then after meshing if they have a combined moment of inertia of $(I_1 + I_2)$ and an angular velocity ω_3:

$$I_1\omega_1 + I_2\omega_2 = (I_1 + I_2)\omega_3$$

Example

Clutch plate A, mass of 40 kg and a radius of gyration of 160 mm, rotates at 10 rev/s. Clutch plate B, mass of 50 kg, a radius of gyration of 120 mm, is stationary. What will be the common angular speed of rotation of the plates when they are engaged?

Applying the principle of conservation of angular momentum, then the angular momentum before engagement equals the angular moment after engagement, i.e. $I_A\omega_A = (I_A + I_B)\omega$. Hence:

$$40 \times 0.16^2 \times 2\pi \times 10 = (40 \times 0.16^2 + 50 \times 0.12^2)2\pi f$$

and the frequency of rotation f = 5.9 rev/s.

29.4 Angular kinetic energy

When a force is used to rotate an object then work is done since the point of application of the force moves through some distance and the work = $T\theta$ (See Section 8.4.1). Consider an object which, starting from rest, is set rotating with an angular acceleration a as a result of a torque T produced by a tangential force F (Figure 29.9) The work done by the force is $T\theta$. But $T = Ia$ and thus the work done = $Ia\theta$. As a result of this rotation from rest through angle θ, the object has an angular velocity ω. Using $\omega^2 = \omega_0^2 + 2a\theta$, we have $\omega^2 = 0 + 2a\theta$ and so the work done = $\frac{1}{2}I\omega^2$. This is the energy transferred to the body as a result of the work done and is called the *angular kinetic energy*. Thus, for angular motion:

angular kinetic energy = $\frac{1}{2}I\omega^2$

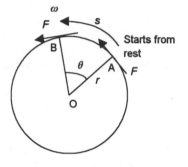

Figure 29.9 *Angular motion*

Example

What is the angular kinetic energy of a clutch plate, mass of 40 kg and a radius of gyration of 160 mm, rotating at 10 rev/s?

The moment of inertia $I = mk^2 = 40 \times 0.160^2$ kg m² and so the angular kinetic energy is $\frac{1}{2} \times 40 \times 0.160^2 \times (2\pi \times 10)^2 = 2021$ J.

Example

A solid cylinder, density of 7000 kg/m³, of diameter 25 mm and length 25 mm is allowed to roll down an inclined plane, the plane being at an elevation of 20° to the horizontal. Determine the linear velocity of the cylinder when it has rolled a distance of 1.2 m.

The loss in potential energy of the cylinder in rolling down the slope must equal the gain in linear kinetic energy plus the gain in rotational kinetic energy. The loss in potential energy is mgh with m being $\frac{1}{4}\pi \times 0.025^2 \times 0.025 \times 7000 = 0.0859$ kg and $h = 1.2 \sin 20° = 0.410$ m. The loss in potential energy is $0.0859 \times 9.81 \times 0.410 = 0.345$ J. The gain in linear kinetic energy is $\frac{1}{2} \times 0.0859v^2$. The gain in rotational kinetic energy is $\frac{1}{2}I\omega^2$, where $I = \frac{1}{2}mr^2 = \frac{1}{2} \times 0.0859 \times 0.0125^2 = 6.71 \times 10^{-6}$ kg m² and $\omega = v/r = v/0.0125$. Thus the gain in rotational energy is $\frac{1}{2} \times 6.71 \times 10^{-6} \times (v/0.0125)^2 = 0.0215v^2$. Since the loss in potential energy equals the total gain in kinetic energy, $0.345 = \frac{1}{2} \times 0.0859v^2 + 0.0215v^2$ and so $v = 2.31$ m/s.

Problems

1 Determine the moment of inertia of an object having a mass of 4 kg and a radius of gyration of 200 mm.

2 Determine the moment of inertia of a sphere about an axis through its centre if it has a radius of 200 mm and a density of 8000 kg/m³.

3 A disc with an outside radius of 250 mm and a mass of 11.78 kg has a moment of inertia of 0.460 kg m² about an axis at right angles to its plane and through its centre. Determine its moment of inertia about an axis at right angles to its plane and passing through a point on its circumference.

4 What is the moment of inertia of a disc, about an axis perpendicular to the face of the disc and through its centre, which has a constant thickness of 50 mm, a diameter of 750 mm and a central hole of diameter 150 mm? The disc material has a density of 7600 kg/m³.

5 A door has a mass of 50 kg and has width of 750 mm. If the door has a radius of gyration of 500 mm, what will be its angular acceleration when the edge opposite the hinged side is given a right-angled push of 24 N?

6 A flywheel has a mass of 360 kg and a radius of gyration of 600 mm and is rotating at 10 rev/s. What uniform torque will be required to bring it to rest in 30 s?

7 A flywheel has a mass of 60 kg and a radius of gyration of 450 mm. What uniform torque will be required to increase its speed from 3.0 to 4.0 rev/s in 20 s?

8 Gear wheel A has a mass of 7 kg and a radius of gyration of 300 mm and is rotating at 200 rev/min. Gear wheel B has a mass of 2.7 kg and a radius of gyration of 500 mm and is stationary. What is the common speed of rotation when the wheels mesh together?

9 A flywheel has a moment of inertia of 8 kg m² about its axis of rotation. What will be the angular kinetic energy stored in the wheel when it is rotating at 4 rev/s?

10 Two coaxial shafts with flywheels are coupled together by a clutch. Initially one with moment of inertia 50 kg m² about the shaft was rotating at 30 rev/s and the other with a moment of inertia of 20 kg m² was at rest. What is the common rate of revolution when they are coupled together?

11 A solid flywheel has a mass of 450 kg and a diameter of 600 mm, with an axle of diameter 100 mm and mass 50 kg. The wheel and axle are set in motion by means of a string wound round the axle and carrying a mass of 10 kg (Figure 29.10). Determine (a) the moment of inertia of the wheel and axle and (b) the kinetic energy of it when the weight reaches the floor 2 m below its starting point.

12 A wheel and axle has a string of length 3 m coiled round the axle and the string is pulled with a constant force of 100 N. When the string leaves the axle, the wheel is rotating at 5 rev/s. What is the moment of inertia of the wheel and axle?

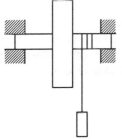

Figure 29.10   *Problem 11*

30 Mechanical power transmission

30.1 Introduction

In Section 2.8 there was an introduction to simple machines with a discussion of basic terms, levers, pulleys, inclined planes and screws. A *machine* was defined as a mechanical device which enables an effort force to be magnified or reduced or applied in a more convenient line of action, or the displacement of the point of action of a force to be magnified or reduced. Also defined was the *force ratio* or *mechanical advantage* MA as load/effort and the *movement ratio* or *velocity ratio* VR as (distance moved by effort)/(distance moved by load) and the efficiency was derived as being MA/VR. This chapter extends that discussion of machines and also discusses mechanical power transmission by means of belt drives and gear trains, these following on from the discussion of angular motion in Chapter 8.

30.2 Machines

The mechanical advantage of a machine depends on the effort applied. For simple machines, the relationship between the effort E and load F is of the form (Figure 30.1):

$$E = aF + b$$

where a and b are constants. This equation is known as the *law of the machine*. In an ideal machine, i.e. where the efficiency is 100%, the effort–load graph passes through the origin and b is zero. However, because of friction, some effort has to be expended to overcome friction and b is the effort needed for this (Figure 30.2).

The above equation gives the MA as:

$$MA = \frac{F}{E} = \frac{F}{aF + b} = \frac{1}{a + b/F}$$

If the load is large then b/F becomes small when compared with a and so the mechanical advantage then becomes approximately $1/a$. This is called the *limiting MA*. The efficiency of a simple machine is MA/VR and so the *limiting efficiency* is:

$$\text{limiting efficiency} = \frac{1/a}{VR} = \frac{1}{a \times VR}$$

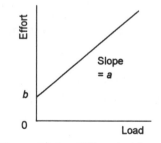

Figure 30.1 *Effort–load graph*

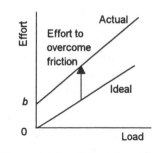

Figure 30.2 *Effort–load graph*

Example

A simple machine is found to require an effort of 21.5 N when the load is 550 N and an effort of 12.5 N when the load is 250 N. What is the law of the machine?

Using $E = aL + b$ we have $21.5 = 550a + b$ and $12.5 = 250a + b$. The first equation gives $b = 21.5 - 550a$ and if we substitute this in the second equation we obtain $12.5 = 250a + 21.5 - 550a$ and so $a = 0.03$. Hence $b = 21.5 - 550 \times 0.03 = 5$. The law of the machine is thus $E = 0.03L + 5$.

Example

A simple machine has the law $E = 0.2F + 0.25$ kN and a VR of 6. Determine (a) the limiting MA, (b) the limiting efficiency, (c) the effort required to overcome friction when the load is zero.

(a) Limiting MA = $1/a$ = $1/0.2$ = 5.
(b) Limiting efficiency = $1/(a \times \text{VR})$ = $1/(0.2 \times 6)$ = 0.83 or 83%.
(c) When $F = 0$ then $E = 0.25$ kN.

30.2.1 Overhauling

With a frictionless machine, if the effort is removed then the load runs back to its original position and the machine is said to *overhaul*. To prevent overhauling, the friction must be large enough for the load to be unable to move by itself. If W is the work done by friction when the effort E moves through a distance x and the load F through a distance y, then applying the conservation of energy gives:

$$Ex = Fy + W$$

If, when the effort is removed, i.e. $E = 0$, the friction is just large enough to prevent the load moving back (a negative value for y), we must have $Fy = W$. Hence, we can write:

$$Ex = 2Fy$$

But x/y is the velocity ratio and F/E the mechanical advantage. Thus we must have MA/VR = ½ and so an efficiency of 50%. To prevent overhauling, a simple machine must have an efficiency less than 50%.

30.2.2 Examples of simple machines

Section 2.8.2 included some examples of machines. Figure 30.3 gives some more.

The *Weston differential pulley* (Figure 30.3(a)) has a fixed upper block which consists of two concentric wheels of different diameters with an endless rope wrapped round them. If a is the diameter of the larger wheel and b the diameter of the smaller wheel, then when the effort moves through a distance x, a length x of the rope is wound on to the

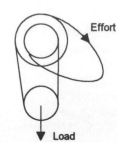

Effort

Load

(a) The differential pulley

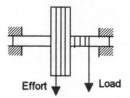

Effort   Load

(b) The wheel and axle

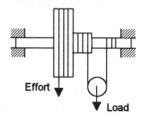

Effort

Load

(c) Differential wheel and axle

Figure 30.3 *Examples of simple machines*

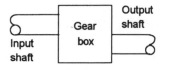

Figure 30.4 *Gear system*

larger wheel while a length bx/a is unwound from the smaller wheel. Thus the load is raised a distance of $\frac{1}{2}(x - bx/a)$ and so the velocity ratio is $2a/(a - b)$.

The *wheel and axle* (Figure 30.3(b)) has the effort applied to a rope wound round a wheel and the load lifted by a rope wound round the axle. If the wheel has a diameter a and the axle a diameter b, then when the effort moves through a distance πa the load rises by πb. Hence the velocity ratio is a/b. If the axle can turn without friction, then the moment of the effort E must equal the moment of the load F and so $Ea = Fb$ and the mechanical advantage is a/b.

The *differential wheel and axle* (Figure 30.3(c)) has an axle consisting of two portions of different diameters. The load is lifted by a pulley which is carried in the loop of a rope wound round the two portions of the axle. If a is the diameter of the wheel, b the diameter of the larger diameter axle and c the diameter of the smaller diameter axle, then when the effort moves through a distance πa we have a length of rope πb wound onto the axle and πc unwound. The load thus rises through a distance $\frac{1}{2}(\pi b - \pi c)$ and so the velocity ratio is $2a/(b - c)$. If b is nearly equal to c then a very large velocity ratio can be produced.

Example

A Weston differential pulley block has pulleys of diameter 150 mm and 200 mm. Determine the velocity ratio.

$$VR = 2a/(a - b) = 2 \times 200/(200 - 150) = 8.$$

Example

A differential wheel and axle has a wheel of diameter 600 mm and axles of diameters 80 mm and 100 mm. Determine the effort required to lift a load of 12 kN if the efficiency is 72%.

$VR = 2a/(b - c) = 2 \times 600/(100 - 80) = 60$. Since efficiency = MA/VR then MA = $0.72 \times 60 = 43.2$. Since MA = load/effort, the effort required is $12 \times 10^3/43.2 = 278$ N.

30.3 Gears

A gear system can be used to change the speed of rotation of a shaft (Figure 30.4). Suppose the input shaft is rotating with an angular velocity ω_i and the output shaft with an angular velocity ω_o. The input power is ω_i, where T_i is the torque on the input shaft. The output power is $T_o\omega_o$, where T_o is the torque on the input shaft. If no power is lost, then the output power equals the input power and so:

$$T_o\omega_o = T_i\omega_i$$

and so, if we define the overall gear ratio of the system, i.e. its *movement ratio* or *velocity ratio* VR = (distance moved by effort)/(distance moved by load), as the ratio of the angular velocity of the input shaft to the angular velocity of the output shaft:

$$\text{gear ratio} = \frac{\omega_i}{\omega_o} = \frac{T_o}{T_i}$$

A reduction gear box is one which reduces the angular velocities, e.g. one with a gear ratio 8 to 1, and thus converts power at a high angular speed and low torque to power at a lower angular speed and high torque.

In the above discussion, the gear system was assumed to be 100% efficient in converting the power of the input shaft to the power of the output shaft. In practice there would be some power loss. The *transmission efficiency* η is defined as:

$$\eta = \frac{\text{output power}}{\text{input power}}$$

Then, output power = $\eta \times$ (input power) and so:

$$T_o\omega_o = \eta T_i\omega_i$$

$$\text{gear ratio} = \frac{\omega_i}{\omega_o} = \frac{T_o}{\eta T_i}$$

Example

A gear box has a gear ratio of 2 to 1. If the input shaft rotates at 20 rev/s when a torque of 200 N m is applied, what will be (a) the number of revolutions per second of the output shaft and (b) the output shaft torque? Assume the system is 100% efficient.

(a) Using gear ratio = ω_i/ω_o, then, since $\omega_i = 2\pi f_i$ and $\omega_o = 2\pi f_o$ we have $f_o = f_i/(\text{gear ratio}) = 20/2 = 10$ rev/s.
(b) Using gear ratio = $\omega_i/\omega_o = T_o/T_1$, then $T_o = $ (gear ratio) $\times T_i = 2 \times 200 = 400$ N m.

Example

If the gear box in the previous example had not been 100% efficient but 85% efficient, what would have been the output torque?

Gear ratio = $\omega_i/\omega_o = T_o/T_1$, so $T_o = 0.85 \times 2 \times 200 = 340$ N m.

30.3.1 Simple gear trains

Figure 30.5 shows a simple gear train; the term *simple* is used when each shaft carries only one wheel. In such a train, the teeth on the wheels are evenly spaced so that they exactly fill the circumference with a whole number of identical teeth and the teeth on the driver and follower mesh without interference. The circumferences are thus proportional to the number of teeth and so:

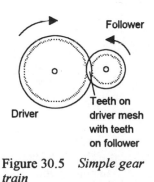

Follower

Teeth on driver mesh with teeth on follower

Driver

Figure 30.5 *Simple gear train*

$$\frac{\text{no of teeth on driver}}{\text{no. of teeth on follower}} = \frac{\text{circumference of driver}}{\text{circumference of follower}}$$

If the driver has, say, twice as many teeth as the follower, then the follower will make two revolutions for each revolution of the driver. Thus, in general:

$$\frac{\text{no. or revs of driver}}{\text{no. of revs of follower}} = \frac{\text{no. of teeth on follower}}{\text{no. of teeth on driver}}$$

Thus, the *velocity ratio* VR or *movement ratio*, i.e. (distance moved by effort)/(distance moved by load) is:

$$\frac{\text{angular vel. of driver}}{\text{angular vel. of follower}} = \frac{\text{no. of teeth on follower}}{\text{no. of teeth on driver}}$$

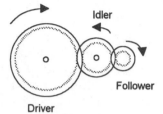

Figure 30.6 *Simple gear train with idler wheel*

With the above arrangement, the two wheels of the train rotate in opposite directions. When the same direction of rotation is required, an idler wheel is included (Figure 30.6). If t_d is the number of teeth on the driver and t_i the number on the idler, then for that pair of wheels $\omega_i/\omega_d = t_d/t_i$, where ω_d is the angular velocity of the driver and ω_i that of the idler. If t_f is the number of teeth on the follower, then for the idler–follower arrangement we have $\omega_i/\omega_f = t_f/t_i$, where ω_f is the angular velocity of the follower. Thus:

$$\frac{\omega_d}{\omega_f} = \frac{\omega_i \dfrac{t_i}{t_d}}{\omega_i \dfrac{t_i}{t_f}} = \frac{t_f}{t_d}$$

The idler thus only changes the direction of rotation; the overall gear ratio is independent of the size of the idler wheel.

30.3.2 Compound gear trains

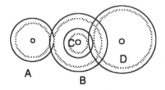

Figure 30.7 *Compound gear trains*

The term *compound* is used for a gear train in which two wheels are mounted on a common shaft. Wheels mounted on a common shaft must rotate with the same angular velocity. Figure 30.7 shows a compound gear train in which wheels B and C are mounted on the same shaft; thus, $\omega_B = \omega_C$. For the pair of wheels A and B we have $\omega_A/\omega_B = t_B/t_A$ and so $\omega_A = (t_B/t_A)\omega_B$, where t_A is the number of teeth on wheel A and t_B the number on wheel B. For the pair of wheels C and D we have $\omega_C/\omega_D = t_D/t_C$, where t_C is the number of teeth on wheel C and t_D the number on wheel D. Since $\omega_B = \omega_C$ we can write $\omega_B/\omega_D = t_D/t_C$ and so $\omega_D = (t_C/t_D)\omega_B$. Thus, the overall velocity ratio is given by:

$$\frac{\omega_A}{\omega_D} = \frac{(t_B/t_A)\omega_B}{(t_C/t_D)\omega_B} = \frac{t_B}{t_A}\frac{t_D}{t_C}$$

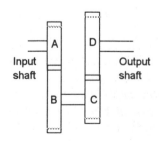

Figure 30.8 *Compound gear train*

Figure 30.8 shows another form of compound gear train; with such an arrangement the output shaft can be brought into line with the input shaft; wheels B and C are on a common shaft.

Example

A compound gear train of the form shown in Figure 30.7 has an input gear A with 40 teeth and A engaging with wheel B having 160 teeth. Wheel B is on the same shaft as wheel C which has 48 teeth and engages with wheel D on the output shaft with 96 teeth. Determine the overall velocity ratio of the system.

$$\frac{\omega_A}{\omega_D} = \frac{t_B}{t_A}\frac{t_D}{t_C} = \frac{160}{40}\frac{96}{48} = 8.$$

30.3.3 Worm gears

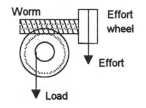

Worm

Effort wheel

Effort

Load

Figure 30.9 *Worm gear*

A gear system which can be used to enable the input and output shafts to be at right angles to each other is shown in Figure 30.9. Rotation of the effort wheel causes the worm to rotate. It engages with the gear teeth on the wheel and hence causes the load shaft to rotate. If the effort wheel has a diameter d_E then one revolution has the effort moving by πd_E. For one revolution of the worm, each tooth on the gear wheel is moved around the circumference of the wheel by an amount equal to the pitch p of the teeth on the worm, assuming it is single threaded. If there are t teeth on the wheel then it will have rotated through $1/t$ for one revolution of the worm. If the load shaft has a diameter d_L then one revolution of the effort wheel will result in the load shaft rotating by $\pi d_L/t$. The velocity ratio VR or movement ratio, i.e. (distance moved by effort)/ (distance moved by load), of the worm gear is thus $(\pi d_L/t)/\pi d_E = t d_L/d_E$. If the worm has n threads then one revolution of the worm causes each tooth on the wheel to rotate by np and so the load shaft to rotate by $n\pi d_L/t$; the velocity ratio is thus $t d_L/n d_E$.

Example

A single threaded worm gear has an effort wheel of diameter 140 mm and a load shaft of diameter 100 mm. If the worm gear has a wheel with 40 teeth, what will be the velocity ratio of the gear?

Velocity ratio = $t d_L/d_E$ = 40 × 100/140 = 28.6.

30.3.4 Gear winches

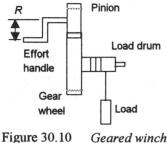

R

Pinion

Effort handle

Load drum

Gear wheel

Load

Figure 30.10 *Geared winch*

The geared winch (Figure 30.10) involves the effort being used to rotate the pinion wheel. This results in the gear wheel rotating and hence winding a cable onto a drum and lifting the load. The gear train used increases the movement ratio. One rotation of the effort handle is a distance moved by the effort of $2\pi R$. If the pinion has n_p teeth and the gear wheel n_g teeth, then one revolution of the effort handle rotates the load drum by (n_p/n_g) revolutions. Thus, if the drum has a diameter D then the distance moved by the load for one revolution of the effort handle is $(n_p/n_g)\pi D$. The velocity ratio is thus $2\pi R/(n_p/n_g)\pi D$.

Example

A geared hand winch has an effort arm with radius 350 mm and a load drum of diameter 100 mm. If the pinion has 12 teeth and the gear wheel 60 teeth, what is the velocity ratio?

$$VR = 2R/(n_p/n_g)D = 2 \times 350/[(12/60) \times 100] = 35.$$

30.4 Belt drives

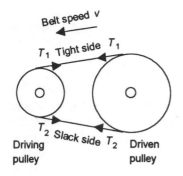

Figure 30.11 *Belt drive*

Power can be transmitted from one shaft to another by means of a continuous belt wrapped round pulleys mounted on the shafts (Figure 30.11). The belt has an initial tension when the shafts are at rest. When the driver pulley starts to rotate, frictional forces between the shafts and the belt cause the tension on one side to increase and on the other side to decrease. It is this difference in tension which is responsible for the transmission of power. If the tension on the 'tighter' side is T_1 and on the 'slacker' side is T_2 then there is a net tangential force of $(T_1 - T_2)$ acting on each pulley. If the belt has a linear speed v, the power transmitted is:

$$power = (T_1 - T_2)v$$

If we assume that the belt is elastic and obeys Hooke's law, the increase in the length of the slack side must equal the reduction in the length of the tight side. The change in length of the belt will be proportional to the tension in the belt and so, if T_0 is the initial tension in the belt when it is not running, then $T_1 - T_0 = T_0 - T_2$ and so:

$$T_0 = \tfrac{1}{2}(T_1 + T_2)$$

If no slipping occurs, the peripheral speed of the driving pulley must be the same as of the driven pulley. Hence, using $v = r\omega$:

driving pulley radius × driving pulley angular velocity
= driven pulley radius × driven pulley angular velocity

Example

A belt is installed with an initial tension of 400 N. If the maximum permissible tight side tension is 560 N, what will be (a) the slack side tension and (b) the maximum power that can be transmitted if the smaller pulley is rotating at 10 rev/s and has a diameter of 100 mm?

(a) Using $T_0 = \tfrac{1}{2}(T_1 + T_2)$, then $400 = \tfrac{1}{2}(560 + T_2)$ and $T_2 = 240$ N.
(b) $v = r\omega = r \times 2\pi f = 0.050 \times 2\pi \times 10 = 3.14$ m/s. Hence, the maximum power that can be transmitted is:

$$max.\ power = (T_1 - T_2)v = (560 - 240) \times 3.14 = 1005\ W.$$

Example

The pulley on a motor shaft has a radius of 40 mm and rotates at 100 rev/s. What will be the number of revolutions per second of a shaft with a pulley of radius 160 mm which is connected to the motor shaft by a belt drive and no slipping occurs?

Since the peripheral speed will be the same for both pulleys we have driving pulley radius × driving pulley angular velocity = driven pulley radius × driven pulley angular velocity and so, as $\omega = 2\pi f$:

$$40 \times 2\pi \times 100 = 160 \times 2\pi \times f_{driven}$$

Hence the driven pulley rotates at 25 rev/s.

Activities

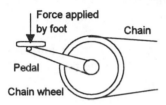

Figure 30.12   *Activity 1*

1 Analyse the elements involved in a rider getting a bicycle to move along a road. Consider how the force applied by the rider's feet is transformed into rotation of the chain wheel (consider the direction of the applied force in relation to the position of the pedal (Figure 30.12) and so how the applied torque varies with angle of rotation) and how this rotation is transformed into rotation of the rear bicycle wheel. What is the effect of changing the number of sprockets on the rear wheel drive, i.e. changing gear?

Problems

1 A simple machine has the law $E = 0.3F + 0.2$ kN and a VR of 7. Determine (a) the limiting MA, (b) the limiting efficiency, (c) the effort required to overcome friction when the load is zero.

2 A simple machine was found to require an effort of 10 N for a load of 30 N and an effort of 68 N for a load of 300 N. Determine the law of the machine.

3 Determine the limiting efficiency for a simple machine with the law $E = 0.25F + 0.6$ kN and a VR of 5.

4 A differential wheel and axle has a wheel of diameter 480 mm and axles of diameters 80 mm and 120 mm. What is the efficiency if it takes an effort of 144 N to raise a load of 2240 N?

5 A Weston differential pulley block has pulleys of diameter 120 mm and 150 mm. Determine the velocity ratio.

6 A Weston differential pulley block has pulleys of diameter 250 mm and 230 mm. Determine the velocity ratio.

7 A differential wheel and axle has a wheel of diameter 200 mm and axles of diameters 60 mm and 70 mm. Determine (a) the velocity ratio, (b) the load that can be lifted by an effort of 120 N if the MA is 24.

8 A simple gear train has a driver gear with 28 teeth meshing with an idler gear of 56 teeth, it in turn meshing with a follower gear of 168 teeth. Determine the speed of the follower when the driver gear rotates at 24 rev/s.

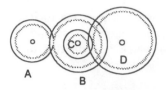

Figure 30.13 *Problem 9*

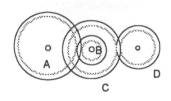

Figure 30.14 *Problem 10*

9 A compound gear train of the form shown in Figure 30.13 has an input gear A with 48 teeth and A engaging with wheel B having 144 teeth. Wheel B is on the same shaft as wheel C which has 80 teeth and engages with wheel D on the output shaft with 160 teeth. Determine the overall velocity ratio of the system.

10 A compound gear train of the form shown in Figure 30.14 has an input gear A with 60 teeth and A engaging with wheel B having 20 teeth. Wheel B is on the same shaft as wheel C which has 40 teeth and engages with wheel D on the output shaft with 50 teeth. Determine the overall velocity ratio of the system and the numbers of revolutions per second that D will make when A revolves at 5 rev/s.

11 A geared hand winch has an effort arm with radius 250 mm and a load drum of diameter 100 mm. If the pinion has 20 teeth and the gear wheel 100 teeth, what is the velocity ratio?

12 A gear box with an overall gear ratio of 5 to 1 has an input shaft with a power of 6 kW at 30 rev/sec. What will be the torque on the input shaft and the torque on the output shaft if the gear box is assumed to be 100% efficient?

13 A gear box with an overall gear ratio of 6 to 1 has a torque of 35 N m applied to the input shaft. What will be the torque on the output shaft if the gear box is 90% efficient?

14 The output shaft of a reduction gear rotates at one-eighth the speed of the input shaft. What will be the efficiency if the output torque is 40 N m when the input torque is 6 N m?

15 Pulleys on two parallel shafts are linked by a belt drive. If the pulley on the drive shaft has a radius of 100 mm and is rotating at 30 rev/s, what will be the belt speed if no slippage occurs?

16 The tensions on the two sides of a belt passing round a pulley are 2000 N and 500 N. If the effective radius of the pulley is 200 mm and it rotates at 10 rev/s, calculate the power transmitted by the belt.

17 A pulley of radius 400 mm is used to drive a belt. When the difference in tension between the tight and slack sides of the belt drive is 2 kN and the pulley is rotating at 5 rev/s, what is (a) the torque applied to the pulley, (b) the power transmitted?

18 An electric motor is used to lift a load of 5 kg by means of a rope wrapped round the a pulley of radius 600 mm driven by the motor. Determine (a) the torque that needs to be exerted and (b) the work done when the pulley has to make 15 revolutions to lift the load.

31 Oscillations

31.1 Introduction

This chapter is about the periodic motion of oscillating bodies. Any sort of motion which repeats itself in equal intervals of time is called *periodic*.

The time between repetitions is called the *periodic time* and the motion occurring within one period is called a *cycle* (Figure 31.1) with the number of cycles per second termed the *frequency*. Since one cycle occurs in the periodic time T, the frequency is $1/T$. The unit of frequency is the hertz (Hz), with 1 Hz equal to one cycle per second. The *amplitude* A of an oscillation is the maximum displacement from the initial rest position, the range of displacement being thus $\pm A$.

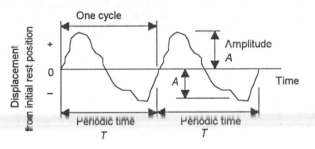

Figure 31.1 *A periodic oscillation*

The term *free oscillation* is used when an elastic system oscillates under the action of forces inherent in the system itself with there being no externally applied forces. The system will oscillate at, what is termed, a *natural frequency*, such frequencies being determined by the properties of the system. In this chapter we are only concerned with free oscillations.

31.2 Simple harmonic motion

The simplest form of periodic motion is *simple harmonic motion* and the following demonstrates the characteristics of such motion. Consider a basic mechanical system of mass which when deflected from its rest position is restored to it by forces arising from elasticity in the system. Figure 31.2 shows such a system as a trolley tethered between two supports by springs. A trolley is used for the mass in order to minimise frictional effects.

When the trolley, the mass of the system, is pulled to one side then one of the springs is compressed and the other stretched and this has the effect of providing a restoring force which is directed in such a direction

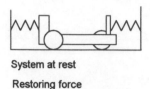

System at rest

Restoring force

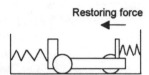

Mass deflected from rest position

Restoring force

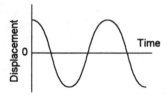

Mass deflected from rest position

Figure 31.2 *A basic mechanical system*

Figure 31.3 *Graph of the displacement with time*

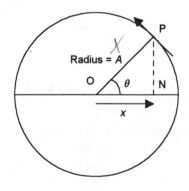

Figure 31.4 *Point P moving in a circular path with a constant angular velocity*

as to endeavour to restore the trolley back to its rest position and is always proportional to its displacement from the rest position. If the trolley is released from this deflected position, the restoring force causes the trolley to move back towards its original rest position and overshoot that position. The restoring force then reverses its direction to still be directed towards the rest position and so oscillations occur. If the displacement from the rest position is measured as a function of time then the result is as shown in Figure 31.3, the displacement variation with time being described by a cosine graph. This form of oscillation is termed simple harmonic motion (SHM).

> *Simple harmonic motion* (SHM) *is said to occur when the motion is under the action of a restoring force which is always directed to a fixed point and has a magnitude which is proportional to the displacement from that point.*

31.2.1 Simple harmonic motion equations

The type of displacement variation with time which is characteristic of simple harmonic motion can be produced by the horizontal displacement x from the centre of a point P rotating in a circular path with a constant angular velocity ω (Figure 31.4), i.e.

$$x = A \cos \theta$$

where A is the amplitude of the oscillation. Since we have $\theta = \omega t$, then we can describe such oscillations by:

$$x = A \cos \omega t$$

The frequency f of the oscillations is the number of cycles completed per second and it is just the frequency multiplied by 2π. Thus:

$$x = A \cos 2\pi f t$$

Since for angular motion the velocity tangential to the circle is $r\omega$, the component of this velocity along the diameter is $r\omega \sin \theta$ (Figure 31.5). Thus the velocity of the displacement of the point N along the x-axis is:

$$v = -A\omega \sin \theta = -A\omega \sin \omega t$$

The minus sign is included to show that the direction of the velocity is towards O and in the opposite direction to the direction in which you would go if the displacement was being increased.

An alternative way of arriving at the same equation is to recognise that the linear velocity v at some instant is the rate of change of displacement dx/dt and differentiate $x = A \cos \omega t$ to give $v = dx/dt = -A\omega \sin \omega t$.

In Figure 31.4 we have $\sin \theta = PN/A$, and since $PN^2 = A^2 - x^2$ then $A \sin \theta = \sqrt{(A^2 - x^2)}$ and so $v = -A\omega \sin \omega t$ can be written as:

$$v = -\omega\sqrt{A^2 - x^2}$$

The maximum velocity is when $x = 0$, i.e. as the mass passes through its rest position, and is:

$$\text{maximum velocity} = -\omega A$$

The centripetal acceleration of P is $\omega^2 A$ (Figure 31.6) and thus the component of this acceleration along x-axis is:

$$a = -\omega^2 A \sin\theta$$

The minus sign is because the direction of the acceleration is towards O and in the opposite direction to which you would go if the displacement was increased. But $x = A \sin\theta$ and so:

$$a = -\omega^2 x$$

Alternatively, since the linear acceleration a at an instant is the rate of change of velocity dv/dt, we can differentiate $v = -A\omega \sin\omega t$ to give $a = -\omega^2 A \cos\omega t$ and hence, as before, $a = -\omega^2 x$.

The acceleration has a maximum value when x equals the maximum displacement A:

$$\text{maximum acceleration} = -\omega^2 A$$

The restoring force $F = ma$ and is thus:

$$F = -m\omega^2 x$$

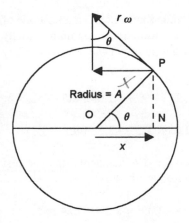

Figure 31.5 *Point P moving in a circular path with a constant angular velocity*

The minus sign indicates that the direction of this restoring force is always in the opposite direction to that for which x increases.

We can write $F = -m\omega^2 x$ as:

$$\omega = \sqrt{\frac{F}{mx}}$$

and thus, since the periodic time $T = 1/f = 2\pi/\omega$:

$$T = 2\pi\sqrt{\frac{\text{mass}}{\text{force per unit displacement}}}$$

$$f = \frac{1}{2\pi}\sqrt{\frac{\text{force per unit displacement}}{\text{mass}}}$$

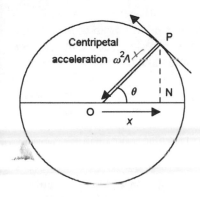

Figure 31.6 *Point P moving in a circular path with a constant angular velocity*

The larger the force needed to produce unit displacement the higher the frequency. Thus a high 'stiffness' mechanical system with its large force per unit displacement will have a high frequency of oscillation.

Example

An object moving with simple harmonic motion has an amplitude of 1.2 m and a periodic time of 3 s. Determine the maximum velocity and maximum acceleration and at what points in the oscillation they occur.

The angular frequency $\omega = 2\pi f = 2\pi/T = 2\pi/3 = 2.09$ rad/s. Thus:

$$\text{maximum velocity} = -\omega A = -2.09 \times 1.2 = -2.51 \text{ m/s}$$

The maximum velocity occurs when the displacement from the rest position is zero. The maximum acceleration is thus:

$$\text{maximum acceleration} = -\omega^2 A = -2.09^2 \times 1.2 = -5.24 \text{ m/s}^2$$

The maximum acceleration occurs when the displacement from the rest position is a maximum.

Example

A body moving with simple harmonic motion has an acceleration of 1.5 m/s^2 when 200 mm from the zero displacement position. Calculate the frequency of the oscillation and the time for one cycle.

Acceleration $a = -\omega^2 x = -(2\pi f^2)x$ and so $f = (1/2\pi)\sqrt{(a/x)} = (1/2\pi)\sqrt{(1.5/0.200)} = 0.44$ Hz. The time for one cycle $= 1/f = 1/0.44 = 2.3$ s.

Example

An object moves with simple harmonic motion and has an amplitude of 500 mm and a frequency of 4 Hz. Determine the velocity and acceleration of the object when it is 200 mm from its rest position.

With $\omega = 2\pi f = 2\pi \times 4 = 25.1$ rad/s, the velocity is:

$$v = -\omega\sqrt{A^2 - x^2} = -25.1\sqrt{0.5^2 - 0.2^2} = -11.5 \text{ m/s}$$

and the acceleration is $a = -\omega^2 x = -25.1^2 \times 0.2 = -126$ m/s.

31.3 Mass on a spring   Consider a mass suspended from a vertical helical spring (Figure 31.7), the mass of the spring being assumed to be negligible. It is made to oscillate in the vertical direction by the mass being pulled down, so extending the spring, and then the mass is released. The spring then exerts a restoring force on the mass. Assuming that the spring obeys Hooke's law, then the restoring force F is proportional to the displacement x of the end of the spring from its rest position. The force is always directed towards the rest position and thus we can write:

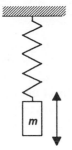

Figure 31.7 *Mass on a spring*

$$F = -kx$$

where k is the spring stiffness. The motion is simple harmonic because F is proportional to $-x$. The magnitude of the force per unit displacement is k and thus:

$$f = \frac{1}{2\pi} \sqrt{\frac{\text{force per unit displacement}}{\text{mass}}} = \frac{1}{2\pi} \sqrt{\frac{k}{m}}$$

Example

A load is suspended from a vertically mounted spring and causes it to extend by 10 mm. With what frequency will the spring oscillate if the load is pulled down further and then released?

The force extending the spring is mg, where m is the mass of the load. Thus $F = mg = kx$ and so $k/m = g/x$. The frequency is thus:

$$f = \frac{1}{2\pi} \sqrt{\frac{k}{m}} = \frac{1}{2\pi} \sqrt{\frac{g}{x}} = \frac{1}{2\pi} \sqrt{\frac{9.8}{0.010}} = 5.0 \text{ Hz.}$$

31.4 Simple pendulum

Consider a simple pendulum in which the mass m of the pendulum acts at a distance L from the point of suspension (Figure 31.8). Note the use of the term 'simple', this is because we will make the assumption that all the mass of the pendulum is concentrated in a small bob on the end of a string of negligible mass; if these assumptions cannot be made the pendulum is termed compound and the analysis of the motion of the pendulum has to be modified from that which follows. When the pendulum is pulled through a small angle θ from the vertical, there is a restoring force acting on the pendulum bob which arises from the component of the weight mg which is tangential to the arc of motion of the bob, i.e. $mg \sin \theta = mg(d/L)$. Since we are only considering small angles, d is approximately the same as x and thus the restoring force is approximately mgx/L and the force per unit displacement is mg/L:

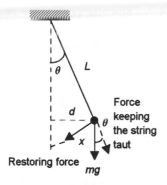

Figure 31.8 *Pendulum*

$$f = \frac{1}{2\pi} \sqrt{\frac{\text{force/unit angular displacement}}{\text{mass}}} = \frac{1}{2\pi} \sqrt{\frac{g}{L}}$$

Example

Calculate the length of cord needed for a simple pendulum to have a periodic time of 2 s.

Using $T = 2\pi\sqrt{(L/g)}$ then $2 = 2\pi\sqrt{(L/9.8)}$ and so $L = 0.99$ m.

31.5 Energy of SHM

The velocity of an object when oscillating with simple harmonic motion and at a displacement x from its central rest position is given by:

$$v = -\omega \sqrt{A^2 - x^2}$$

Thus the kinetic energy at displacement x is:

$$\text{kinetic energy} = \tfrac{1}{2}mv^2 = \tfrac{1}{2}m\omega^2(A^2 - x^2)$$

Because the restoring force is proportional to the displacement, the work done to move the object from its central rest position to a displacement x is given by the average force acting over that displacement multiplied by the displacement. The force is zero at the central position and $m\omega^2 x$ at displacement x. Thus:

$$\text{work done} = \tfrac{1}{2}m\omega^2 x^2$$

Thus the potential energy of the object when displaced by x is:

$$\text{potential energy} = \tfrac{1}{2}m\omega^2 x^2$$

The total energy at this displacement is the sum of the potential and kinetic energies thus is:

$$\text{energy} = m\omega^2 A^2$$

The total energy is thus constant at all displacements, depending only on the amplitude of the oscillation. When the oscillating body is passing through its rest position, i.e. $x = 0$, all the energy is kinetic energy and there is no potential energy; when the oscillating body is at the extreme position of its oscillation then all the energy is potential energy and there is no kinetic energy. Thus, as an object oscillates there is a continual changing of potential energy to kinetic energy and vice versa with, in the absence of losses or inputs to the system, the sum remaining constant.

Example

A piston of mass 0.2 kg moves with simple harmonic motion. If the amplitude of the piston oscillation is 70 mm and its frequency is 10 Hz, calculate (a) the maximum acceleration, (b) the maximum velocity, (c) the maximum kinetic energy.

(a) Since $a = -\omega^2 x = -(2\pi f)^2 x$, the maximum acceleration occurs at the maximum value of x, i.e. when $x = $ the amplitude A, and thus maximum $a = -(2\pi f)^2 A = -(2\pi \times 10^2) \times 0.070 = -280$ m/s^2.
(b) Since $v = -\omega\sqrt{(A^2 - x^2)}$, the maximum velocity occurs when $x = 0$ and thus maximum $v = -2\pi \times 10 \times 0.070 = 4.40$ m/s.
(c) The maximum kinetic energy occurs when the velocity is a maximum and so maximum KE $= \tfrac{1}{2}mv^2 = \tfrac{1}{2} \times 0.2 \times 4.40^2 = 1.93$ J.

Activities

1 Determine the acceleration due to gravity in your locality by constructing a simple pendulum with a length of thread and a small compact bob and then measuring the time taken for a number of oscillations, e.g. 20, and hence obtain a value for the periodic time.

In order that the point of support of the pendulum can be accurately fixed and not vary during the oscillations, clamp the end between two flat pieces of wood or metal.

Problems

1 A body oscillates with simple harmonic motion. The frequency of the oscillation is 12 Hz and the amplitude is 200 mm. What is the maximum acceleration and the maximum velocity attained and at what points in the path of the oscillating body are these maximums obtained?

2 A valve moves with simple harmonic motion. If the amplitude of the movement is 5 mm and the periodic time 0.005 s, what will be the maximum velocity and the maximum acceleration?

3 An object oscillates with simple harmonic motion. At a displacement of 1 m from its central rest position the acceleration is 200 m/s². Determine the frequency of the oscillation.

4 An object oscillates with simple harmonic motion of frequency 4 Hz and amplitude 150 mm. Determine the velocity and acceleration of the particle at a displacement of 90 mm.

5 An object oscillates with simple harmonic motion. Determine the periodic time if the acceleration is 4 m/s² when the displacement from the centre of the oscillation is 2 m.

6 An object oscillates with simple harmonic motion and has a periodic time of 8 s and amplitude 4 m. Determine the maximum velocity and the velocity when the object is 2 m from its central rest position.

7 An object of mass 0.1 kg oscillates with simple harmonic motion of frequency 15 Hz. Determine the restoring force acting on the object when it is at a displacement of 30 mm.

8 A vertical helical spring has a stiffness of 2 kN/m, calculate the frequency of the oscillation that will be produced when a mass of 3 kg is supported by the spring.

9 A 2.5 kg mass when attached to the lower end of a vertical helical spring causes it to extend by 20 mm. What will be the frequency of oscillation of the system?

10 A mass of 5 kg is suspended from a fixed support by a vertical spring. If the natural frequency of oscillation is 2 Hz for vertical oscillations, what is the stiffness of the spring?

11 A simple pendulum has a periodic time of 1.000 s where the acceleration due to gravity is 9.81 m/s². What will be its periodic time when it is taken to a place where the acceleration due to gravity is 9.85 m/s²?

12 What is the length of a simple pendulum with periodic time 3.0 s?

13 A compact object of mass 0.20 kg is suspended by a cord of length 1.0 mm from a fixed support and allowed to swing back-and-forth like a simple pendulum. What will be (a) the periodic time and (b) the maximum velocity of the object if the amplitude of the oscillation is 100 mm?

32 Heat transfer

32.1 Introduction

Thermodynamics is the science dealing with the relations between the properties of substances and the quantities termed work and heat. *Work* is defined as the transfer of energy that occurs when the point of application of a force moves through some distance in the direction of the force. *Heat* is defined as the transfer of energy that occurs between two systems when there is a temperature difference between them. As with other forms of energy, the SI unit for work or heat is the joule (J). This chapter is a basic introduction to heat transfer, outlining the various modes of heat transfer and then concentrating on heat transfer by conduction.

32.2 Heat transfer processes

There are three basic modes of heat transfer: *conduction*, *convection* and *radiation*. These may occur separately or simultaneously. Separate equations can be written for each mode of heat transfer and generally, when heat transfer is occurring by more than one mode, we can just consider each mode separately and then sum the results to give the total heat transfer.

1 *Conduction*

This can basically be considered as being the mode of transfer when energy is just passed from one atom or molecule to another through the material without there being bodily movement of the material. If you think of a crowd of people, the analogue of conduction is that energy is transferred through the crowd by one person jostling the next who in turn jostles the next one and so on, the jostling thus being passed through the crowd. An example of conduction is when one end of a metal bar is heated and, with the other end cool, heat energy is transferred through the bar from the hot end to the cool end.

2 *Convection*

This occurs as a result of the gross motion of parts of the material. The analogue of convection with the crowd is that people start moving through the crowd and so a disturbance in one region of the crowd is transferred to another region by people moving to that region. An example of convection is when water is heated in a pan. The hot water at the bottom of the pan rises through the cooler water and so transfers heat from the hotter part of the water to the cooler part.

3 *Radiation*

Unlike conduction and convection, this method of transferring heat does not require matter for the transfer medium. *Radiation* is the transfer of heat energy by electromagnetic waves. We are all familiar with the transfer of heat energy by radiation from the sun to the earth through the vacuum of space.

32.3 Conduction

For conduction to occur there must be a temperature difference between different parts of the material. Conduction occurs in solids where it might be the sole method of heat transfer. It can also occur in liquids and gases but is then often occurring with other modes of heat transfer.

The rate of heat transfer by conduction is proportional to the temperature gradient in the direction of the heat flow and to the area of the material perpendicular to the direction of the heat flow. This is known as the *Fourier law of heat conduction*. It can be expressed as:

heat transfer rate per unit area $q = -k$(temperature gradient)

where k is a constant called the *thermal conductivity*. Because the temperature decreases in the direction of the heat flow, the equation includes a minus sign so that such a negative temperature gradient gives a positive heat flow. Typical values of thermal conductivity are given in Table 32.1. Metals have much higher thermal conductivities than non-metals.

Table 32.1 *Typical values of thermal conductivities*

Material	k W/m K	Material	k W/m K
Copper	380	Fibreboard	0.11
Mild steel	54	Cork	0.04
Concrete	1.4	Glass wool blanket	0.04
Common brick	1.2	Expanded polystyrene	0.03
Softwood	0.13	Air	0.029

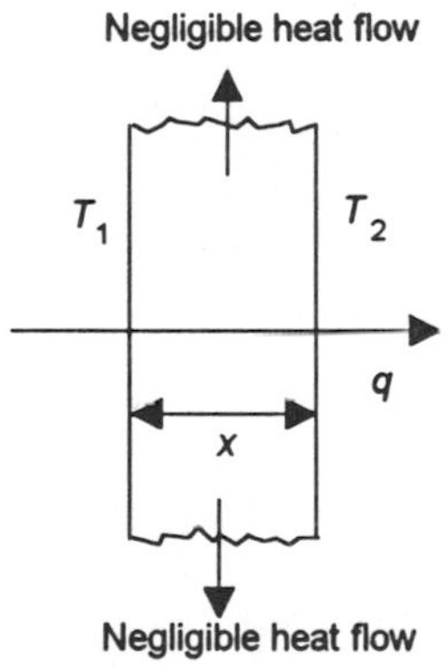

Figure 32.1 *Heat conduction through a plane wall*

32.3.1 Heat conduction through plane layers

Consider uniaxial heat conduction through a plane wall; heat transfer in the transverse direction is assumed to be negligible (Figure 32.1). If the wall surfaces are at temperatures T_1 and T_2 and the wall thickness is x, the temperature gradient in the direction of the heat flow is $(T_1 - T_2)/x$ and so:

$$\text{heat transfer rate per unit area } q = -k\frac{T_1 - T_2}{x}$$

For a plane wall of surface area A, the rate of heat transfer by conduction $Q = qA$ and thus:

heat transfer rate $Q = -kA\dfrac{T_1 - T_2}{x}$

Example

Calculate the rate of heat loss by conduction through the wall of an industrial refrigerator if the wall has an area of 6 m² and consists of a layer of cork 200 mm thick sandwiched between metal sheets. The temperatures of the inner and outer surfaces of the cork are −15°C and 20°C. Cork has a thermal conductivity of 0.04 W/m K.

Heat transfer rate $= -0.04 \times 6 \times \dfrac{-15 - 20}{0.200} = 42$ W.

32.3.2 Conduction through composite materials

In many situations involving heat conduction we have the heat being conducted through a series of layers, e.g. a cavity brick wall consisting of two layers of brick separated by an air cavity. Consider a plane wall which consists of three layers (Figure 32.2), each layer having a different thermal conductivity and thickness. We will assume that all the heat transfer takes place through the wall at right angles to the wall. Thus we have the same rate of energy input Q to each layer. For the first layer:

$$Q = -k_1 A\dfrac{T_2 - T_1}{x_1}$$

and for the second layer:

$$Q = -k_2 A\dfrac{T_3 - T_2}{x_2}$$

and for the third layer:

$$Q = -k_3 A\dfrac{T_4 - T_3}{x_3}$$

Rearranging these equations gives:

$$T_2 - T_1 = -\dfrac{Q}{A}\dfrac{x_1}{k_1}$$

$$T_3 - T_2 = -\dfrac{Q}{A}\dfrac{x_2}{k_2}$$

$$T_4 - T_3 = -\dfrac{Q}{A}\dfrac{x_3}{k_3}$$

Adding the above three equations gives:

$$T_4 - T_1 = -\dfrac{Q}{A}\left[\dfrac{x_1}{k_1} + \dfrac{x_2}{k_2} + \dfrac{x_3}{k_3}\right]$$

and so we can write:

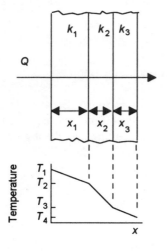

Figure 32.2 *Composite wall*

$$Q = -\frac{A}{\frac{x_1}{k_1} + \frac{x_2}{k_2} + \frac{x_3}{k_3}}(T_4 - T_1)$$

$T_4 - T_1$ is the overall temperature difference across the composite wall; we thus do not need to know the temperatures of the intermediate surfaces in order to calculate the heat transferred.

The term *overall heat transfer coefficient U* or, as it is often called, the *U-value* is defined as:

$$U = \frac{1}{\frac{x_1}{k_1} + \frac{x_2}{k_2} + \frac{x_3}{k_3}}$$

and relates the rate of heat transfer per unit area to the temperature difference and so we can write:

$$Q = -UA(T_4 - T_1)$$

Table 32.2 shows some typical U-values for composite plane layers used in buildings.

Table 32.2 *U-values*

Element	Composition	U-value, W/m² K
Solid wall	Brickwork 215 mm, plaster 15 mm	2.3
Cavity wall	Brickwork 102.5 mm, cavity 50 mm, brickwork 102.5 mm	1.6
Cavity wall	Brickwork 102.5 mm, cavity 25 mm, polystyrene board, 25 mm, aerated concrete block 100 mm, plaster 13 mm	0.58
Window	Single glazing	5.7
Window	Double glazing, airspace 20 mm	2.8

Example

The wall of a furnace consists of a layer of firebrick of thickness 200 mm with an outer layer of insulation of thickness 10 mm. The firebrick has a thermal conductivity of 0.7 W/m K and the insulator a thermal conductivity of 0.1 W/m K. Determine the rate of heat transfer per square metre of wall when the internal surface temperature is 600°C and the external surface temperature is 50°C.

$$Q = -\frac{A}{\frac{x_1}{k_1} + \frac{x_2}{k_2}}(T_3 - T_1) = -\frac{1}{\frac{0.200}{0.7} + \frac{0.010}{0.1}}(50 - 600) = 1426 \text{ W}$$

Example

Calculate the heat loss per square metre through a wall with a U-value of 1.6 W/m² K when the internal surface temperature of the wall is 20°C and the external surface temperature is 0°C.

$$Q = -UA(T_4 - T_1) = -1.6 \times 1 \times (0 - 20) = 32 \text{ W}.$$

Activities

1 Using the apparatus available in your laboratory, e.g. the apparatus called Searle's bar, determine the thermal conductivity of a metal.

Problems

1 Calculate the rate of heat loss by conduction though a single brick wall if the wall has an area of 20 m² and a thickness of 100 mm. The temperatures of the inner and outer surfaces of the wall are 20°C and 5°C. The brick has a thermal conductivity of 1.2 W/m K.

2 Determine the heat transferred by conduction per square metre of surface area through a wall of thickness 300 mm if one side of the wall is at a temperature of 230°C and the other at 15°C. The wall has a thermal conductivity of 0.9 W/m K.

3 Calculate the rate of heat input required to maintain a temperature difference of 2°C between the two sides of a metal sheet of area 2 m² and thickness 10 mm if the sheet has a thermal conductivity of 50 W/m K.

4 Determine the rate at which heat is transferred along a lagged copper rod of length 500 mm and cross-sectional area 1000 mm² if the temperature difference between the two ends is 100°C. Take the thermal conductivity of copper as 380 W/m K.

5 A double-glazed window consists of two panes of glass of thickness 5 mm separated by an air space of thickness 20 mm. If the thermal conductivity of glass is 0.9 W/m K and that of air 0.03 W/m K, determine the heat loss per square metre of glazing when the internal temperature is 20°C and the external temperature 0°C.

6 The cavity wall of a house consists of brick of thickness 120 mm, a 50 mm air cavity and then another brick section of thickness 120 mm. What is the rate at which heat is conducted through a cavity wall of area 15 m² when the difference in temperature between the inner surface and outer surfaces of the wall is 20°C. Take the thermal conductivity of the brick to be 1.0 W/m K and the air 0.03 W/m K.

7 Determine the rate of heat transfer through a wall of area 100 m² when there is a temperature difference of 20°C between the inner and outer surfaces of the wall and the wall has a U-value of 1.5 W/m² K.

8 A double-glazed window has a U-value of 2.8 W/m² K. Determine the heat loss per square metre of glazing when the internal temperature is 30°C and the external temperature is 0°C.

33 Fluid mechanics

33.1 Introduction

The term *fluid* is used for substances that can flow and have no particular shape of their own but can assume the shape of a containing vessel. Thus the term fluids encompasses both liquids and gases; gases can be compressed and show significant volume changes whereas liquids do not show noticeable changes in volume when subject to pressure and can generally be regarded as incompressible. This chapter is about the basic properties of incompressible fluids and includes a consideration of the pressure and thrust relationship to depth of immersion in a fluid and the energy changes involved in the flow of such fluids.

33.2 Pressure

The term *pressure p* is defined as being the force F per unit area A, i.e. $p = F/A$, the basic unit being the pascal (Pa) with 1 Pa = 1 N/m². For a fluid:

1   The pressure exerted on any point on a surface in the fluid is always at right angles to the surface.

2   The pressure at any point in a static fluid is the same in all directions at that point. If this was not the case then there would be a net pressure in some particular direction and so a force causing the fluid to move in that direction.

33.2.1 Pressure due to a column of liquid

The pressure at some depth in a fluid at rest due to the fluid above it is the weight of the fluid above it divided by the area (Figure 33.1). The weight is mg, where the mass m is the volume multiplied by the density ρ. When the height of the fluid is h and the area over which it is acting is A, then the volume is Ah and so:

$$p = \frac{F}{A} = \frac{mg}{A} = \frac{hA\rho g}{A} = h\rho g$$

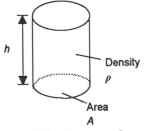

h

Density
ρ

Area
A

Figure 33.1 *Pressure due to a column of fluid*

Example

What is the pressure at the base of a column of liquid of height 20 cm and density 1000 kg/m³ arising from the liquid?

$p = h\rho g = 0.20 \times 1000 \times 9.8 = 1960$ Pa = 1.96 kPa.

33.2.2 Atmospheric pressure

The earth's atmosphere is a fluid and exerts a pressure at the surface of the earth due to the height of air above the surface. The atmospheric pressure at the surface of the earth is about 100 kPa, this sometimes being referred to as a pressure of 1 bar.

The pressure due to a fluid which is measured relative to zero pressure is termed the *absolute pressure*. However, pressures are often measured relative to the prevailing atmospheric pressure and are then said to be *gauge pressure*. Hence:

absolute pressure = gauge pressure + atmospheric pressure

33.3 Measurement of pressure

A common method of measuring the pressure due to a gas is to use the *U-tube manometer*. This consists of a U-tube containing a liquid (Figure 33.2) and the pressure difference between the gases above the liquid in the two limbs produces a difference h in vertical heights of the liquid in the two limbs. For a liquid at rest, the pressure at the base of each limb must be the same. Thus we must have:

$$P_1 + h_1\rho g = P_2 + h_2\rho g$$

where ρ is the density of the manometric liquid. Hence the pressure difference between the gases above the two limbs is:

$$P_1 - P_2 = (h_2 - h_1)\rho g = h\rho g$$

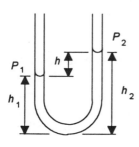

Figure 33.2   *U-tube manometer*

where h is the vertical difference in heights of the liquid in the two limbs. If one of the limbs is open to the atmosphere then the pressure difference is the gauge pressure. Pressures measured by means of such a manometer are sometimes expressed in terms of the difference in heights of the liquid level, .e.g. as 20 mm of mercury, rather than in units of pressure.

Water, alcohol and mercury are commonly used manometric liquids. U-tube manometers are simple and cheap and can be used for pressure differences in the range 20 Pa to 140 kPa. Errors can arise due to the height measured not being truly vertical, the effects of temperature on the density of the liquid, and incorrect values of the acceleration due to gravity being used. The accuracy is also affected by difficulties in obtaining an accurate reading of the level of the manometric liquid in a tube due to the meniscus. The accuracy is thus typically about ±1%.

The *inclined tube manometer* (Figure 33.3) is a U-tube manometer with one limb having a larger cross-section than the other and the narrower limb being inclined at some angle θ to the horizontal. It is generally used for the measurement of small pressure differences and gives greater accuracy than the conventional U-tube manometer. The vertical displacement d of the liquid level in the inclined limb is related to the movement x of the liquid along the tube by $d = x \sin \theta$. The displacement x is the measured quantity, thus:

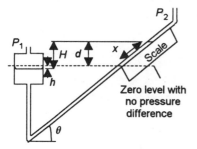

Figure 33.3   *Inclined tube manometer*

$$P_1 - P_2 = \left(\frac{A_2}{A_1} + 1\right) d\rho g = \left(\frac{A_2}{A_1} + 1\right) \rho g x \sin \theta$$

Since A_2 is much greater than A_1, the equation approximates to:

pressure difference = $\rho g x \sin \theta$

Thus for a particular angle and manometer liquid, the movement x of the liquid along the tube is proportional to the pressure difference. Since this movement is greater than would occur with the conventional U-tube manometer, greater accuracy is possible.

33.3.1 Diaphragms

With diaphragm pressure gauges (Figure 33.4), a difference in pressure between two sides of a diaphragm results in it bowing out to one side or the other. Diaphragms may be flat, corrugated or dished, the form determining the amount of displacement produced and hence the pressure range which can be measured. It also determines the degree of non-linearity. If the fluid for which the pressure is required is admitted to one side of the diaphragm and the other side is open to the atmosphere, the diaphragm gauge gives the gauge pressure. If fluids at different pressures are admitted to the two sides of the diaphragm, the diaphragm gauge gives the pressure difference.

There are a number of methods used to detect and give a measure of the deformation of the diaphragm. One form is the *capacitance diaphragm gauge* (Figure 33.5(a)) in which the diaphragm is between two fixed plates and its movement thus increases the capacitance with respect to one fixed plate and decreases it with respect to the other. Such a gauge is generally used with the two capacitors in opposite arms of an a.c. bridge, the out-of-balance voltage then being related to the pressure difference across the diaphragm. The range of capacitance diaphragm gauges is generally about 1 kPa to 200 kPa with an accuracy of about ±0.1% and a bandwidth up to 1 kHz. Another form of diaphragm pressure gauge is the *strain gauge diaphragm gauge*. This generally involves strain gauges being stuck on the diaphragm, or a silicon sheet used for the diaphragm with the strain gauges being formed in it by the introduction of doping material (Figure 33.5(b)). Whatever the form used, four strain gauges are generally used with two of the gauges being subject to tension and two to compression. These form the arms of a Wheatstone bridge and the out-of-balance voltage is then taken as a measure of the pressure difference across the diaphragm. Typically metal wire strain gauge instruments are used over the range 100 kPa to 100 MPa, with the integrated semiconductor gauge instrument over the range 0 to about 100 kPa. The accuracy is about ±0.1%.

Another form of diaphragm gauge is the *piezoelectric diaphragm gauge* (Figure 33.6). This consists essentially of a diaphragm which presses against a piezoelectric crystal. Movement of the diaphragm causes the crystal to be compressed and a potential difference is consequentially produced across its faces. Such a gauge can only be used

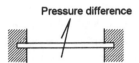

Pressure difference

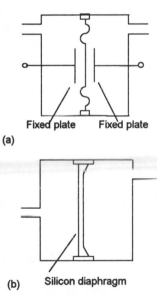

Figure 33.4 *Diaphragm pressure gauge*

Fixed plate   Fixed plate

(a)

(b)   Silicon diaphragm

Figure 33.5 *Diaphragm pressure gauges*

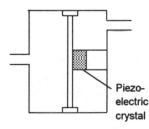

Piezo-electric crystal

Figure 33.6 *Piezoelectric pressure gauges*

for dynamic pressures and typically for pressures in the range 200 kPa to 100 MPa with a bandwidth of 5 Hz to 500 kHz.

33.3.2 Bourdon tubes

The *Bourdon tube* is a pressure gauge where an elastic deformation is produced as a result of pressure and is a tube which may be in the form of a spiral (Figure 33.7(a)) or a 'C' (Figure 33.7(b)). In all forms, an increase in the pressure in the tube causes the tube to straighten out to an extent which depends on the pressure. This displacement may be monitored in a variety of ways, e.g. to directly move a pointer across a scale, via gearing to move a pointer across a scale, to move the slider of a potentiometer. Bourdon tube instruments typically operate in the range 10 kPa to 100 MPa, the range depending on the form of the tube and on the material from which it is made. C-shaped tubes made from brass or phosphor bronze have a pressure range from about 35 kPa to 100 MPa. Spiral tubes are more expensive and have a greater sensitivity but as a consequence a lower maximum pressure that can be measured, typically about 50 MPa. Bourdon tubes are robust with an accuracy of about ±1% of full-scale reading. The main sources of error are hysteresis, changes in sensitivity due to temperature changes and frictional effects with linkages and pointers.

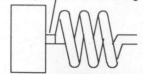
End of spiral attached to sliding contact

Rotary potentiometer
(a)

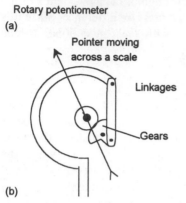
Pointer moving across a scale

Linkages

Gears

(b)

Figure 33.7   *Bourdon tubes*

33.4 Archimedes' principle

When an object is immersed in a fluid of density ρ, then the pressure p_2 on its bottom face must be greater than the pressure p_1 on its top face since it is at a greater depth h in the fluid, e.g. the cube shown in Figure 33.8, and thus $P_2 - P_1 = h\rho g$. Hence:

$$\frac{F_2}{A} - \frac{F_1}{A} = h\rho g$$

$$F_2 - F_1 = Ah\rho g$$

But Ah is the volume of the cube and thus $Ah\rho g$ is the weight of the fluid displaced by the cube. Thus there is an upthrust acting on an immersed object equal to the weight of fluid displaced. This is known as *Archimedes' Principle* and applies to all objects immersed in fluids, regardless of their shape.

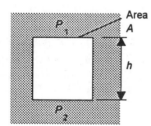
Area
A

P_1

h

P_2

Figure 33.8   *Upthrust*

Example

What will be the upthrust acting on an object of volume 100 cm³ when immersed in a liquid of density 950 kg/m³?

The upthrust is the weight of fluid displaced by the object and is thus given by $V\rho g = 100 \times 10^{-6} \times 950 \times 9.8 = 0.93$ N.

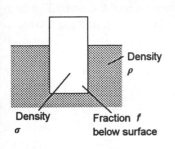

Density
ρ

Density
σ

Fraction f
below surface

Figure 33.9 *Floating*

33.4.1 Floating

When objects float in a fluid then the weight of the object is just balanced by the upthrust. Thus if an object of volume V floats in a liquid with a fraction f of its volume below the surface of the liquid (Figure 33.9), then the upthrust is $fV\rho g$, where ρ is the density of the liquid. If the object has a density r then its weight is $V\sigma g$ and $fV\rho g = V\sigma g$. Hence the fraction immersed is $f = \sigma/\rho$.

The density of ice is about 920 kg/m³ and that of sea water about 1030 kg/m³. Thus icebergs float with the fraction immersed of 920/1030 = 0.89. Nearly 90% of an iceberg is below the sea surface.

The simple *hydrometer* (Figure 33.10) is used for the measurement of the density if liquids. The hydrometer is floated in a liquid and the depth to which it floats is a measure of the density of the liquid. A scale on the stem enables the density to be directly read.

33.5 Thrust on an immersed surface

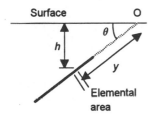

Scale

Weighted
bulb

Figure 33.10 *Hydrometer*

The thrust exerted on any surface in a fluid is always at right angles to that surface. Consider a plane surface immersed in a fluid (Figure 33.11), the surface being at an angle θ to the horizontal. For an elemental strip of area δA a distance y along the extended line of the plane from the surface, the pressure on the element is $\rho gh = \rho gy \sin \theta$, with ρ being the density of the fluid. The thrust on the element is thus:

$$\text{thrust on element} = \rho gy \sin \theta \, \delta A$$

The total thrust exerted on the plate will be the sum of the thrust on each elemental area:

$$\text{total thrust} = \rho g \sin \theta \times \text{sum of all the } y \, \delta A \text{ terms}$$

But the sum of all the $y \, \delta A$ terms is the first moment of area of the plate about the axis through O. Thus the sum is equal to the product of the entire plate area A and the distance of the plate centroid from O. Hence:

$$\text{total thrust} = \rho g \sin \theta \times \overline{y} A$$

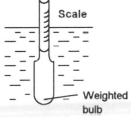

Surface   O

h

y

Elemental
area

Figure 33.11 *Immersed plane surface*

Taking the depth of the centroid below the surface to be $\overline{h} = \overline{y} \sin \theta$:

$$\text{total thrust } F = \rho g A \overline{h}$$

The total thrust is thus the product of the pressure at the centroid and the total area of the surface.

Example

Determine the resultant thrust acting on a vertical wall if the height of water one side is 3 m and on the other side is 8 m, the wall having a width of 6 m (Figure 33.12). The density of water is 1000 kg/m³.

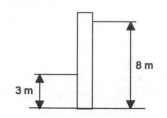

Figure 33.12 *Example*

Consider first the thrust on the left-hand side of the wall. The total thrust is the product of the pressure at the centroid and the total area of the surface and thus, as the centroid is at half the immersed height, is $\rho gAh/2 = 1000 \times 9.8 \times 3 \times 6 \times 3/2 = 2.65 \times 10^5$ N. The thrust on the right-hand side of the wall is $1000 \times 9.8 \times 8 \times 6 \times 8/2 = 18.82 \times 10^5$ N. Therefore the resultant horizontal thrust on the wall is $18.82 \times 10^5 - 2.65 \times 10^5 = 16.17 \times 10^5$ N.

33.5.1 Centre of pressure

The position at which the total thrust on an immersed surface can be considered to act is termed the *centre of pressure* (rather like the centre of gravity resulting from the weight of an object). The thrust on an elemental area is $\rho g y \sin \theta \, \delta A$ and so the moment of this thrust about O is $\rho g y \sin \theta \, \delta A \times y$. Hence the total moment of the thrust about O is:

total moment = $\rho g \sin \theta \times$ sum of all the $y^2 \delta A$ terms

But the sum of all the $y^2 \delta A$ terms is the second moment of area I_O of the surface about O. Hence:

total moment = $\rho g I_O \sin \theta$

If we consider the entire thrust F to be acting at a distance y_c from O, then:

$F y_c = \rho g I_O \sin \theta$

But, total thrust $F = \rho g A \overline{h}$, hence:

$$y_c = \frac{I_O}{A \overline{y}}$$

Thus the distance of the centre of pressure from O is the second moment of the area about O divided by the first moment of the area about O.

Example

Determine the position of the centre of pressure of a vertical rectangular plate of width b and depth d with one edge at the free surface (Figure 33.13).

The second moment of area of a rectangular plate about an axis through its centroid is $bd^3/12$, the centroid being at a depth $\frac{1}{2}d$. The theorem of parallel axes enables us to determine the second moment of area about O as $I_O = bd^3/12 + (bd)(\frac{1}{2}d)^2 = bd^3/3$. Hence:

$$y_c = \frac{I_O}{A \overline{y}} = \frac{bd^3/3}{bd \times \dfrac{d}{2}} = \frac{2d}{3}$$

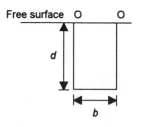

Figure 33.13 *Example*

33.6 Fluid flow

The term *steady flow* is used when, at every point across the cross-section of the flow, the same velocity is assumed to occur. The following analysis refers to conditions of steady flow.

33.6.1 Equation of continuity

When an incompressible fluid flows through a pipe, the volume of fluid passing through a section of the pipe is termed the *volume rate of flow*. If the cross-sectional area of the pipe is A and the fluid advances a distance s in time t then the volume moved in time t is As and the volume rate of flow is As/t. Thus, if steady flow velocity is v then the volume flow rate is Av. The *mass flow rate* is ρAv, where ρ is the fluid density.

Consider flow with an incompressible fluid through a pipe for which the cross-sectional area changes (Figure 33.14). With the steady flow, the mass entering a section of the pipe per second is constant and the mass entering a section must equal the mass leaving. Since the fluid is incompressible, the density is constant and so the volume entering a section per second must equal the volume leaving. The volume entering a section XX with cross-sectional area A_1 in time t at velocity v_1 is $A_1 v_1 t$. The volume leaving through YY in the same time is $A_2 v_2 t$, where A_2 is the cross-sectional area at YY and v_2 the velocity. Thus:

$$A_1 v_1 = A_2 v_2$$

This is termed the *equation of continuity*.

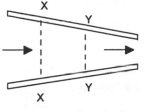

Figure 33.14 *Steady flow through a pipe*

33.6.2 Bernoulli's equation

Consider the steady flow of an incompressible fluid. The various forms of energy that the fluid can have are:

1 *Potential energy*

   This is the energy arising from a mass m of fluid being at some height z above some datum line. The potential energy is mgz and thus:

   potential energy per unit mass $= gz$

   Thus if a fluid enters a system at height z_1 it has a potential energy per unit mass of gz_1 and if it leaves at height z_2 a potential energy per unit mass of gz_2.

2 *Kinetic energy*

   This is the energy arising from the motion of a fluid. The kinetic energy of a mass m of fluid moving with velocity v is $\frac{1}{2}mv^2$ and thus:

   kinetic energy per unit mass $= \frac{1}{2}v^2$

   Thus if a fluid enters a system with velocity v_1 it has a kinetic energy per unit mass of $\frac{1}{2}v_1^2$ and if it leaves with velocity v_2 a kinetic energy per unit mass of $\frac{1}{2}v_2^2$.

3   *Displacement energy*

Displacement energy (or *pressure energy* as it is often termed) is the work done by a volume of fluid entering a system having to displace the volume ahead of it to progress. Work has to be done to move fluid through a section of a pipe against the pressure existing in the pipe at that section. For a cross-sectional area A at pressure p, the force needed is pA. If the fluid is to be pushed through a distance s against the fluid pressure then the work that has to be done is pAs and, since As is the volume V of liquid moved, then the work done is pV. If the volume flow rate is Q then the work done per second is pQ. The mass flow rate is ρQ and so in one second the mass flowing is ρQ and:

$$\text{work done per unit mass} = \frac{p}{\rho}$$

In general, the energy possessed by unit mass of a fluid will be the sum of its potential, kinetic and displacement energies:

$$\text{total energy per unit mass} = gz + \frac{v^2}{2} + \frac{p}{\rho}$$

Each of the energy terms has the unit of J/kg.

For steady flow of an incompressible fluid through a system, e.g. that shown in Figure 33.15, where the height and cross-sectional area may change, then as a consequence of the conservation of energy we must have the total energy at section 1 equal to the total energy at section 2. If there are no losses due to friction:

$$gz_1 + \frac{v_1^2}{2} + \frac{p_1}{\rho} = gz_2 + \frac{v_2^2}{2} + \frac{p_2}{\rho}$$

This is known as Bernoulli's equation. The equation can only be applied to real fluids if there is no significant amount of energy transferred to the fluid by heating, no external work is done, the flow can be assumed to be steady and the fluid is incompressible.

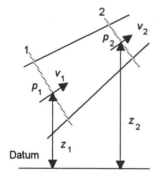

Figure 33.15   *Flow through a system*

Example

A horizontal pipe tapers uniformly from a diameter of 140 mm to 80 mm and carries oil of density 850 kg/m³. The pressure at the wider diameter section is measured as 80 kPa and that at the smaller diameter section as 50 kPa. Calculate the velocities at both sections and the volumetric rate of flow. Neglect any losses.

The equation of continuity $A_1v_1 = A_2v_2$ gives $\frac{1}{4}\pi \times 0.140^2 v_1 = \frac{1}{4}\pi \times 0.080^2 v_2$ and hence $v_1 = 0.327v_2$. As the height z is the same for both sections, the Bernoulli equation becomes:

$$\frac{v_1^2}{2} + \frac{p_1}{\rho} = \frac{v_2^2}{2} + \frac{p_2}{\rho}$$

$$\frac{(0.327v_2)^2}{2} + \frac{80 \times 10^3}{850} = \frac{v_2^2}{2} + \frac{50 \times 10^3}{850}$$

Thus $v_2 = 8.89$ m/s and so $v_1 = 2.91$ m/s. The volume rate of flow is A_1v_1, or A_2v_2, and hence $\frac{1}{4}\pi \times 0.140^2 \times 2.91 = 44.7 \times 10^{-3}$ m³/s.

Example

A large diameter open water tank discharges water to the atmosphere through a pipe at its base. The height of the water in the tank above the discharge pipe is 5.0 m. With what velocity and at volume per second will the water emerge through the discharge pipe if it has a diameter of 100 mm? Neglect any losses.

The pressures at the open surface in the tank and the outlet can be taken as being the same since they are open to the atmosphere and the difference in atmospheric pressure between the two will be negligible. The velocity at the upper surface in the tank can be assumed to be insignificant since it has a much larger diameter than the discharge pipe. Thus the Bernoulli equation becomes::

$$gz_1 = gz_2 + \frac{v_2^2}{2}$$

and so $v_2^2 = 2 \times 5.0 \times 9.8$ and $v_2 = 9.9$ m/s. The volume rate of flow $= A_2v_2 = \frac{1}{4}\pi \times 0.100^2 \times 9.9 = 0.078$ m³/s.

33.6.3 Measurement of flow rate

There are a range of methods that can be used to determine the flow rate of a fluid, a group of these involve the application of Bernoulli's equation. For a fluid flowing through a constriction in a pipe (Figure 33.16), since $A_1v_1 = A_2v_2 =$ the volume flow rate Q then a decrease in cross-sectional area from A_1 to A_2 means that v_2 is greater than v_1. Thus there is a gain in kinetic energy. For a horizontal pipe with energy conserved, this gain in energy can only occur from a drop in the displacement energy. But the displacement energy per unit mass is the pressure divided by the density. Thus there is a pressure drop at the constriction from P_1 to P_2.

For an incompressible fluid we can apply Bernoulli's equation, and since we assume a horizontal pipe and $z_1 = z_2$:

$$\frac{1}{2}v_1^2 + \frac{p_1}{\rho} = \frac{1}{2}v_2^2 + \frac{p_2}{\rho}$$

This can be rearranged to give:

$$\frac{v_2^2 - v_1^2}{2} = \frac{p_1 - p_2}{\rho}$$

Thus, using $A_1v_1 = A_2v_2 = Q$:

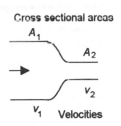

Cross sectional areas

A_1

A_2

v_2

v_1 Velocities

Figure 33.16 *Flow through a constriction*

$$Q = \frac{A_2}{\sqrt{1 - \left(\frac{A_2}{A_1}\right)^2}} \sqrt{\frac{2(p_2 - p_1)}{\rho}}$$

Thus a measurement of the pressure difference ($p_2 - p_1$) enables the volume flow rate Q to be determined. Note that the relationship between the pressure and the volume rate of flow is non-linear. There are a number of forms of differential pressure devices based on the above equation and involving constant size constrictions, e.g. the venturi tube, nozzles, Dall tube and orifice plate. In addition there are other devices involving variable size constrictions, e.g. the rotameter. The following are discussions of the characteristics of the above devices.

The *venturi tube* (Figure 33.17) has a gradual tapering of the pipe from the full diameter to the constricted diameter. The pressure loss occurring as a result of the presence of the venturi tube is about 10 to 15%, a comparatively low value. The pressure difference between the flow prior to the constriction and the constriction can be measured with a simple U-tube manometer or a differential diaphragm pressure cell. The instrument can be used with liquids containing particles, is simple in operation, capable of accuracy of about ±0.5%, has a long-term reliability, but is comparatively expensive and has a non-linear relationship between pressure and the volume rate of flow.

A cheaper form of venturi is provided by the *nozzle flow meter* (Figure 33.18). Two types of nozzle are used, the venturi nozzle and the flow nozzle. The venturi nozzle (Figure 33.18(a)) is effectively a venturi tube with an inlet which is considerably shortened. The flow nozzle (Figure 28.18(b)) is even shorter. Nozzles produce pressure losses of the order of 40 to 60%. Nozzles are cheaper than venturi tubes, give similar pressure differences, and have an accuracy of about ±0.5%. They have the same non-linear relationship between the pressure and the volume rate of flow.

The *Dall tube* (Figure 33.19) is another variation of the venturi tube. It gives a higher differential pressure and a lower pressure drop. The Dall tube is only about two pipe diameters long and is often used where space does not permit the use of a venturi tube.

The *orifice plate* (Figure 33.20) is simply a disc with a hole. The effect of introducing it is to constrict the flow to the orifice opening and the flow channel to an even narrower region downstream of the orifice. The narrowest section of the flow is not through the orifice but downstream of it and is referred to as the *vena contracta*. The pressure difference is measured between a point equal to the diameter of the tube upstream of the orifice and a point equal to half the diameter downstream. The orifice plate has the usual non-linear relationship between the pressure difference and the volume rate of flow. It is simple, reliable, produces a greater pressure difference than the venturi tube and is cheaper but less accurate, about ±1.5%. It also produces a greater pressure drop. Problems of silting and clogging can occur if particles are present in liquids.

The rotameter (Figure 33.21) is an example of a *variable area flow meter*; a constant pressure difference is maintained between the main

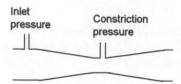

Figure 33.17 *Venturi tube*

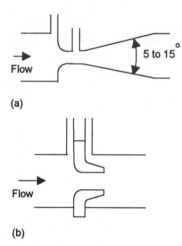

(a)

(b)

Figure 33.18 *Nozzles: (a) venturi, (b) flow*

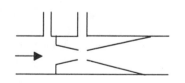

Figure 33.19 *Dall tube*

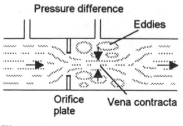

Figure 33.20 *Orifice plate*

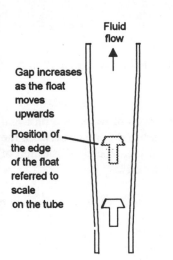

Fluid flow

Gap increases as the float moves upwards

Position of the edge of the float referred to scale on the tube

Figure 33.21 *Rotameter*

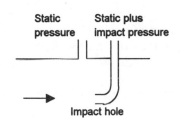

Static pressure

Static plus impact pressure

Impact hole

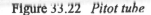

Figure 33.22 *Pitot tube*

flow and that at the constriction by changing the area of the constriction. The rotameter has a float in a tapered vertical tube with the fluid flow pushing the float upwards. The fluid has to flow through the constriction which is the gap between the float and the walls of the tube and so there is a pressure drop at that point. Since the gap between the float and the tube walls increases as the float moves upwards, the pressure drop decreases. The float moves up the tube until the fluid pressure is just sufficient to balance the weight of the float. The greater the flow rate the greater the pressure difference for a particular gap and so the higher up the tube the float moves. A scale alongside the tube can thus be calibrated to read directly the flow rate corresponding to a particular height of the float. The rotameter is cheap, reliable, has an accuracy of about ±1% and can be used to measure flow rates from about 30 × 10^{-6} m³/s to 1 m³/s.

The *Pitot tube* can be used to directly measure the velocity of flow of a fluid, rather than the volume rate of flow and consists essentially of just a small tube inserted into the fluid with an opening pointing directly upstream (Figure 33.22). The fluid impinging on the open end of the tube is brought to rest and the pressure difference measured between this point and the pressure in the fluid at full flow. The difference in pressure between where the fluid is in full flow and the point where it is stopped is due to the kinetic energy of the fluid being transformed to potential energy. The displacement energy, i.e. pressure energy, per unit mass for the fluid at the impact point is p_I/ρ, where p_I is the pressure at that point and ρ the density. In full flow at a point just prior to the impact where the fluid has a pressure p_s and a velocity v, the fluid has a displacement energy per unit mass p_s/ρ and kinetic energy per unit mass of $\frac{1}{2}v^2$. Thus, applying the principle of conservation of energy:

$$\frac{p_I}{\rho} = \frac{p_s}{\rho} + \frac{1}{2}v^2$$

$$v = \sqrt{\frac{2(p_I - p_s)}{\rho}}$$

The velocity is proportional to the square root of the pressure difference.

Example

A Venturi meter is used to measure the flow of oil, of density 820 kg/m³, in a horizontal pipe of diameter 120 mm. If the throat of the meter is 40 mm diameter, what will be the flow rate of oil if a manometer connected between the throat and the pipe indicates a pressure difference of 800 Pa. Assume there are no losses.

Since $A_1v_1 = A_2v_2$ then $d_1^2v_1 = d_2^2v_2$ and so $120^2v_1 = 40^2v_2$ and so, since $z_1 = z_2$, Bernoulli's equation becomes:

$$\frac{v_1^2}{2} + \frac{p_1}{\rho} = \frac{v_2^2}{2} + \frac{p_2}{\rho}$$

$$\frac{p_1 - p_2}{\rho g} = \frac{(120^2 v_1)^2}{2g(40^2)^2} - \frac{v_1^2}{2g}$$

Hence $v_1 = 0.16$ m/s and the volumetric rate of flow $= 0.0018$ m³/s.

Example

A Pitot tube has one outlet facing the stream of water in a pipe and the other at right angles to it. If the pressure difference between the two outlets gives a height difference in a manometer of 100 mm of water, what is the velocity of the water stream?

$$\frac{p_I}{\rho} = \frac{p_s}{\rho} + \tfrac{1}{2}v^2$$

$$v^2 = \frac{2(p_I - p_s)}{\rho} = \frac{2\rho g \Delta h}{\rho} = 2 \times 9.8 \times 0.100$$

and so the velocity is 1.4 m/s.

Activities

1 Put some lead shot or sand in the bottom of a test tube and a rolled strip of graph paper against the inner surface of the upper half of the test tube (Figure 33.23). Then, using liquids of different density, calibrate the arrangement as an hydrometer.
2 Use a flow meter to determine the volumetric rate of flow of water through a hose pipe.

Problems

1 What is the pressure at the base of a column of liquid of height 25 cm and density 1000 kg/m³ arising from the liquid?
2 A U-tube manometer has water, density 1000 kg/m³ as the manometric liquid. What will be the pressure difference between the two limbs of the instrument when there is a difference in water levels in the two limbs of 55 mm?
3 A rectangular tank has a base 100 cm by 40 cm and contains water to a depth of 30 cm. What will be (a) the pressure and (b) the force on the base due to the water? Water has a density of 1000 kg/m³.
4 What is the absolute pressure at a depth of 100 m in the sea, density 1030 kg/m³, when the atmospheric pressure is 101 kPa?
5 A U-tube manometer has mercury, density 13 600 kg/m³ as the manometric liquid. One of the limbs is open to the atmosphere. If the level of the mercury in the limb open to the atmosphere is 100 mm higher than that in the other limb, what is (a) the gauge pressure, (b) the absolute pressure in the closed limb. The atmospheric pressure is 101 kPa.
6 What will be the force required to keep a block of wood of volume 0.10 m³ below the surface of water of density 1000 kg/m³?
7 What will be the upthrust acting on an object of volume 0.10 m³ when immersed in a liquid of density 980 kg/m³?

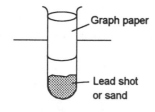

Figure 33.23 *Activity 1*

8 A hollow spherical ball is made of copper sheet and has an external diameter of 80 mm. If it just floats completely immersed in water, what is the thickness of the sheet? Copper has a density of 8800 kg/m³ and water 1000 kg/m³.

9 A rectangular dock gate has a width of 4 m. What will be the resultant horizontal thrust acting on the gate when the water is at a height of 5.5 m on one side and 3.5 m on the other? The density of the water can be taken as 1000 kg/m³.

10 A rectangular dock gate has a width of 12 m. What will be the resultant horizontal thrust acting on the gate when the sea is at a depth of 9 m on one side of the gate and 4.5 m on the other side? The density of sea water can be taken as 1025 kg/m³.

11 Determine the position of the centre of pressure of a vertical circular plane surface of diameter d when it is completely immersed with its upper edge at the free surface.

12 An open tank with vertical walls and a height of 3 m has an opening of diameter 1 m in a vertical side covered by a plate, the centre of the plate being 2 m below the top of the tank. If the tank contains water to a depth of 2.5 m, what will be the resultant force acting on the plate and the point at it which it may be considered to act? The density of the water can be taken as 1000 kg/m³.

13 Water, density 1000 kg/m³, flows down a pipe which tapers from 150 mm diameter to 100 mm diameter at the lower end. The difference in height between the two ends of the pipe is 3.0 m and the rate of flow is 0.80 m³/min. Calculate the pressure difference between the upper and lower ends of the pipe.

14 A horizontal pipe carries water, density 1000 kg/m³. The pipe tapers from 80 mm diameter to 50 mm diameter, the pressure at the wider end being 100 kPa and the other end 70 kPa. Calculate the velocity of the water at each end of the pipe and the volume passing through the pipe per second.

15 An open water tank discharges water, density 1000 kg/m³, to the atmosphere through a pipe that falls a distance of 12 m from the water level in the tank. The discharge pipe has a uniform diameter of 50 mm. What will be (a) the discharge velocity, (b) the volume discharged per second?

16 An open water tank discharges water, density 1000 kg/m³, to the atmosphere through a horizontal pipe of diameter 20 mm and 5 mm below the water surface. What will be the velocity of discharge and the volumetric rate of flow? Neglect any losses.

17 Water, density 1000 kg/m³, passes through a horizontal venturi meter which has an inlet diameter of 140 mm and throat diameter of 30 mm. If the pressure difference between the inlet and throat is 2.0 kPa, what is the velocity of the water entering the meter? Neglect any losses.

34 Engineering systems

34.1 Introduction

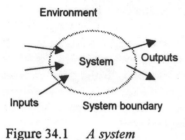

Figure 34.1   *A system*

This chapter is an introduction to a systems approach to engineering. If you want to use an amplifier then you might not be interested in the internal working of the amplifier but what output you can obtain for a particular input. In such a situation we talk of the amplifier being a system. A *system* can be defined as an arrangement of parts within some boundary which work together to provide some form of output from a specified input or inputs. The boundary divides the system from the environment and the system interacts with the environment by means of signals crossing the boundary from the environment to the system, i.e. inputs, and signals crossing the boundary from the system to the environment, i.e. outputs (Figure 34.1). With an engineering system an engineer is more interested in the inputs and outputs of a system than the internal workings of the component elements of that system.

34.2 Block diagrams

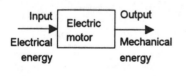

Figure 34.2   *Electric motor system*

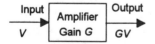

Figure 34.3   *Amplifier system*

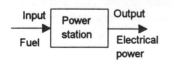

Figure 34.4   *Power station system*

A useful way of representing a system is as a *block diagram*. Within the boundary described by the box outline is the system and inputs to the system are shown by arrows entering the box and outputs by arrows leaving the box. Figure 34.2 illustrates this for an electric motor system; there is an input of electrical energy and an output of mechanical energy, though you might consider there is also an output of waste heat. The interest is in the relationship between the output and the input rather then the internal science of the motor and how it operates. It is convenient to think of the system in the box operating on the input to produce the output. Thus, in the case of an amplifier system (Figure 34.3) we can think of the system multiplying the input V by some factor G, i.e. the amplifier gain, to give the output GV.

34.2.1 Connected systems

We can consider a power station as a system which has an input of fuel and an output of electrical power (Figure 34.4). However, it can be more useful to consider the power station as a number of linked systems (Figure 34.5). Thus we can have the boiler system which has an input of fuel and an output of steam pressure. The steam pressure then becomes the input to the turbine to give an output of rotational mechanical power. This in turn becomes the input to the electrical generator system which gives an output of electrical power.

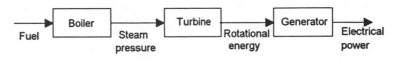

Figure 34.5 *Power station*

In drawing a system as a series of interconnected blocks, it is necessary to recognise that the lines drawn to connect boxes indicate a flow of information in the direction indicated by the arrow and not necessarily physical connections. As a further illustration, Figure 34.6 shows the basic form of a radio communication involving analogue signals. The input signal is the input to the modulator system which puts it into a suitable form for transmission. The signal is then transmitted before becoming the input to the receiver where it passes through the demodulator system to be put into a suitable form for reception. Because the input to the demodulator system is likely to be not only the transmitted signal but also noise and interference, the output signal from the modulator has added to it, at a *summing junction*, the noise and interference signal. A summing junction is represented by a circle with the inputs to quadrants of the circle given + or − signs to indicate whether we are summing two positive quantities or a summing a positive quantity and a negative quantity and so subtracting signals.

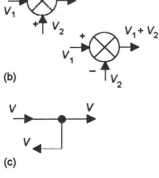

(a)

(b)

(c)

Figure 34.7 *Block diagram elements*

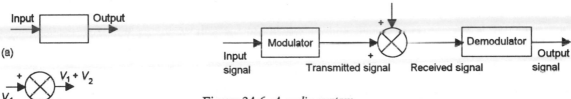

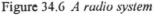

Figure 34.6 *A radio system*

Another element we will encounter in block diagram representations of systems is the *take-off point*. This allows a signal to be 'tapped' and used elsewhere in the system. For example, in the case of a central heating system, the overall output is the temperature of a room. But this temperature signal is also tapped off to become an input to the thermostat system where it is compared with the required temperature signal.

Figure 34.7 shows the three elements involved in block diagrams; these are the basic block (Figure 34.7(a)), a summing junction (Figure 34.7(b)) and a take-off point (Figure 34.7(c)).

34.3 Measurement systems

The purpose of a *measurement system* (Figure 34.8) is to give the user a numerical value corresponding to the variable being measured, e.g. a thermometer may be used to give a numerical value for a temperature.

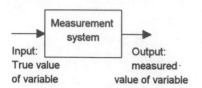

Figure 34.8 *A measurement system*

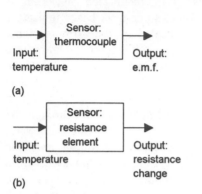

Figure 34.9 *Sensors: (a) thermocouple, (b) resistance thermometer element*

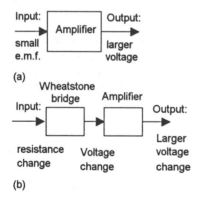

Figure 34.10 *Signal processing*

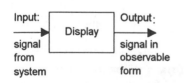

Figure 34.11 *Data presentation element*

A measurement system consists of three basic elemental systems which are used to carry out particular functions. These functional elements are:

1   *Sensor*
    This is the element of the system which is effectively in contact with the process for which a variable is being measured and gives an output which depends on its value, taking information about the variable being measured and changing it into some form which enables the rest of the measurement system to give a value to it. For example, a thermocouple is a sensor which has an input of temperature and an output of a small e.m.f. (Figure 34.9(a)). Another example is a resistance thermometer element which has an input of temperature and an output of a resistance change (Figure 34.9(b)).

2   *Signal processor*
    This element takes the output from the sensor and converts it into a form which is suitable for display or onward transmission in some control system. In the case of the thermocouple this may be an amplifier to make the e.m.f. big enough to register on a meter (Figure 34.10(a)). There often may be more than one item, perhaps an element which puts the output from the sensor into a suitable condition for further processing and then an element which processes the signal so that it can be displayed. Thus in the case of the resistance thermometer there might be a Wheatstone bridge, which transforms the resistance change into a voltage change, then an amplifier to make the voltage big enough for display (Figure 34.10(b)).

3   *Data presentation*
    This presents the measured value in a form which enables an observer to recognise it (Figure 34.11). This may be via a display, e.g. a pointer moving across the scale of a meter or perhaps information on a visual display unit (VDU). Alternatively, or additionally, the signal may be recorded, e.g. on the paper of a chart recorder or perhaps on magnetic disc.

Figure 34.12 shows how these basic functional elements form a measurement system.

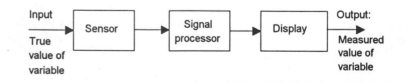

Figure 34.12 *Measurement system elements*

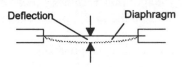

Diaphragm deflects because
pressure greater on other side
than this side

Figure 34.13 *Bending sensor*

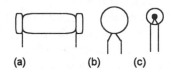

Figure 34.14 *Thermistors:
(a) rod, (b) disc, (c) bead*

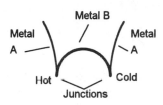

Figure 34.15 *Thermocouple*

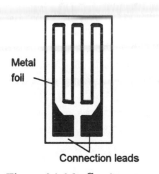

Figure 34.16 *Strain gauge*

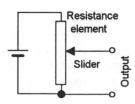

Figure 34.17 *Potentiometer*

34.3.1 Examples of system elements

The following are examples of commonly used sensors:

1 *Displacement of a rigid element*
   For example, in a length comparator an element, a rod or plunger, rests against an object and senses changes in position, the change in position being transformed into a movement of the rigid element.

2 *Extension or contraction of a spring*
   For example, a spring balance employs a spring to sense changes in force, the changes being transformed into a change in length of the spring.

3 *Bending or twisting of an elastic element*
   For example, some pressure gauges employ a diaphragm which bends as a result of a pressure difference between its two faces (Figure 34.13).

4 *Dimensional changes as a result of temperature changes*
   For example, the mercury-in-glass thermometer gives a volume of mercury as a result of a temperature change.

5 *Electrical resistance changes as a result of temperature changes*
   The sensor might be a metal coil or a semiconductor (thermistor, Figure 34.14) which changes in resistance when the temperature changes.

6 *Generation of an e.m.f. as a result of temperature changes*
   The thermocouple, a junction between two dissimilar metals, generates an e.m.f. related to the temperature of its junction. Figure 34.15 shows a thermocouple where the e.m.f. is related to the temperature difference between its two junctions.

7 *Electrical resistance changes due to strain changes*
   The electrical resistance strain gauge changes its resistance when subject to strain (Figure 34.16). It is stuck to the surface for which the strain is to be measured.

8 *Voltage changes due to movement of a slider*
   A potentiometer consists of a resistance element with a sliding contact which can be moved over the length of the element (Figure 34.17). With a constant supply voltage between the ends of the element, the voltage between the slider and one end is a measure of the position of the slider.

9 *Electrical inductance changes as a result of movement*
   For example, the linear variable differential transformer (LVDT) (Figure 34.18) is essentially a transformer with its primary and two secondary coils symmetrically spaced along a tube. When the central rod of magnetic material moves, the amount of magnetic material linking the primary coil with each of the two secondaries differs and thus the voltages in the two secondaries will be different. The

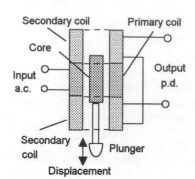

Figure 34.18 *LVDT*

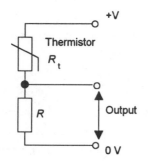

Figure 34.19 *Resistance to voltage conversion for a thermistor*

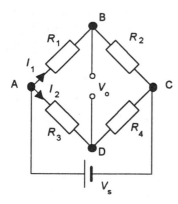

Figure 34.20 *Wheatstone bridge*

difference between these secondary voltages thus becomes a measure of the position of the central rod.

The following are examples of commonly used signal processors:

1 *Electronic amplification*
This is used to amplify small electrical signals, often for further processing.

2 *Conversion of a resistance change to a voltage change*
Figure 34.19 shows how a *potential divider circuit* can be used to convert the resistance change produced by a thermistor when subject to a temperature change into a voltage change. A constant voltage V is applied across the thermistor and another resistor in series. When the resistance of the thermistor changes, the fraction of the voltage across the fixed resistor changes. The output voltage is proportional to the fraction of the total resistance which is between the output terminals. Thus:

$$\text{output} = \frac{R}{R + R_t} V$$

Another method is the *Wheatstone bridge* Figure 34.20. The resistance element being monitored forms one of the arms of the bridge. When the output voltage V_o is zero, then there is no potential difference between B and D and so the potential at B must equal that at D. The potential difference across R_1, i.e. V_{AB}, must then equal that across R_3, i.e. V_{AD}, and so $I_1 R_1 = I_2 R_2$. We also must have the potential difference across R_2, i.e. V_{BC}, equal to that across R_4, i.e. V_{DC}. Since there is no current through BD then the current through R_2 must be the same as that through R_1 and the current through R_4 the same as that through R_3 and so $I_1 R_2 = I_2 R_4$. Dividing these two equations gives:

$$\frac{R_1}{R_2} = \frac{R_3}{R_4}$$

If R_1 is the sensor and the others are fixed resistances, then when R_1 changes so the bridge goes out-of-balance and produces an output voltage V_0 which is, for small changes, proportional to the change in resistance.

34.4 Control systems

A *control system* can be thought of as a system which for some particular input or inputs is used to control its output to some particular value or give a particular sequence of events or give an event if certain conditions are met. In this book we will consider just control systems being used to control outputs to particular values.

34.4.1 Open- and closed-loop control systems

Consider two alternative ways of heating a room to some required temperature. In the first instance there is an electric fire which has a selection switch which allows a 1 kW or a 2 kW heating element to be selected. The decision might be made that to obtain the required temperature it is only necessary to switch on the 1 kW element. The room will heat up and reach a temperature which is determined by the fact the 1 kW element is switched on. The temperature of the room is thus controlled by an initial decision and no further adjustments are made. This is an example of *open-loop control*. Figure 34.21 illustrates this. If there are changes in the conditions, perhaps someone opening a window, no adjustments are made to the heat output from the fire to compensate for the change. There is no information *fed back* to the fire to adjust it and maintain a constant temperature.

Now consider the electric fire heating system with a difference. To obtain the required temperature, a person stands in the room with a thermometer and switches the 1 kW and 2 kW elements on or off, according to the difference between the actual room temperature and the required temperature in order to maintain the temperature of the room at the required temperature. There is a constant comparison of the actual and required temperatures. In this situation there is *feedback*, information being fed back from the output to modify the input to the system. Thus if a window is opened and there is a sudden cold blast of air, the feedback signal changes because the room temperature changes and is fed back to modify the input to the system. This type of system is called *closed-loop*. The input to the heating process depends on the deviation of the actual temperature fed back from the output of the system from the required temperature initially set Figure 34.22 illustrates this type of system.

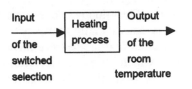

Figure 34.21 *Open-loop system*

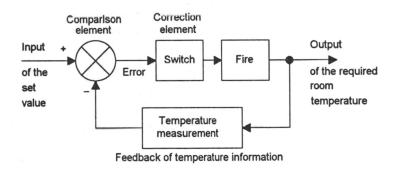

Figure 34.22 *Closed-loop system*

The comparison element in the closed-loop control system is represented by the summing symbol with a + opposite the set value input and a − opposite the feedback signal to give the sum

+set value − feedback value = error

This difference between the set value and feedback value, the so-called error, is the signal used to control the process. Because the feedback signal is subtracted from the set value signal, the system is said to have *negative feedback*.

In an open-loop control system the output from the system has no effect on the input signal to the plant or process. The output is determined solely by the initial setting. In a closed-loop control system the output does have an effect on the input signal, modifying it to maintain an output signal at the required value. Open-loop systems have the advantage of being relatively simple and consequently cheap with generally good reliability. However, they are often inaccurate since there is no correction for errors in the output which might result from extraneous disturbances. Closed-loop systems have the advantage of being relatively accurate in matching the actual to the required values. They are, however, more complex and so more costly with a greater chance of breakdown as a consequence of the greater number of components.

34.4.2 Basic elements of a closed-loop system

Figure 34.23 shows the general form of a basic closed-loop system, the following being the functions of the constituent elements:

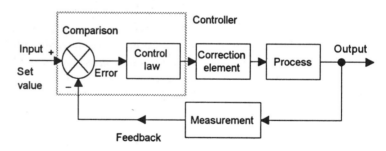

Figure 34.23 *Basic elements of a closed-loop control system*

1   *Comparison element*
    This element compares the required value of the variable being controlled with the measured value of what is being achieved and produces an error signal.

    Error = reference value signal − measured value signal

2   *Control law implementation element*
    The control law element determines what action to take when an error signal is received. The control law used by the element may be just to supply a signal which switches on or off when there is an error, as in a room thermostat, or perhaps a signal which is proportional to the size of the error and thus proportionally opens or closes a valve according to the size of the error. The term *control unit* is often used for the combination of the comparison element and

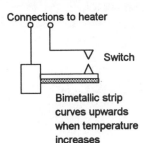

Figure 34.24 *Bimetallic thermostat switch*

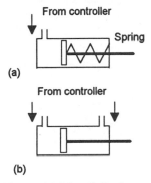

Figure 34.25 *Cylinders: (a) single acting, (b) double acting*

the control law implementation element. Thus it might be a thermostat switch which is used to compare the set value with the actual value and give an on–off control signal (Figure 34.24).

3   *Correction element*
The correction element or, as it is often called, the *final control element*, produces a change in the process which aims to correct or change the controlled condition. The term *actuator* is used for the element of a correction unit that provides the power to carry out the control action. An example of an actuator is a motor to correct the rotational speed of a shaft or perhaps, via possibly a screw, rotate and correct the position of a workpiece. Another example of actuator is an hydraulic or pneumatic cylinder (Figure 34.25). The cylinder has a piston which can be moved along the cylinder depending on a pressure signal from the controller. With a single acting cylinder, the pressure signal is used to move the piston against a spring; with the double acting cylinder the controller can increase the pressure on one side or the other of the piston and so move it in the required direction.

4   *Process element*
The process is what is being controlled, e.g. it might be a room in a house with its temperature being controlled.

5   *Measurement element*
The measurement element produces a signal related to the variable condition of the process that is being controlled. It might be a temperature sensor with suitable signal processing.

A *feedback loop* is a means whereby a signal related to the actual condition being achieved is fed back to modify the input signal to a process. The feedback is said to be *negative feedback* when the signal which is fed back subtracts from the input value. It is negative feedback that is required to control a system. *Positive feedback* occurs when the signal fed back adds to the input signal.

As a a example of a control system involving feedback, consider the motor system shown in Figure 34.26 for the control of the speed of rotation of the motor shaft. The input of the required speed value is by means of the setting of the position of the movable contact of the potentiometer. This determines what voltage is supplied to the comparison element, i.e. the differential amplifier, as indicative of the required speed of rotation. The differential amplifier produces an amplified output which is proportional to the difference between its two inputs. When there is no difference then the output is zero. The differential amplifier is thus used to both compare and implement the control law. The resulting control signal is then fed to a motor which adjusts the speed of the rotating shaft according to the size of the control signal. The speed of the rotating shaft is measured using a tachogenerator, this being connected to the rotating shaft by means of a pair of bevel gears. The signal from the tachogenerator gives the feedback signal which is then fed back to the differential amplifier.

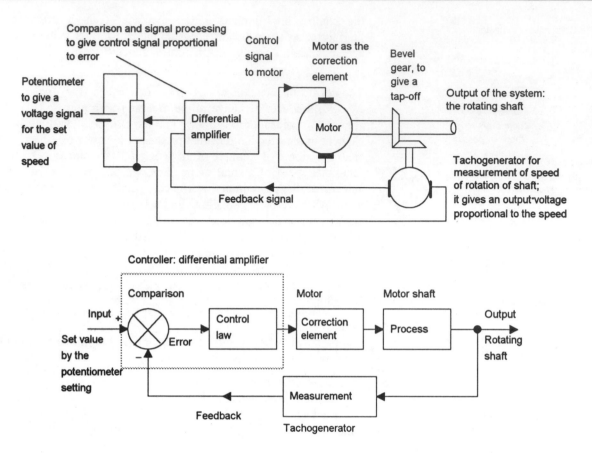

Figure 34.26 *Control of the speed of rotation of a shaft*

Figure 34.27 shows how the above system can be modified to enable it to be used to control the position of, say, a workpiece. The correction element is a motor with a gear and screw, the resulting rotation of the screw being used to give a displacement of the workpiece. The sensor that can be used to give a measurement signal is a rotary potentiometer coupled to the drive shaft of the screw, as the screw rotates it moves a slider over the resistance of the potentiometer and so gives a voltage related to the position of the workpiece.

Figure 34.28 shows a position control system using a hydraulic or pneumatic cylinder as an actuator to control the position of a workpiece. The inputs to the controller are the required position voltage and a voltage giving a measure of the position of the workpiece, this being provided by a potentiometer being used as a position sensor. The output from the controller is an electrical signal which depends on the error between the required and actual positions and is used, probably after some amplification, to operate a solenoid valve. When there is a current to the solenoid of the valve it switches pressure to one side or other of the cylinder and causes its piston to move and hence move the workpiece in the required direction.

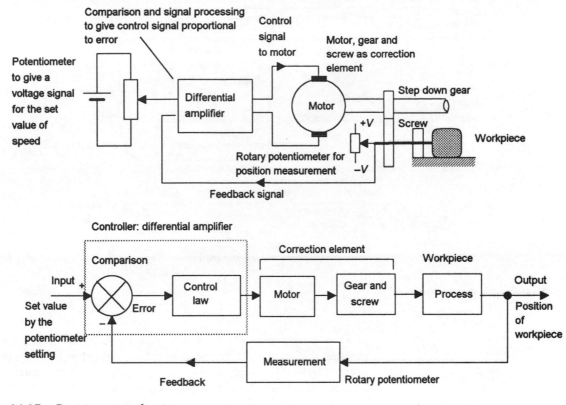

Figure 34.27 *Position control system*

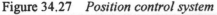

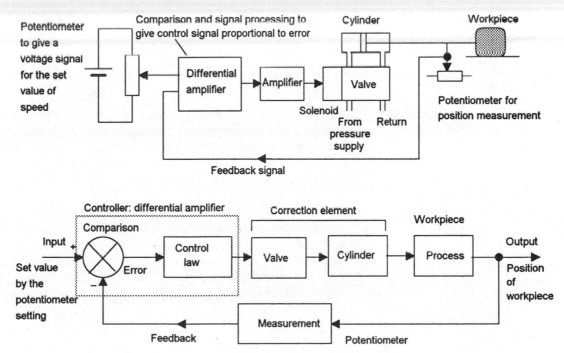

Figure 34.28 *Position control system*

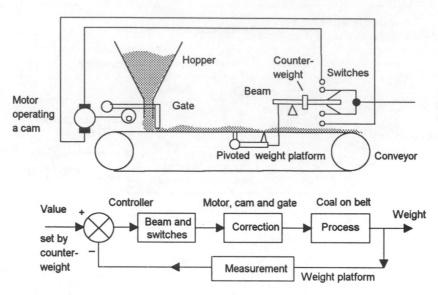

Figure 34.29 *Conveyor belt system*

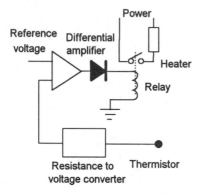

Figure 34.30 *Problem 4*

As a further example of a control system, Figure 34.29 shows a constant speed conveyor belt used to deliver a fixed weight of, say, coal per unit time. The weight of the coal on a segment of the conveyor belt applies a downward force on a platform and hence a beam. The beam is balanced, by means of a counterweight on the beam, to give a zero error signal when the required weight is passing over the platform. An error causes the beam to deflect and so trip on–off switches which cause a motor to operate and rotate a cam in such a direction as to either raise or lower the gate on the hopper and change the amount of coal being conveyed by the belt.

Problems

1 Identify the sensor, signal processor and display elements in the following systems: (a) pressure measurement with a Bourdon gauge, (b) temperature measurement with a mercury-in-glass thermometer, (c) temperature measurement with a resistance thermometer.

2 Explain the difference between open- and closed-loop control systems.

3 Identify the basic functional elements that might be used in the closed-loop control systems involved in (a) a temperature-controlled water bath, (b) a speed-controlled electric motor, (c) rollers in a steel strip mill being used to maintain a constant thickness of strip steel.

4 Figure 34.30 shows a temperature control system and Figure 34.31 a water level control system. Identify the basic functional elements of the systems.

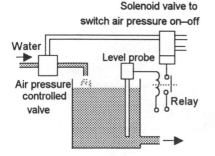

Figure 34.31 *Problem 4*

Answers

Chapter 1

1. ±0.25°C
2. ±0.02 A
3. +0.05 units
4. ±0.01%
5. (a) 50.4 s, (b) 2.13 mm, (c) 50.5 cm³
6. (a) 20×10^{-3} V, (b) 15×10^{-6} m³, (c) 230×10^{-6} A, (d) 20×10^{-3} m³, (e) 15×10^{-12} F, (f) 210×10^{9} Pa, (g) 1×10^{6} V
7. (a) 1.2 kV, (b) 0.2 MPa, (c) 200 dm³, (d) 2.4 pF, (e) 3 mA, (f) 12 GHz
8. (a) 13, (b) 0.21, (c) 0.014, (d) 19×10^{2}, (e) 0.0013
9. 31.6 g
10. 250 kg/m³, 0.250
11. 0.20 N

Chapter 2

1. (a) Chemical energy to heat energy to KE, (b) PE to KE, (c) chemical energy to KE to PE, (d) PE to KE, (e) PE to KE to electrical energy
2. 19.6 kJ
3. 6 000 J
4. 50 J
5. 110 250 J
6. 25 J
7. 150 m in direction of force
8. 50 J
9. 9.7 kJ
10. 1.98 kJ
11. 1.47 MJ
12. 7.84 J
13. 41.16 J
14. 960 N
15. (a) 7.2 MJ, (b) 18 kN
16. (a) 80 J, (b) 4 m/s
17. 83.3 kJ
18. 16 N
19. 1.25 m
20. 8 kN
21. 3.7 m/s
22. 50 W
23. 1470 kW
24. 24 kW
25. 1286 N
26. 90%
27. 147 kW
28. 0.32 N
29. 254.8 N, 254.8 N
30. (a) 2, (b) 3
31. 235 N

32. 10, 7.35, 73.5%
33. 8.3
34. (a) 24, (b) 16, (c) 67%

Chapter 3

1. 1.8 K
2. 2.3 K
3. 1.30×10^{7} J
4. 3.35 MJ
5. 430 kJ
6. 0.053 kg
7. 0.012 m
8. 0.0672 m
9. 167.5 mm³
10. 1.394 mm²
11. 7.997 m
12. 294 mm³
13. 44 mm³
14. 0.04668 m³
15. 0.70 m³
16. 0.4 m³
17. 0.012 m³
18. 0.56 m³
19. 0.022 m³
20. 1.13 m³
21. 0.32 m³
22. 84°C
23. 0.0010 m³
24. 0.819 kg
25. 125 kPa, 181 kPa
26. 0.118 kg
27. 442 kPa

Chapter 4

1. 80 cm
2. 38.7°
3. 15 cm
4. 19.5°
5. 1.4, 38.2°
6. 0.67
7. (a) 20 cm, real, (b) 3.8 cm, virtual, (c) 40 cm, real, (d) 24 cm, virtual, (e) 9.3 cm, virtual, (f) 6 cm, virtual, (g) 4 cm, virtual
8. (a) 0.4 cm, real, (b) 0.67 cm, virtual, (c) 6 cm, virtual
9. 90 cm
10. 15.6 cm, 25.6
11. 24.4 cm
12. 1.33 m
13. Sound path bends upwards over the sea

Chapter 5

1. (a) 6.08 N at 34.7° to the horizontal, (b) 6.71 N at 26.6° west of north, (c) 7.61 N at 28.5° to 5 N force, (d) 10.2 N at 13.0° to 8 N force, (e) 11.6 N at 12° to 7 N force, (f) 17.9 N at 18.9° to 10 N force, (g) 13.5 N at 45.7° to 12 N force
2. 3.17 kN at 9.4° to 1.2 kN force
3. 21 N, 16 N
4. 14.0 N
5. 7.3 N, 16.3 N
6. $F_1 = 1.2$ kN, $F_2 = 2.3$ kN
7. 35.4°
8. 44.7 kN at 26.6° to the 40 kN force
9. 58° and 48° with respect to the 90 N force
10. 22.8 kN at 70.3° to the horizontal
11. 197 N
12. (a) 7.7 N, 6.4 N, (b) 5.1 kN, 14.1 kN, (c) 11.3 N, 4.1 N, (d) 5.2 kN, 29.5 kN
13. 28.2 N, 10.3 N
14. 12.3 kN, 8.6 kN
15. 5.0 N, 8.7 N
16. 2.0 m from the 2 kg mass, 58.9 N
17. 1.3 kN and 3.7 kN
18. 125 N, 120 N
19. 105 N
20. 100 mm from one end, in centre of the section
21. 300 mm from the thicker end
22. (a) 10 mm from the base along line bisecting base and joined to the opposite apex, (b) 217 mm along central axis from left end, (c) 19 mm above base and 17 mm from the right end, (d) 250 mm above base on central axis
23. 4.6 m
24. 1.6 m from load pivot
25. 2.72 N, 2.18 N
26. 20 N, 30 N
27. 3.6 m from 35 kg mass
28. 48.75 N, 16.25 N
29. 27.5 N, 17.5 N
30. 225 kN, 135 kN

Chapter 6

1. 20 MPa
2. 3.1 MPa
3. $6.7 \times 10^{-3} = 0.67\%$
4. (a) 127.3 MPa, (b) 0.0006, (c) 212 GPa
5. 2.9×10^{-6}
6. 0.955 mm
7. 204 GPa

8   50 kN
9   (a) 456 MPa, (b) 410 MPa,
     (c) 200 GPa
10   (a) 480 MPa, (b) 167 GPa
11   79 GPa, 145 MPa
12   31.1%
13   1 MN
14   48.6 mm
15   12%
16   Diameter 143 mm, thickness 32 mm
17   9.5×10^{-6} m, 0.0048 J
18   38.1 kJ

Chapter 7

1   30 m/s
2   400 m
3   20 s
4   (a) 7.5 s, (b) 112.5 m
5   3 m/s
6   0.2 m/s^2
7   12 m/s
8   500 m
9   20 s
10   10.1 m, 5.2 m/s
11   0.51 s
12   31.3 m/s
13   9.8 m/s
14   1.6 s
15   9.9 m/s
16   4.1 m, 0.92 s
17   22.5 m
18   176 m
19   (a) 34 mm/s, (b) 50 mm/s, (c) 66 mm/s
20   (a) 8 m/s, (b) 11 m/s, (c) 16 m/s
21   (a) 10 m/s^2, (b) 10 m/s^2, (c) 11.25 m
22   1.2 m/s^2, 12.6 m
23   -4 m/s^2, 54 m
24   (a) Distance–time non-zero gradient
     straight line, velocity–time straight line
     with zero gradient, (b) distance–time
     curved, velocity–time non-zero gradient
     straight line, (c) as (a), (d) as (b)
25   300 m

Chapter 8

1   (a) 628.7 rad, (b) 95.5 rev, (c)
     94.2 rad/s, (d) 1.27 rev/s, (e) 9.42 rad/s
2   0.63 rad.s
3   3.49 rad/s
4   8.64 rad/s^2, 25
5   0.21 rad/s^2, 5
6   0.84 rad/s^2, 19.2 rev
7   0.31 rad/s^2, 7.5 revs
8   1.26 rad/s^2
9   1.05 m/s
10   (a) 0.57 rad/s^2, (b) 0.10 m/s^2
11   2.1 m/s
12   72.7 rad.s
13   7.54 m/s
14   20 rad/s
15   20 rad/s^2
16   20 N m
17   60 N m
18   5.3 rev/s

19   754 kW

Chapter 9

1   28 kg
2   1.31 N, 3.27 m/s^2
3   1248 N
4   (a) 5.4 m/s^2, (b) 10.9 N
5   0.24 m/s^2
6   3.14 kN
7   6.4 m/s
8   3 m/s in opposite direction
9   3.1 m/s
10   (a) 4.0×10^3 kg m/s, (b) 2.0×10^5 N
11   134 N
12   0.58
13   (a) 58.8 N, (b) 64.8 N
14   3.0 m/s
15   9 N
16   1.1 m

Chapter 10

1   No
2   (a) 200 V, (b) 2000 Ω
3   (a) 19 to 21 Ω, (b) 61.2 to 74.8 Ω, (c)
     37.6 to 56.4 Ω
4   (a) 680 Ω, 5%, (b) 473 Ω, 10%, (c)
     685 Ω, 5%
5   (a) 4k7, 5R6, (b) 6.8 MΩ, 4.7 Ω
6   50 mA
7   11.6 V
8   >1.6 W
9   0.20 A, 4.0 V, 6.0 V, 10.0 V
10   750 Ω
11   500 Ω and 1900 Ω in series
12   (a) 0.41 W, 1.02 W, 1.43 W, (b) 5 W,
     2 W, 7 W
13   (a) 30 Ω, (b) 2.73 Ω
14   (a) 6.32 kΩ, (b) 0.0911 W
15   6 Ω
16   (a) 0.25 A, (b) 2 V, 3 V
17   (a) 1 A, 0.33 A, (b) 5.33 W
18   6.9 mA
19   2.03 V, 9.97 V
20   0.57, 0.29, 0.14
21   (a) 22 Ω, 0.55 A, (b) 6 Ω, 2 A, (c)
     12.8 Ω, 0.94 A
22   1.875 A
23   2.22 A
24   (a) 0.22 A, (b) -0.09 A, (c) 0.22 A
25   (a) 0.51 Ω, (b) 9995 Ω
26   (a) 0.0200 Ω, (b) 9990 Ω
27   29.4 V
28   1966.7 Ω, 100 Ω
29   0.0125 Ω, 4.0×10^7 Ω m
30   31.2 m

Chapter 11

1   (a) 1.0 T, (b) 1.4 T
2   0.20 T
3   4×10^{-4} Wb
4   1.5 V
5   8000 V
6   (a) 10 V, (b) 5 V
7   0.52 mV

8   (a) 20 V, (b) 10 V
9   60 V
10   250 V, 4.0 A
11   (a) 1.0 N/m, (b) 0.87 N/m, (c) zero
12   0.21 N
13   0.28 N m
14   21

Chapter 12

1   (a) 30 Ω, 0.5 A, (b) 10.8 Ω, 1.11 A, (c)
     24 Ω, 0.5 A, (d) 20 Ω, 0.5 A, (e)
     4.25 Ω, 2.8 A, (f) 1.2 Ω, 10 A
2   2/5 V_s
3   (a) 1 A, 5 V; 0.5 A, 5 V; 0.5 A, 5 V, (b)
     1 mA, 8 V; 0.67 mA, 4 V; 0.33 mA,
     1.3 V; 0.33 A, 2.7 V, (c) 0.2 A, 12 V;
     0.2 A, 12 V; 0.1 A, 10 V; 0.1 A, 10 V;
     0.2 A, 10 V
4   2.5 V
5   2.5 A
6   0.8 mA
7   (a) 0.22 A, (b) -0.09 A, (c) 0.22 A
8   0.353 A, 0.078 A, 0.4 A, 0.431 A,
     0.322 A
9   6.0 V, 0.4 Ω
10   4.0 V
11   3.33 V, 0.17 Ω
12   2.0 V, 0.05 Ω
13   (a) 0.32 A, 0.09 A, 0.23 A, (b) 1.25 A,
     0.63 A, 0.68 A, (c) 0.89 A, 2.07 A,
     1.18 A, (d) 0.73 A, 0.11 A, 0.62 A
14   (a) 3 A, (b) -1.0 A
15   (a) 16 V, 140 Ω, (b) 8 V, 6.6 Ω
16   (a) 0.5 A, (b) 0.71 A
17   (a) 0.11 A, 140 Ω, (b) 1.2 A, 6.6 Ω
18   (a) 0.5 A, (b) 0.71 A
19   (a) 12 mA, (b) 10 mA

Chapter 13

1   4×10^4 V/m
2    $+60$ μC, -60 μC
3   4 μF
4   4 V
5   (a) $+16$ μC, -16 μC, (b) 8 V, 4 V
6   (a) 1.1 μF, (b) 14 μF
7   (a) 6 μF, (b) $+10$ μC, -10 μC, $+20$ μC,
     -20 μC, $+30$ μC, -30 μC, (d) 10 V
8   (a) 10 μF, (b) 360 μC, (c) 18 V, 12 V,
     6 V
9   (a) 370 pF, (b) 3700 pC, (c) 10 V
10   885 pF
11   3.77 m^2
12   2.1×10^{-7} m
13   7.5 m
14   44.25 pF
15   0.0531 μF
16   0.021 mm
17   1.6 kV
18   6 V
19   7.2 mJ
20   20 μF
21   (a) 1 μF, (b) 10 μC, (c) 5 V, 10/3 V,
     10/6 V, (d) 25 μJ, 16.7 μJ, 8.3 μJ
22   125 μJ, 89.3 μJ

23   (a) 88.5 pF, (b) 4.425×10^{-5} J

Chapter 14

1   (a) 2 mA, 1.86 mA, (b) 0, 0
2   (a) 6.7 Ω, (b) 1.2 Ω
3   ±0.1 V
4   8.3 Ω, 1.3 Ω
5   See Section 14.4; base is lightly doped and thin so few electrons lost by recombining with holes
6   (a) 0.995, (b) 199
7   8.64 mA
8   0.99
9   26 μA, 1.3 mA, 1.27 mA
10   1.004 mA
11   (a) 0.998, (b) 82.3
12   (a) 10 V, (b) 9.9 mA, (c) 0.1 V
13   10 kΩ
14   5 kΩ

Chapter 15

1   1000 ampere-turns
2   2000 A/m, 1.26 T, 37.7 mWb/m²
3   0.66 A
4   348
5   1.27×10^6 A/Wb, 5.1 A
6   2.65×10^6 A/Wb
7   11.9 mWb
8   5.96 A
9   0.77 T
10   4456, 5730, 4377, 3395
11   800 A/m
12   503 ampere-turns
13   1.75 A
14   1751, 1421, 1074, 769, 597
15   350 ampere-turns
16   3 A

Chapter 16

1   2 H
2   0.4 V
3   0.01 H
4   0.02 H
5   90 μWb
6   3.35 mH
7   9.87 mH
8   0.18 mJ
9   0.0225 J
10   −0.5 V
11   −0.5 V
12   1 mH
13   60%
14   0.29
15   0.15 V
16   0.24 V

Chapter 17

1   (a) $v = 10 \sin 2\pi 50t$ V, (b) (i) 5.88 V, (ii) 9.51 V, (iii) −5.88 V
2   (a) $i = 50 \sin 2\pi \times 2 \times 10^3 t$ mA, (b) (i) −47.6 mA, (ii) −29.4 mA, (iii) 47.6 mA
3   (a) 0, (b) 10 V, (c) 1.99 V

4   0, 0.20, 0.39, 0.56, 0.72, 0.84, 0.93, 0.99, 1.0
5   (a) 1 V, (b) 0, (c) 1 A, (d) 1 A, (e) 0.5 V, (f) 1 V
6   As given in the problem
7   (a) 1.27 A, (b) 0
8   (a) 5 V, (b) 0
9   As given in the problem
10   As given in the problem
11   2.83 V
12   15 mA
13   8.0 V
14   As given in the problem
15   7.5 mA, 19 mA, 2.53
16   4 A
17   (a) 6.50 V, (b) 7.14 V, (c) 1.10
18   (a) 15.0 mA, (b) 17.1 mA, (c) 1.14
19   As given in the problem
20   As given in the problem
21   5 V, 7.07 V, 1.4
22   1 V
23   (a) 1.4 V, (b) 7.07 V
24   (a) 7.1 mA, (b) 2.12 A
25   (a) 141.4 mA, (b) 5.66 A
26   10 V
27   See text
28   0.25 V
29   11.25 V
30   156.25 Hz

Chapter 18

1

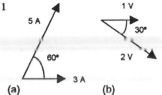

(a)   (b)

2   (a) 628 Ω, (b) 15.9 mA
3    $i = 0.05 \sin (200t - 90°)$
4   (a) 0.157 Ω, 3.14 Ω
5   (a) 314.2 Ω, (b) 0.080 A
6   750 sin (150t + 90°) V
7   (a) 79.6 Ω, (b) 126 mA
8   1.41 A
9   (a) 31 831 Ω, (b) 1592 Ω
10   (a) 5.31 Ω, (b) 3.77 A
11   252.1 V, 85.4° leading current
12   1.46 A, 72.3° lagging
13   25 Ω, 373 mH
14   21.7 Ω, 11.0 A, 46.4° current lagging the voltage
15   17 Ω, 28.1° voltage leading current
16   (a) 75 Ω, (b) 1.33 A, (c) 59.9 V, 79.8 V
17   (a) 24.1 Ω, (b) 4.15 A at 51.5° lagging the voltage
18   (a) 50 Ω, (b) 79.6 μF
19   148 μF
20   (a) 75.2 Ω, (b) 3.19 A, 57.9° leading the voltage

21   (a) 48.0 Ω, (b) 0.5 A, 33.6° leading the voltage
22   (a) 70.7 V, 45° current lagging voltage, (b) 50 V, 100 V, 50 V
23   (a) 0.86 A, 49.9° lagging the voltage, (b) 174 V, (c) 86 V
24   356 Hz
25   (a) 8.4 μF, (b) 9.4
26   (a) 118.6 Hz, (b) 11.2
27   9.28 A, 860 W
28   13.6 mW
29   (a) 0.53, (b) 8.15 W, (c) 15.3 V A, (d) 10.8 V Ar
30   133.3 kV A, 106.7 kV Ar
31   0.8
32   1536 W
33   1.5 kW, 1.2 kW

Chapter 19

1   28.4 A, 32.5° lagging the voltage
2   5 A, 53.1° lagging the voltage
3   15.5 mA, 30.8° lagging
4   200 Ω, 5.3 μF
5   (a) 1.73 A, 1.0 A, (b) 57.7 Ω, 100 Ω
6   (a) 167 mA, (b) 100 mA, (c) 195 mA
7   7118 Hz
8   0.85 to 2.3 MHz
9   2251 Hz, 100 kΩ, 1 mA, 141
10   59.3 Hz, 200 Ω, 0.5 A, 4.6
11   (a) 38.7 mA, 83.1° lagging, (b) 9.12, (c) 1.11 W, (d) 9.28 VA, (e) 9.21 VAr
12   93.3 W
13   7.3 μF
14   15.4 μF
15   (a) 8.04, (b) 58.0 μF

Chapter 20

1   20 000 V/s
2   30 V/s
3   17.3 V
4   (a) 0.4 mA, (b) 0.147 mA
5   (a) 0.3 ms, (b) 0.736 mA, (c) 0.446 mA, (d) 0
6   (a) 30 V, (b) 11.0 V, (c) 1.07 V
7   (a) 0.25 s, (b) 6.3 V, (c) 8.6 V
8   (a) 1.29 s, (b) 1.07 mA
9   (a) 5 μA, (b) 4.6 s
10   (a) 0.767 V, (b) −0.767 V
11   (a) 0.05 s, (b) 40 A/s, (c) 1.73 A, (d) 2 A
12   0.35 A
13   (a) 0.1 s, (b) 40 A/s, (c) 3.1 A, (d) 4 A
14   73 mA
15   (a) 5 A, (b) 0.092 A
16   4.6 ms
17   (a) 80 V, (b) 2.75 A/s
18   (a) 0.95 A, (b) 0.69 s
19   0.1 H
20   20 H

Chapter 21

1   (a) 254 V, (b) 440 V

2 (i) (a) 240 V, (b) 8 A, (c) 8 A, (ii) (a) 240 V, (b) 10 A, (c) 10 A, (iii) (a) 240 V, (b) 20.4 A, (c) 20.4 A
3 (i) (a) 415 V, (b) 23.0 A, (c) 39.9 A, (ii) (a) 415 V, (b) 8.3 A, (c) 14.4 A
4 (a) 312 V, (b) 1.2 A, (c) 1.2 A
5 4.4 A lagging V_s by 143°
6 (i) 14.4 kW, (ii) 10.95 kW
7 (i) 3.2 kW, (ii) 34.6 kW, (iii) 3.2 kW
8 1955 W, 56%
9 (a) 254 V, (b) 5.08 A, (c) 5.08 A, (d) 3090 W, (e) 0.8
10 (a) 440 V, (b) 5.46 A, (c) 9.46 A, (d) 4480 W, (e) 0.62
11 53.5 A
12 8626 W
13 9.9 A
14 4.4 kW, 0.28
15 400 kW, 0.76
16 −584 W, 2180 W

Chapter 22

1 840 V
2 (a) 1.33, (b) 2.25 A
3 (a) 75 V, (b) 5 A, (c) 1.5 A, (d) 375 W
4 (a) 1.7 A, 20.8 A, (b) 240 V
5 1920 V
6 (a) 25 V, (b) 600 ampere-turns
7 0.42 A, 8.33 A
8 (a) 91%, (b) 90%
9 (a) 98%, (b) 97%
10 287 W, 530 W
11 97%
12 100 A
13 (a) 50 A, (b) 5 A
14 953 V, 69.4 A
15 1/4.47
16 1/4.0
17 11
18 250 Ω
19 (a) 0.375 W, (b) 0.5 W, 1/3
20 25 Ω, 10 mW

Chapter 23

1 1.2 N m
2 94 V
3 228 V
4 Large start-up current
5 As given in the problem
6 (a) 50 A, (b) 1.1 Ω
7 (a) 50 A, (b) 0.9 Ω
8 8 Ω
9 6.3 Ω
10 750 rev/min
11 3.7 Ω
12 0.033
13 39.9 rev/s
14 1450 rev/min
15 (a) 25 rev/s, (b) 23.75 rev/s
16 (a) 12.5 rev/s, (b) 12.2 rev/s
17 Higher resistance-to-reactance ratio than main winding, see Section 21.4
18 To produce phase difference, see Section 21.4

19 To give phase difference between the two parts of the field, see Section 21.4.1

Chapter 24

1 102 MPa, 144 MPa
2 0.15 mm
3 150 MPa, 270 kN
4 3.6 MPa, 51.5 MPa
5 34 MPa, 3.4 MPa
6 3.5 MN
7 144 MPa compression
8 2.4 MPa
9 14.7 MPa compression, 18.3 MPa tension
10 53 MPa compression, 34 MPa tension

Chapter 25

1 (a) 62.5 MPa, (b) 8.2×10^{-4}
2 8.0 mm
3 140 kN
4 (a) 162 kN, (b) 324 kN
5 4.97 MPa
6 25.1 kN
7 42 mm
8 34.5 kN m
9 1.7 kN m
10 1.8 kN m
11 0.025 rad
12 5.7 MPa
13 165 MPa
14 140 mm
15 150 mm
16 73 MPa
17 (a) 65.2 MPa, (b) 149.4 GPa, (c) 4.36 $\times 10^{-4}$
18 33.3 MPa, 46.9 MPa
19 10 kN m
20 4.7 MPa
21 15.8 MPa
22 501 kW
23 17.7 rev/s
24 4.4 mm
25 127 mm, 0.57°

Chapter 26

1 About 6.0 N at 53.5°
2 7.2 N at 9° east of north
3 45.6 N at 61.6° from the horizontal
4 (a) 2.06 N at 14° south of east, (b) 4.17 N at 4.5° north of east, (c) 6.85 N at 49.9° south of east
5 34.3 N at 27.8° west of north
6 5.3 N at 33.8° east of south
7 4.1 N at 5.8° north of west
8 (a) F_{CD} +8.7 kN, F_{AD} −17.3 kN, F_{BD} +10.0 kN, (b) F_{FE} −16.4 kN, F_{BF} −20 kN, F_{AF} −28.3 kN, F_{EC} −41.7 kN, F_{ED} +34.9 kN
9 (a) F_{BD} − 5.2 kN, F_{AD} −3.0 kN, F_{CD} +2.6 kN, (b) F_{AD} +250 N, F_{CD} +600 N, F_{BD} −650 N, (c) F_{AE} −14.4 kN, F_{EF} +2.9 kN, F_{DE} +7.3 kN, F_{DG} +10.1 kN, F_{BF} −8.7 kN, F_{FG} −20.2 kN, F_{CG}

−20.2 kN, (d) F_{AF} −54.8 kN, F_{EF} +26.4 kN, F_{FG} +43.3 kN, F_{BG} −64.9 kN, F_{DH} +36.1 kN, F_{CH} −72.2 kN, F_{OH} +26.0 kN, (e) F_{AD} −17.3 kN, F_{CD} +8.7 kN, F_{BE} −10 kN, F_{CE} +8.7 kN, F_{DE} 0
10 (a) −25.5 kN, (b) +8.5 kN, (c) −23.1 kN
11 (a) Stable, (b) unstable, (c) redundancy, (d) unstable

Chapter 27

1 (a) −250 N, +250 N m, (b) −250 N, +375 N m
2 (a) + 8 kN, −12 kN m, (b) +8 kN, −8 kN m
3 +20 kN, −20 kN m
4 (a) +0.75 kN m, −1.5 kN, (b) +1.5 kN m, −1.5 kN, (c) +2.25 kN m, 0
5 (a) 24 kN m fixed end, (b) 12 kN entire length
6 (a) 90 kN m at fixed end, (b) 60 kN at fixed end
7 97 mm from base
8 89.5 mm from base
9 14.0×10^6 mm^4
10 4.17×10^6 mm^4
11 (a) 325.5×10^6 mm^4, (b) 6.25×10^6 mm^4, (c) 3.07×10^5 mm^4, (d) 1.25×10^6 mm^4, (e) 130.3×10^6 mm^4
12 0.95 mm
13 350 MPa
14 4.2 m
15 5.9 N m, 79 MPa
16 15 kN m
17 205 MPa
18 268 mm
19 5.25 kN
20 3.57×10^{-3} m^3
21 44 kN/m
22 40 MPa
23 61 MPa

Chapter 28

1 178 N, increased by a factor of 4
2 80 N
3 1.4 rev/s
4 205 N, 216 N
5 8.85 m/s
6 17.1 m/s
7 64°
8 0.16 m
9 12.1 m/s
10 14 N
11 0.39
12 1.1
13 1.49 m, 53 N
14 35.5 N, 24.8 cm

Chapter 29

1 0.16 kg m^2
2 4.29 kg m^2

3   1.196 kg m^2
4   25.65 kg m^2
5   1.44 rad/s^2
6   271 N m
7   3.8 N m
8   96.6 rev/min
9   2527 J
10   21.4 rev/s
11   (a) 81.25 kg m^2, (b) 196 J
12   0.6 kg m^2

Chapter 30

1   (a) 3.3, (b) 48%, (c) 0.2 kN
2    $E = 0.21F + 3.6$ N
3   80%
4   65%
5   10
6   25
7   (a) 40, (b0 2880 N
8   4 rev/s
9   6
10   0.42, 12 rev/s
11   25
12   31.8 N m, 159.2 N m
13   189 N m
14   83.3%
15   18.8 m/s
16   18.9 kW
17   (a) 800 N m, (b) 25.1 kW
18   (a) 29.4 N m, (b) 2.77 kJ

Chapter 31

1   1137 m/s^2 at $x = A$, 15 m/s at $x = 0$
2   0.5 m/s, 7030 m/s$^?$
3   2.25 Hz
4   3.0 m/s, 56.8 m/s^2
5   4.44 s

6   3.1 m/s, 2.7 m/s
7   26.6 N
8   4.1 Hz
9   3.5 Hz
10   790 N/m
11   1.002 s
12   2.2 m
13   (a) 2.0 s, (b) 0.31 m/s

Chapter 32

1   3.6 kW
2   645 W
3   20 kW
4   76 W
5   29.5 W
6   157 W
7   3 kW
8   84 W

Chapter 33

1   2.45 kPa
2   539 kPa
3   (a) 2.94 kPa, (b) 11.76 N
4   1.11 MPa
5   (a) 13.33 kPa, (b) 114.33 kPa
6   980 N
7   960 N
8   1.5 mm
9   3.53×10^5 N
10   3.67 MN
11   $5d/8$
12   11.5 kN, 1/24 m below centre of plate
13   29 kPa
14   0.0 m/s, 0.1 m/s, 0.016 m/s
15   (a) 13 m/s, (b) 0.03 m^3/s
16   9.9 m/s, 7.8×10^{-4} m^3/s
18   0.00845 m/s

Chapter 34

1   (a) Sensor - Bourdon tube, signal processor - gears, display - pointer and scale, (b) sensor - volume of mercury, signal processor - tube, display - mercury level against scale, (c) sensor - resistance coil, signal processor - bridge, display - meter
2   Open - no feedback, closed - feedback, see text
3   (a) Measurement - temperature sensor, controller - thermostat, correction - heater, process - water bath, (b) measurement - rotary speed sensor, controller - differential amplifier, correction - motor, process - shaft, (c) measurement - sensor of thickness, e.g. LVDT, controller - differential amplifier, correction - rollers, process - steel strip
4   Measurement - thermistor with resistance-to-voltage convertor, comparison - differential amplifier, correction - relay and heater, process - the enclosure being controlled; measurement - level probe, controller - relay and solenoid valve, correction - flow control valve, process - water tank, measurement - level probe

Index